IT INNOVATION FOR ADAPTABILITY AND COMPETITIVENESS

IFIP – The International Federation for Information Processing

IFIP was founded in 1960 under the auspices of UNESCO, following the First World Computer Congress held in Paris the previous year. An umbrella organization for societies working in information processing, IFIP's aim is two-fold: to support information processing within its member countries and to encourage technology transfer to developing nations. As its mission statement clearly states,

> *IFIP's mission is to be the leading, truly international, apolitical organization which encourages and assists in the development, exploitation and application of information technology for the benefit of all people.*

IFIP is a non-profitmaking organization, run almost solely by 2500 volunteers. It operates through a number of technical committees, which organize events and publications. IFIP's events range from an international congress to local seminars, but the most important are:

- The IFIP World Computer Congress, held every second year;
- Open conferences;
- Working conferences.

The flagship event is the IFIP World Computer Congress, at which both invited and contributed papers are presented. Contributed papers are rigorously refereed and the rejection rate is high.

As with the Congress, participation in the open conferences is open to all and papers may be invited or submitted. Again, submitted papers are stringently refereed.

The working conferences are structured differently. They are usually run by a working group and attendance is small and by invitation only. Their purpose is to create an atmosphere conducive to innovation and development. Refereeing is less rigorous and papers are subjected to extensive group discussion.

Publications arising from IFIP events vary. The papers presented at the IFIP World Computer Congress and at open conferences are published as conference proceedings, while the results of the working conferences are often published as collections of selected and edited papers.

Any national society whose primary activity is in information may apply to become a full member of IFIP, although full membership is restricted to one society per country. Full members are entitled to vote at the annual General Assembly, National societies preferring a less committed involvement may apply for associate or corresponding membership. Associate members enjoy the same benefits as full members, but without voting rights. Corresponding members are not represented in IFIP bodies. Affiliated membership is open to non-national societies, and individual and honorary membership schemes are also offered.

IT INNOVATION FOR ADAPTABILITY AND COMPETITIVENESS

IFIP TC8 / WG8.6
Seventh Working Conference on
IT Innovation for Adaptability and Competitiveness
May 30–June 2, 2004, Leixlip, Ireland

Edited by

Brian Fitzgerald
University of Limerick
Ireland

Eleanor Wynn
Intel Corporation
USA

Springer Science+Business Media, LLC

Electronic Services <http://www.wkap.nl>

Library of Congress Cataloging-in-Publication Data

A C.I.P. Catalogue record for this book is available from the Library of Congress.

IT Innovation for Adaptability and Competitiveness
Edited by Brian Fitzgerald and Eleanor Wynn

ISBN 978-1-4757-8027-7   ISBN 978-1-4020-8000-5 (eBook)
DOI 10.1007/978-1-4020-8000-5

Copyright © 2004 IFIP International Federation for Information Processing
Originally published by Kluwer Academic Publishers in 2004
Softcover reprint of the hardcover 1st edition 2004

Printed on acid-free paper.

Contents

Conference Chairs

GENERAL CHAIR:

Karl Kautz
Copenhagen Business School, Denmark

ORGANIZING CHAIR:

Martin Curley
Intel Corporation IT Innovation, Ireland

PROGRAM CO-CHAIRS:

Brian Fitzgerald
University of Limerick, Ireland

Eleanor Wynn
Intel Corporation, USA

Program Committee

Pär Ågerfalk
University of Limerick, Ireland

David Avison
ESSEC, France

Richard Baskerville
Georgia State University, USA

Carol Brooke
Lincoln University, UK

Deborah Bunker
University of New South Wales, Australia

Claudio Ciborra
London School of Economics, UK

Siobhán Clarke
Trinity College Dublin, Ireland

Kieran Conboy
NUI Galway, Ireland

Kevin Crowston
Syracuse University, USA

Elizabeth Davidson
University of Hawaii, USA

Jan Damsgaard
Copenhagen Business School, Denmark

Brian Donnellan
Analog Devices, Ireland

Sponsors

Intel Corporation
University of Limerick
Science Foundation Ireland

Contributing Authors

Ivan Aaen
Aalborg University, Denmark

Ritu Agarwal
University of Maryland, USA

Ghada Alaa
Brunel University, UK

J. P. Allen
University of San Francisco, USA

Steven Alter
University of San Francisco, USA

Kenneth T. Anderson
Intel Corporation, USA

Ulf Ärlig
Combitech Systems AB, Sweden

Esther Baldwin
Intel Corporation, USA

Chris Barry
National University of Ireland, Galway, Ireland

Piero Bassetti
Bassetti Foundation on Innovation and Responsibility, Italy

Keith Beggs
Intel Corporation, USA

Pernille Bjørn
Roskilde University, Denmark

Anna Börjesson
Ericsson and IT University of Gothenborg, Sweden

Jim Brown
Draeger Safety UK Ltd, Northumbria University, UK

Tom Butler
University College Cork, Ireland

Claudio Ciborra
London School of Economics, UK

Brian Donnellan
Analog Devices B.V

Helga Drummond
Liverpool University, UK

Audrey Dunne
University College Cork, Ireland

Guy Fitzgerald
Brunel University, UK

Helle Damborg Frederiksen
Aalborg University, Denmark

Steve Furnell
Plymouth University, UK

Helle Zinner Henriksen
Copenhagen Business School, Denmark

Joachim Höck
Polizei Oberbayern, Germany

Charles House
Intel Corporation

Edoardo Jacucci
University of Oslo, Norway

Björn Johansson
Jönköping University, Sweden

Jannis Kallinikos
London School of Economics, UK

Karlheinz Kautz
Copenhagen Business School, Denmark

Jeffrey Kim
University of Washington, USA

Frank Land
London School of Economics, UK

Linda Levine
Carnegie Mellon University, USA

Brian Lings
University of Exeter, UK

Björn Lundell
University of Skövde, Sweden

Sabine Madsen
Copenhagen Business School, Denmark

Lars Mathiassen
Georgia State University, USA

Anders Mattsson
Combitech Systems AB, Sweden

Tom McMaster
University of Salford, UK

Shaila Miranda
University of Oklahoma, USA

Michael Ney
TU München, Germany

Malvina Nisman
Intel Corporation, USA

Joyce O'Connor
National College of Ireland, Ireland

Carl Magnus Olsson
Viktoria Institute, Gothenburg, Sweden

Cindy Pickering
Intel Corporation, USA

Jan Pries-Heje
The IT-University of Copenhagen, Denmark

Nancy L. Russo
Northern Illinois University USA

Tony Salvador
Intel Corporation, USA

Christian Salzmann
BMW Car IT GmbH, Germany

V. Sambamurthy
Michigan State University, USA

Jeff Sampler
London Business School, UK and IESE Business School, Spain

Kurt M. Saunders
California State University, USA

Bernhard Schätz
TU München, Germany

Ada Scupola
Roskilde University, Denmark

E. Burton Swanson
UCLA, USA

Prodromos Tsiavo
London School of Economics, UK

Richard Vidgen
University of Bath, UK

Philip Vos Fellman
University of Southern New Hampshire, USA

David Wastell
University of Manchester Institute of Science and Technology, UK

Robert Zmud
University of Oklahoma, USA

Message from the Organizing Chair

IT Innovation for Competitive Advantage and Adaptiveness

Martin Curley,
Director, IT Innovation, Intel Corporation

Achieving competitive advantage from Information Technology or at least proving the business value of IT has long been a holy grail for both CIO's and academic researchers. The statement by Robert Solow in 1987 "I see computers everywhere except in the productivity statistics" initiated a more than decade long debate on the business value of IT. This became known as the "IT productivity paradox" which stated that despite enormous improvements in the underlying technology, the benefits of IT spending have not been found in aggregate output spending. A summary report of all related research in this area, published by the Centre of Information Technology and Organizations (CRITO) at UC Irvine (Dedrick et al, 2003), came to the conclusion that the Productivity Paradox had at last been refuted and that investment in IT leads to increased value and improved productivity. Indeed increasingly evidence is available to show that when viewed over a longer period, investments in IT can significantly outperform other kinds of investments. (Brynjolfsson 2002).

In a study from the University of Groningen (2002) on ICT and Productivity, van Ark et al linked the slower adoption of ICT in Europe (compared to the US), to the productivity gap between the US and Europe. This was particularly prominent in the ICT intensive industries where the US saw a rapid acceleration of productivity growth in the second of the last decade, whilst growth in Europe in general stagnated. There is a consensus growing that investment in ICT leads to productivity growth elsewhere in the economy, particularly in the service sectors.

Innovation is crucial to growth and survival of national economies. In this context IT Innovation is emerging as a substantive approach and tool for driving productivity and growth. The combination of IT enabled business process re-engineering coupled with the increasing flexibility of IT solutions development enabled by web services, means that transformational IT solutions which can transform a firm, industry or indeed a country are becoming more commonplace.

Additionally the ever improving economics of IT infrastructure performance driven by Moore's law, means that IT Innovation as a sub-discipline of information technology will become more substantial and compelling. Who would have imagined in 1976, when a Cray C1 computer costing $5million delivered 0.16 Gigaflops, that desktop PC's many times more powerful would be commonplace in 2004. Today a PC based on a 3GHZ Pentium ® 4 microprocessor delivers computing power of 6 Gigaflops at a price of approx $1400. With this kind of power available to millions of users worldwide, the sweet spot for IT innovation has forever shifted from the mainframe to the PC client. Dale Jorgenson (2001) summarized the impact of Moore's Law when he said "Despite differences in methodology and data sources, a consensus is building that the remarkable behavior of IT prices provides the key to the surge in economic growth!"

IT innovation really means IT *enabled* innovation as any innovation requires the co-evolution of the concept, the IT solution, the business processes and the organization. Transformational success is achieved when these four entities are co-evolved in parallel. However when dissonance occurs between the evolution paths across a major transition then significant problems occur. Organizations that succeed at a major IT enabled transformation typically have a compelling vision, a determined credible champion, a well developed IT capability and momentum which is built through early quick wins.

Rapid Solutions prototyping is a key experimentation process for furthering innovation as new or modified concepts are rapidly made real in a solution or environment that can be experimented with. Fast iteration of the rapidly developed prototypes can lead to order of magnitude improvements in functionality and capability and decreased time-to-market.

Within Intel we have used IT enabled Innovation and rapid solution prototyping to deliver new capabilities. For example in our engineering computing activity, we rapidly migrated a suite of design tools from a Unix/Risc platform to a Linux/Intel Architecture platform and have achieved more than $500 million savings in capital avoidance in three years while meeting computing demand which is growing by more than 100% annually.

Another example of IT Innovation is using individual PCs for caching of rich media content to deliver new capabilities such as eLearning and video to the desktop to tens of thousands of employees worldwide at almost zero incremental cost.

One way of describing the impact of IT innovation are improvements in efficiency, effectiveness or transformation. Typically efficiency and effectiveness improvements drive incremental business improvements, however IT enabled transformation can drive structural changes and advances. Let's look at some public sector examples.

At Westminster City Council in the UK, Peter Rogers the CEO and the council Leader Simon Mallet have developed a vision of how the city could be transformed using wireless technology, enabling delivery of better services to citizens at lower cost – for example the use of wireless WiMAX technology with IP camera technology can reduce CCTV installation cost by 80% dramatically advancing the crime-free agenda of the city.

In Portugal each third level campus is being unwired using WiFI technology and the government, working with private industry is promoting the adoption of wireless notebooks by all third level students, helped by low interest loans provided by the major banks. In this way the Portuguese government hopes to transform learning in Portugal and ensure the Portuguese information society has one of the fastest learning velocities in Europe.

At the National Health Service in the UK, more than £9 billion is being invested in ICT to transform the UK health service. Against a backdrop of a mission "saving lives, cost effectively" these ICT investments will introduce better solutions such as decision support systems for doctors, improved administration systems to enable easier appointment booking and mobile point of care solutions, based on wireless tablet technology to in-hospital staff and district nurses.

All of these solutions are transformational, involving a lofty vision and elements of public-private partnership. In an increasingly complex world with pervasive computing looming in the horizon, those countries which embrace IT enabled innovation will lead as the transition from the resource based economy to the knowledge based economy continues unabated.

This conference discusses the many aspects of IT innovation, including high technology adoption, innovation diffusion in firms and industry/public sector and the business value of IT Innovation. I hope it contributes to the evolution of IT Innovation as a discipline and improved solutions for citizens and customers everywhere.

References

Brynjolfsson, E and L.M. Hitt. 2003. Computing Productivity: Firm Level Evidence. MIT Sloan, Center for Information Systems Research (CISR). Working paper No. 4210-01

Curley, M. 2004. "Managing Information Technology for Business Value", Intel Press. January.

Dedrick, J., V. Gurbaxani, and K.L. Kraemer. 2002. "Information Technology and economic performance: A critical review of the empirical evidence". University of California, Irvine, Center for Research in IT and Organizations (CRITO). November.

Jorgenson, D. 2001. "Information Technology and the US economy." American Economic Review, 91:1, 1-32.

Van Ark, B. Inklaar, R. McGuckan, R. 2002. "Changing Gear, Productivity, ICT and Services, Europe and United States – Research Memorandum GD-60". University of Groningen. December.

Preface

IT Innovation for Adaptability and Competitiveness

Eleanor Wynn

Brian Fitzgerald

IFIP WG 8.6 has as its focus diffusion of technological innovation. In this conference we have solicited papers on the topic of IT innovations that can further an organization's ability to adapt and be competitive. Thus we address the problem at an earlier starting point, that is, the emergence of something innovative in an organization, applied to that organization, and its process of being diffused and accepted internally.

A further extension of this would be the propagation of a successful innovation outside the originating organization as a product, service or example of technology use that builds the firm's markets. In this discussion we are supposing that said innovations are indeed a contribution. In reality, an idea is only labeled an innovation once it is accepted. Before that time, it can be just an idea, a crackpot idea, a disturbance, obsession, distraction or dissatisfaction with the status quo. Many innovations are of course deliberately cultivated in research labs, but again their success is the determinant of their eventual designation as "innovative".

Conversely, some ideas really are crackpot concoctions or technologies in search of a use that linger in the environment as potential innovations long after their use is discredited. Case in point: voice recognition software, which does have some applications but has been over hyped and over applied for about 20 years. Today some call centers won't let users punch a single button on their telephone sets; they MUST tell the voice recognition program what they want. Some of these systems will revert to an operator if the voice recognition system doesn't understand, while others will just hang up. We were relieved to note the following title in the March 5 *Financial Times*: "To speak to an operator, start swearing now." Someone has developed an innovation to recognize user frustration and bypass the prior innovation of persistent automated voice "response"!

It is the matching of a capability to a need that is the innovation, and the uptake of this match that is the adoption or diffusion. In a large organization, this process can be long, challenging, and fraught with possibilities for failure, frustration and financial loss.

What makes something an innovation is its eventual utility. In IT, the case is even stronger. Innovation in IT is what helps the firm to survive, adapt and compete on operating costs, on production, on coordination of resources and in the marketplace. Necessity is said to be the mother of invention. As Chesbrough (2003) starkly declares: "Most innovations fail. And companies that don't innovate die". With this in mind, we suggest that innovation in the organization is not a luxury, but a critical means of keeping up with changing circumstances and opportunities. The organization that doesn't innovate at least in parallel with its industry or markets, can be doomed.

Let's take the example of the American steel industry (Tiffany, 2001; Christensen; 2003). Japanese steel makers began using highly efficient production technologies in the 1970s. They also focused on particular markets for steel products utilizing "mini-mill" technologies that could be efficient using scrap rather than ore and in smaller production batches. Meanwhile the American steel industry, with its installed base of foundry equipment, could not see the rationale for paying the price of upgrading their technology. By the time they did see the rationale, they could no longer afford to make the purchases. The markets had been undercut by superior Japanese products that cost less. Had they considered innovation as essential to survival, or conceived that the day would arrive when this major US industry would even see foreign competition within their own markets, they would have acted differently.

Innovation in industry and in technology are "nothing new". Technology innovations have revolutionized civilizations, trade and economies for millennia. Iron implements, gunpowder and antibiotics all made indelible marks on history and culture. There is a proposed parallel with adaptation in species, in the sense that adaptation to the environment that make individuals more successful become adaptations to the gene pool. However, environments in nature do not stay static, and so adaptation continues, given normal cycles of natural change. Cycles of historical change are potentially more turbulent than change in nature, and we are in a particularly turbulent historical period now, both socially and technologically.

So, innovation really is part of the normal life cycle or life process of a business or an industry. Innovations arise as responses either to new needs or to perceived failures, inefficiencies or obstacles in the current process. Innovations tend to beget more innovations. This is especially true of information technology. Indeed, we can go so far as to state that information technology innovation is insufficient unless leveraged successfully in a business context, either for adaptability or for improved competitiveness. As computing power increases and computing devices shrink, more can be done. Large mainframes gave way to minicomputers, which led to desktop computing. The "real" origin of the Internet was a patch to a network set up by two computer scientists at Stanford and UCLA. A professor in Santa Barbara wanted to be on the file transfer system and he "invented" TCP/IP as a way to avoid a "party line" effect (everyone talking at the same time) when he tapped into the wire that went through Santa Barbara on its way from Stanford to UCLA. The next phase in an innovative process like that one is of course to refine, to begin to see new possibilities arising from the leap that has just been made, and successful examples then beget exponential growth.

This is one reason why we are cautious about the idea that diffusion of innovation is a problem to be solved independently of the contextual validity of the innovation.. Innovations that make an impact, providing they are made within a context of immediate application, tend to be self-propelling to some extent.

However, there is something that stands in the way of the adoption even of valuable innovations and that is the worldview or formative context of the environment (Weick, 2001). Innovations are made in the context of institutional embeddedness. That is, the object of innovation does not stand alone, but is set within an economy, a set of cultural and business practices, a set of values and perhaps most important, a set of interests. Some innovations do in fact defy all of the above and succeed in spite of circumstances. Other inventions need only contravene one of the embedding conditions, let's say interests, in order to meet with failure or to be delayed by decades until the use is absolutely compelling. The case of the American steel industry should be a lesson in that regard: when the use becomes compelling, will it still be possible to employ the innovation, or is there a window of opportunity beyond which too many changes in the environment make adoption, though necessary, impractical?

Hence the focus on innovation and innovation processes as a value in and of themselves. The tendency to entropy is as endemic as the tendency to change and adapt. Nature doesn't really care whether species adapt to climate changes or other conditions. It is up to the species to permute themselves accordingly. Within social systems, then innovation must be conscious, even self-conscious. It requires an ontological reflection as to who are we, why are we here and how do we plan to survive given uncertain futures? None of this is imperative day-to-day, so it is possible to go for some time without innovating until a point of crisis is reached.

A coherent innovation strategy would anticipate many possible scenarios and have innovations available that can be tailored to meet the needs. If innovation only comes up when the crisis is at hand, it is likely to take too long. This is partly because of the reflective process we referred to earlier. The process of organization reflection, suggestion, problem-setting, differentiation, concept testing and then product testing can take three years from start to finish, with the best of intentions. Innovation by definition is not the familiar, but the unfamiliar. Many stakeholders have a hard time telling a viable innovation from frivolity or waste; the problem must be perceived in order for a solution to be apparent as such, and so forth, in a fairly deep and emergent social process (Nonaka and Takeuchi, 2001).

Planning in an organization can easily be bounded by the familiar. It can be based on assumptions that are linear with today's environment, e.g. assume a certain growth rate in the market, assume a certain amount of incremental change in the environment, and prepare for that Weick, 2001). But history tends not to be smooth and linear but to contain major disjunctures (can also be referred to conversely as "conjunctures" per Fernand Braudel (1995) and discontinuities. It is safe to say that the present period is exemplary of such a disjuncture. Therefore, nonlinearity in the rate and kind of change should be expected, and multiple innovations should be encouraged to meet a range of possible near term future needs.

An IT department, which after all is the focus of the conference, tends to be pulled strongly towards first of all, stability and reliability, which can be seen as contradictory to innovation. IT organizations in many corporations today are still seen as commodity functions that constitute a necessary cost of doing business, but not as a strategic option for radically increasing profits. There are many contrary examples in the literature (Jelassi and Ciborra 1994), but the fact remains that most chief executives are as happy to raise IT costs as the ordinary householder would be to have his or her electricity bill increase. There is no perception that an increased electricity bill would change the quality of life (unless it is feeding a hot tub).Similarly, IT costs are seen as something that needs to be "kept down". In addition, nobody wants to take risks with something as basic as IT. No electricity is a huge disruption and network downtime can bring a company to its knees. Risk aversion is therefore endemic to the concept of IT. Innovation is constantly needed but also threatens to disrupt, and being innovation, the return on the risk is usually uncertain. Indeed, given the two constraints above, innovations in IT tend to be incremental rather than radical. But let's take a look at the steps that have already happened within the domain of IT.

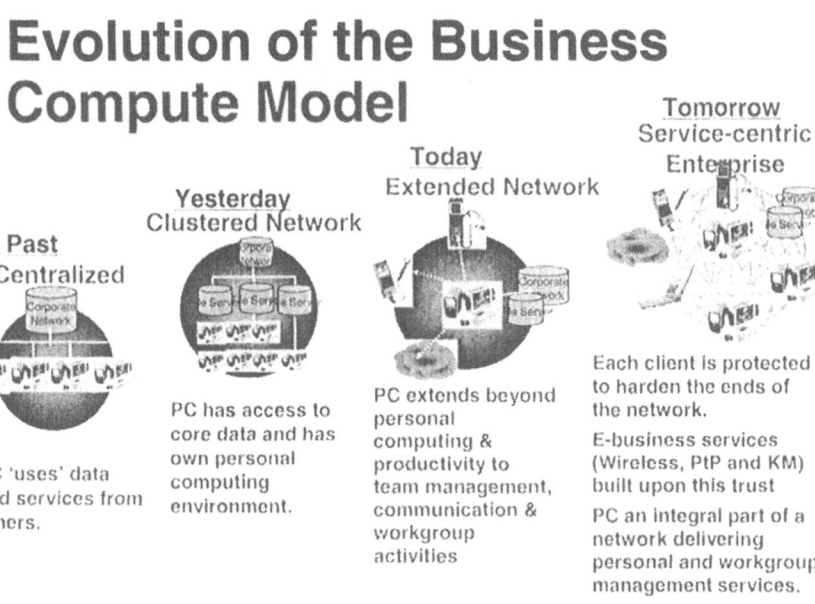

Graphic courtesy Chuck House, Intel IT

These are order of magnitude changes in capability. Costs are discontinuous with a maintenance or gradual improvement approach. Benefits tend to be exponential, though given that new capabilities are involved, baselines often don't exist. Opportunities, however, become obvious after the fact.

Often the actual opportunities that emerge are different from the opportunities that were anticipated. As we know, many software projects fail or are abandoned. There is a constant process pushing toward innovation, not always successfully. Part of the problem is the mismatch between vested technical expertise and the ability to envision the organization as a social environment. Users tend to be held treated as a mystery, ignored, force-fitted or indulged with superficial adjustments. Some innovations are not in fact really innovative, precisely because, although they address one piece of a corporate problem, they ignore the institutional context. They serve only one stakeholder, not the chain of stakeholders that are impacted. They represent a technological capability for which there is no use at the moment, or whose use has not yet been married to the capability. Actually fostering, creating and implementing innovation requires a systemic view of the organization and how it works.

This systemic view can be the most difficult part of the innovation process or innovation artifacts.

In his book Hidden Order, Holland (1996) described the features of a self-organizing system in nature or in any system that takes on properties of self-organization, i.e. where there is a certain degree of agency, autonomy, complexity and interdependence, as follows:
- Aggregation
- Nonlinearity
- Flows
- Diversity
- Tagging
- Internal models
- Building blocks

This is not the place to discuss the features at length. Suffice it to say that a self-organizing system consists of a certain amount of mass, has emergent properties that are non-linear (ie can "take off" or "collapse" depending on key elements), has independently interacting elements, is diverse in its forms, possesses some kind of communication mechanism, has type consistencies with expected behaviors, and can grow organically by means of higher level entities than the units of each type.

The feature to be called out here is diversity. Diversity would work in nature by means of sufficient variety in an ecosystem that the system can rebuild itself in case of a collapse. For instance, trees are cut, birds have no more habitat, insects proliferate, etc. as one set of cascading effects. In fact, this system can eventually regenerate itself by the same mechanisms that enabled it to exist in the first place, a property of robustness that depends on diversity. Seedlings of trees sprout, weeds provide some cover for water retention (assuming they don't choke out the seedlings), growth creates shade, shade helps retain water, trees get bigger, leaves or needles compost, and eventually, if they aren't extinct, birds can come back. It isn't guaranteed that an ecosystem can rebuild, but if it does rebuild it does so by means of diversity. Not every species or element is equally affected by environmental circumstances. Some can survive to begin a process of regeneration.

Organizations try to build some of this robustness into network systems. Indeed part of that robustness can be generated by inadvertent diversity in the accumulation of historical artifacts that make it less than a perfectly rational system. New systems for network security against viruses may rely on a biological model of immunity, where there is sufficient slight diversity in operating systems to slow a virus down, just as nature protects against extinction through biodiversity. .

So, diversity is a survival strategy and an adaptation strategy. And innovation in technology provides ample diversity. Neil Smelser (1995) in his economic sociology has referred to stages of innovation during a given historical period. For instance in the early stages of the industrial revolution there was a high degree of diversity in the types of inventions devised for a particular usage. This was also evident in medieval science as Kuhn describes it, with individual scholars inventing whole nomenclatures and models for systems that eventually became defined as optics or chemistry once established. At the point of paradigm convergence or stability, whole lines of research fell by the wayside that once had flourished when no common agreement existed about how to define this realm.

However, even though conventional wisdom suggests that diversity is an important survival and adaptation strategy for innovation, the IT sector is one in which there are no absolutes. For example, when MicroPro who produced the once-dominant Wordstar word processing package sought to diversify into other product offerings, this allowed their main competitor, WordPerfect, to usurp their dominant position. Yet, several other companies failed to survive because of a lack of innovation and diversification. RCA, once the dominant pioneer in consumer electronics failed precisely because of their lack of diversity as they bet all on vacuum tube technology which was completely superceded by transistor and solid state electronic technology.

Likewise, the future trajectory and potential of innovation is by definition unpredictable. Gordon Moore, founding CEO of Intel, has admitted that when IBM awarded Intel the design win for their 8088 processor in the IBM PC, it was not seen as one of the top 50 market applications for the 8088 product. Yet, today most of Intel's revenue and profits stem from the Pentium microprocessor range descended from the 8088 used in the IBM PC. Hidden within novelty, therefore, are different models, even though they appear to perform similar functions. Models carry implications and have more or less extendibility or scalability. When technologies are new, it is less likely that any one observer will be able both to understand the underlying technological model and to understand the social model implied in it.

During the computer era, change, including paradigm change, has happened so rapidly that invention seems not only normal but obligatory. It is an engine of economies and as new possibilities become apparent on the horizon of current capabilities, innovation continues to be spurred forward towards that next thing. Innovation makes gaps apparent, gaps that never existed because no solution was at hand, no basis for noticing an absence was evident.

What we know as the current era of globalization (Friedman, 2004) is a product of convergence or *conjoncture* of a combination of technological capabilities that together add up to a critical mass phenomenon resulting in a state change in world labor markets. In a recent New York Times editorial, the columnist Thomas Friedman wrote from Bangalore, India about the circumstances that enabled the current economic vibrancy of that city. Although there are downsides to be noted (Madon & Sahay, 2001), as well. Friedman writes:

> Globalization 3.0 was produced by three forces:…first, massive installation of undersea fiber optic cable and the increased bandwidth (thanks to the dotcom bubble), that have made it possible to globally transmit and store huge amounts of data for practically nothing. Second is the diffusion of PCs around the world…third is the variety of software applications….that when combined with all those PCs and bandwidth, made it possible to create global "workflow platforms".

Thus a series of innovations and breakthroughs in separate technology areas combined with geopolitical and economic circumstances to create a large threshold effect of offshore outsourcing of knowledge work.

Let's take the example of distributed collaboration within an organization. Again, first came the mainframe, then came dedicated lines between mainframes, then came minicomputers and their networks, then came the desktop and networked computing, giving rise to e-mail, e-mail attachments, then the web with graphics, etc. etc. Some of these changes early on made it more possible for a company to operate with remote offices and still have some kind of real-time coordination, or near-real time. That ability led to the need to communicate different kinds of material remotely, not just data but documents, documents with graphics, documents from different operating systems and so on. Most of us have lived through these transitions. So, technology enabled corporations to operate remotely, which led to more remote office, which led to overseas outsourcing, people not having to move to follow a job within the company (a major population mover for the middle classes during the 50s and 60s). But the more distributed the more desire for something like real time communication and the "experience" of collocation. So many elements go into this movement of being distributed because you can, then wanting it to be better, then relying more on remote communications therefore needing better tools, that the solution is still in the process of being devised. Remote collaboration takes a combination of key technologies, interoperability among them, a compelling social experience and graphics and usability stickiness at the top level. Many different "solutions" are individually invented around these needs.

Such a complex area of inquiry and invention practically creates an ecosystem of its own, in terms of infrastructure, technical capability and choices (e.g. objects, content management, instant messaging, and imaging of members), connection speeds, standards, graphics elements, design, selection among key features, all hanging together without overcomplicating the interface. In a case like this, eventually one small innovation can become the enabling factor; or else, one small perception of the situation. In our recent work at Intel (Lu, Wynn, Chudoba and Watson-Mannheim, 2003), we discovered that 2/3 of employees, including all job types and geographic regions, work on three or more teams at any given time. This observation had not previously been embodied in any well known collaboration suite on the market. As a result in existing tools, each team is allocated a partitioned space.

The growth in number of separate teams or projects requires more and more overhead on the part of the user to manage across these teams. Indeed, it becomes almost impossible to see a roll-up of all one's responsibilities.

Hence the single observation of multi-teaming could lead to a key innovation in collaboration environments. At the same time, once we conceive of teams as interlocking or clustering in like-minded networks, the requirement for security and permissions on the individual team sites becomes a more complex problem. Indeed when a group at Intel working on the development of new enterprise collaboration tools showed our collaboration concept at an internal invention fair, the most ardent of the admirers were two security gurus who saw at last a user model for technical work they had developed over the years. Thus innovations interact and encounter each other in a cumulative process. That is what happens assuming the "flows" in the self-organizing system framework are fluid enough for innovations to encounter one another in a social or professional network.

In recent years, the open source software (OSS) phenomenon is one where emergent properties reveal innovations beyond those planned or intended. While the concept of free software had been around for years, the coining of the term "open source" didn't radically change the core definition. At heart, free software and open source software are equivalent concepts. However, the free, as in unfettered rather than no cost, access to source code is not what makes the open source phenomenon so innovative. Indeed for many organizations, it is the free as in beer that is attractive, and they are as likely to deploy closed software provided it is zero cost (Fitzgerald & Kenny, 2004). However, the open source term is far more business-friendly and less ambiguous than free software. Certainly, Wall Street and the stock market embraced it enthusiastically, with VA Linux and Red Hat achieving record-breaking IPOs. Also, despite some claims, access to the source code is not the key factor in itself, but rather how this facilitates the collaborative development model of truly independent pan-globally located developers which allows for a rapid evolution of 'killer applications'. The OSS phenomenon has also elicited a great deal of interest beyond the domain of software. It provides new business models and innovative modes of work organisation which have been extremely successful. For example, rather than stifling a local software industry as has been suggested, it appears that small companies and SMEs (small to medium sized enterprises) can treat OSS products as an infrastructure, akin to the rail, highway, electricity networks, and bootstrap a lucrative service model on top.

Furthermore, it has been suggested that the OSS principles of openness and inclusiveness, provide an exemplar for the future of society, and help point the way to addressing the 'Digital Divide'. However, this picture is not so straightforward as it seems, as attempts to stimulate open source communities in Africa have failed to take off there, largely due it seems to a basic mistrust that there can be any value in initiatives that are provided for free.

In brief, the selection process, otherwise known as adoption or diffusion, exists in a complex environment of prior inventions, known needs, encounters among participants in different disciplines, and ultimately in the perception of the need within the context of the new capability. Without some kind of self-examination, in the one case an internal survey, arising from self-reflection (professionals asking themselves: "do other Intel employees work the same way we, the professionals looking at the situation, do? If so, what does that imply for return on investment in exploring collaboration tools from the point of view of our needs rather than from the point of view of what is out there?")

What does this imply for corporate governance, control systems or infrastructures that would support innovation? One provocative writer (Obstfeld, in press) has drawn out a relationship between types of personal networks supported within an organization and the propensity to innovate. Pursuant to Watts' description of the two extreme types of networks, completely closed and redundant vs. completely open and random, Obstfeld describes network bridges that enclose "holes" in a network as particularly fruitful patterns for the creation of innovation. While it is beyond the scope of this preface to fully describe the type of corporate structure that would support innovation, we feel that it lies in such a direction: internal communications systems and methods of social signaling across boundaries where self-organizing networks produce adaptive systems. Many of the papers for the conference support similar scenarios for the support of information technology innovation.

In earlier paragraphs we discussed the diversity of innovation as an indicator of a time of rapid innovation, before a pattern of usage and institutionalization sets in to a new type of technology. Our conference papers are reflective of this diversity. While information systems innovation has been proceeding apace for about fifty years, we still find ourselves in formative stages of new capabilities, as well as new circumstances. Currently, wireless technology and globalization in the capabilities and conditions sectors respectively, are driving a large array of invention. The papers submitted to the conference reflect that branching.

We have divided the papers into the following sections. It was not initially obvious to us what the clustering pattern was. The papers seemed so diverse. Brian Fitzgerald took one cut at clustering them, which gave us a structure. After that, Eleanor Wynn re-sorted them and then renamed the clusters. We believe this may resemble the pattern for innovations in the corporation in the marketplace. At first, innovations defy classification. Or, they are placed in the wrong category and compared on the wrong terms. This has happened recently in the so-named knowledge management sector. Nonaka and Takeuchi 2001) brought us a very robust definition of corporate knowledge and how it is co-created. Then a consulting industry arose. In that process, many approaches aggregated based on some kind of relationship to the concept of knowledge. But underlying approaches varied widely in both their technological and their sociological assumptions.

Library and information science professionals and academics eagerly undertook the complex problem of classifying, tracking and understanding the cognition of knowledge-seeking. People with a social science or interactionist bent, whether academics or practitioners, focused on social networks and how to utilize them. In the middle many sophisticated technologies arose that went around the problem of classification and subtly addressed the sociological side using patterning matching and inferencing technologies like collaborative filtering or Bayesian networks. Looking at the situation from a participant observer or action research perspective, it became clear that the field had divided itself into "technology solutions" and "people-to-people solutions". This division is inherently spurious but it comes easily to hand in many environments. (Bloomfield & Vurdubakis, 1994)

Interimly the result of that was to overlook those sophisticated technologies that did not match any prior conceptions of "knowledge" or "management", that instead were quite risk-taking in terms of where they penetrated the problem: by trying to make sense of tracking the behavior of participants and objects in a knowledge-based environment. The risk on the technology side was the confrontation with Protestant ethic of "management must be orderly" or the Cartesian ethic of reduction to basic terms. Alas, those basic "terms" are unstable in an organization just at the point where their content becomes interesting. Organizational "knowledge" is unlike scientific knowledge in its volatility and time-dependency for relevance. In other words, organizational knowledge is actually highly innovative, but very hard to keep up with. Trying to box into a classification scheme, unless a natural classification already exists or the field is defined by its classifications and terms, e.g. natural science, software, etc. is a guarantee of instant irrelevance. Also classification simply cannot anticipate what will happen in a turbulent environment. Does this mean that classification and taxonomic systems for content are wrong? No it does not. But they do not keep pace with the dynamics of language that actually drive innovative thinking in the organization, in an industry and in the policy and political environments in which these exist.

The tracking process, which comes in various forms, but notably Bayesian systems and collaborative filtering, does not anticipate the content, terminology or behavior, but by various means clusters it into statistical patterns that are then interpreted and labeled by people, who understand through recognition when a relevant relationship has been made. This is especially important for quality filtering, Collaborative filtering simply points out what others who chose one object in common with a user, also chose.. It uses the object of knowledge or the choice as the basis of comparison. This choice then can predict other choices across domains based on similarities implied in the users just because they made these choices. It is incredibly efficient. It does not rely on labeling or classifying but tracks far more subtle evaluations made by individuals as they act.

The "people" vs. "technology" polarity breaks down completely here because the technology is sophisticated but reflective rather than predictive. We believe that a key aspect of innovation is to break down older dichotomies, to search for new frameworks and to implement those frameworks into the adaptive organization. In this process "who are we?" – *that* type of ontological question-- is just as important as "what shall we do/how shall we proceed?"—the epistemological question.

In short, thanks to our illustrious authors, who defy classification, we have discovered a clustering of the conference papers along the following lines and have labeled and relabeled them accordingly. We were very pleased with both the quantity and the quality of papers received. Given that our call for papers on adaptability and competitiveness was, we hope, not squarely in any one conventional topic area, we were gratified that authors found a way to match their interests with our theme in a way that we see as productive and imaginative.

THE ROLE OF IT IN ORGANIZATIONAL INNOVATION:

INNOVATING SYSTEMS DEVELOPMENT & PROCESS

ASSESSING INNOVATION DRIVERS

INNOVATION ADOPTION

NEW ENVIRONMENTS, NEW INNOVATION PRACTICES

J. P. Allen and Jeffrey Kim, Digital Gaming: Organizing for Sustainable Innovation

Michael Ney, Bernhard Schätz, Joachim Höck, Christian Salzmann, Introducing Mobility: The mPolice Project,

Tony Salvador, Kenneth T. Anderson, Supporting the Re-emergence of Human Agency in the Workplace

Audrey Dunne and Tom Butler, Learning Management Systems: A New Opportunity

Chris Barry, Web-Based Information Systems - Innovation or Re-Spun Emperor's Clothing

PANELS

Piero Bassetti, ICT Innovation: From Control to Risk and Responsibility

Frank Land et al., PANEL TITLE: The Darker Side of Innovation,

V. Sambamurthy, Panel: IT as a Platform for Competitive agility,

Esther Aleman: Innovation in academe

REFERENCES

Bloomfield, B. P. & Vurdubakis, T. Boundary Disputes: Negotiating the Boundary Between the Technical and the Social in the Development of IT Systems, *Information Technology & People*, Vol. 7(1), 1994, 9-24.

Braudel, Fernand (1995) *A History of Civilizations*. Penguin USA.

Chesbrough, Henry. (2003) *Open Innovation: The new Imperative for Creating and Profiting from Technology*. Harvard Business School Press, Cambridge, MA. Christensen, Clayton M. (2003) *The Innovator's Solution: Creating and Sustaining Successful Growth*. Harvard Business School Press, Cambridge, MA.

Fitzgerald, B. and Kenny, T. (2004) Developing an Information Systems Infrastructure with Open Source Software, *IEEE Software*, February 2004, pp.50-55.

Friedman, Thomas (2004) While you were sleeping, 'the third great era of globalization' began. *New York Times* OpEd section, Friday March 5.

Holland, John (1996) Hidden Order: How Adaptation Builds Complexity. Perseus Publishing, USA.

Jelassi, Tawfik and Claudio Ciborra (1994) *Strategic Information Systems*. John Wiley & Sons, Chichester

Lu, Mei, Eleanor Wynn, Kathy Chudoba and Mary Beth Watson-Mannheim (2003) Understanding virtuality in a global organization: towards a virtuality index. *Proceedings of the International Conference on Information Systems*, Seattle WA.

Madon, Shirin & Sundeep Sahay (2001) ICTs and cities in the developing world: A network of flows, Information Technology & People, 14, 3

Moingeon, B and A. Edmondson (1996) *Organizational Learning and Competitive Advantage*. Sage Publications, London, New Delhi, Thousand Oaks, CA.

Nonaka, Ikujiro and Toshihiro Nichigushi (2001) *Knowledge Emergence: Social, Technical and Evolutionary Dimensions of Knowledge Creation*.

Obstfeld, David (in press) unpublished manuscript. University of California, Irvine, Graduate School of Management. Title withheld pending author's changes. Working title: Knowledge creation and social networks.

Rigby, Rhymer (2004) To speak to an operator, start swearing now. Financial Times March 5, 2004. http://search.ft.com/search/article.html?id=040305000891

Smelser, Neil J & Richard Swedberg (1995) *Handbook of Economic Sociology*. Princeton University Press US.

Tiffany, Paul A. (2001)The Decline of American Steel: How Management, Labor and Government Went Wrong. Replica Books.

Watts, Duncan (2002) *Six Degrees: The Science of a Connected Age*. WW Norton & Co, New York.

Weick, Karl E. (2001) *Making Sense of the Organization*. Blackwell Business, Malden MA and Cambridge, UK

Acknowledgements

Organising and hosting a conference such as this requires an enormous amount of dedicated effort on the part of a great number of individuals and organizations. We thank the latter sincerely, and wish to record our appreciation specifically to the following:

Lorraine Casey
Kathy Driehaus
David Fleming
Lorraine Morgan
Intel Corporation
Irish Chapter of the Association for Information Systems (IAIS)
Limerick Travel
Science Foundation Ireland
University of Limerick

The original version of this book was revised.
An erratum to this book can be found at DOI 10.1007/978-1-4020-8000-5_31

PART I

THE ROLE OF IT IN ORGANIZATIONAL INNOVATION

EVOLVING SELF-ORGANIZING ACTIVITIES
Addressing Innovation and Unpredictable Environments

Ghada Alaa and Guy Fitzgerald
Department of Information Systems and Computing, Brunel University, Uxbridge, Middlesex, UK

Abstract: Traditional development methods are systematic, prescriptive and plan-driven, and can reduce creativity and the ability to respond in situations of rapid change. In the ere of the Internet and increased uncertainty it is argued that development teams need to be self-organizing, i.e. able to reflect and adapt freely to specific problem situation instead of following rigid methods. A way of operationalizing these concepts is presented and this paper elaborates a case study experience, a business-business (B-B) e-marketplace relating to the pharmaceutical sector, where self-organizing activities have been evolved in modeling upfront requirements for the portal. It is based on brainstorming sessions and the use of a modeling tool, the 'e-Business Issues Roadmap' that categorizes possible issues that such an e-business project might encounter and hence provides a representation or a global 'picture' of the problem situation. The outcome is the trigger of business innovative ideas that are unlikely to have been generated by traditional modeling techniques. By re-visiting this exercise the 'picture' becomes up-to-date and accordingly stakeholders will self-organize to address the necessities of the specific situation. The paper concludes with the introduction of the concept of the 'Practice' that is defined as 'a discourse during which requirements for software development activities and responses evolve'. It helps with the habits and dynamics required to instill self-organization within development teams rather than leaving it to evolve by chance.

Key words: innovation, emergence, complex adaptive systems, self-organization, practice

1.   INTRODUCTION

The turbulent nature of the business environment (Turban et al., 2000) requires organizations to react quickly and creatively to make the most of new opportunities and business models. The globalization effect of the Internet creates pressure on organizations to raise the quality of their products and services as well as react more creatively and quickly to be able to compete internationally. The new development environment is thus characterized by high competitive pressures, large amounts of change, uncertainty and increased time pressures.

Avison and Fitzgerald (2003) point to the current debate and questioning of formal methodologies and characterize this as a 'backlash against methodology'. One element of this backlash is that some developers are rejecting the use of methodologies and turning to less formal more 'ad hoc' development. Highsmith (1998) suggests that the discipline of software engineering is based on rigorous concepts like predictable events, deterministic patterns and linear construction of objects, and that this is insufficient to cope with the high frequency of change and the speed of the market. The limitations of traditional development methods for software innovation and adaptability are discussed in section 2, and the requirements of innovative and adaptive development processes are addressed in section 3.

Alaa and Fitzgerald (2004) identify the importance of fostering innovative development environments and raising the ability of developers and stakeholders to react and adapt freely to specific problem situations. It is argued that innovation and adaptability can be achieved through self-governance and self-organization of their activities. These arguments are based on the theory of complex adaptive systems and the phenomenon of organization emergence that sees the world as a living system, and will be discussed further in section 4 and 5.

Experience with a B-B e-marketplace case directed to the pharmaceutical industry in Egypt will be elaborated where self-organizing activities in modeling upfront requirements for the portal have been evolved (see section 6). The outcome is the trigger of business-value ideas as a response to specific problem situations. Hence self-organization builds in the ability to respond and adapt freely according to the requirements of the specific situation. The paper concludes with the necessity to instil self-organization within development teams instead of leaving it to evolve by chance, as suggested through the introduction of the concept of the 'Practice'. This is defined as "a discourse during which software development activities and responses evolve " (see section 7).

2. LIMITATIONS OF TRADITIONAL METHODOLOGIES FOR INNOVATION AND ADAPTABILITY

According to Baskerville et al. (1992) the traditional thinking of methodical information systems development is unsuitable for the new software development environment, they argue that new businesses are unpredictable and non-linear whereas traditional development methods are systematic, ordered, regular and regimented and therefore are unable to respond to frequently changing and shifting environments.

De Macro and Lister (1987) state that as traditional methods are rigorous and lessen the use of imagination and creativity. They argue that strict use of methods lead people to concentrate on procedures and documentation rather than the actual product or service. Avison and Fitzgerald (2003) refer to this as 'strict adherence' to the methodology rules that they describe as slavish and inhibiting of creative thinking together with an inability to respond to change.

Traditional methodical thinking is typically oriented towards large-scale, long-term, software projects that have long-term objectives (Baskerville et al., 1992) but in a world involving frequent, radical changes, having processes as repeatable and optimized is like a 'Dinosaur' (Highsmith, 2000) and will lead to software failure. Therefore Highsmith (2000) suggests in order to succeed in the post-modern globalization of commerce and industry rapid change-tolerant development processes aiming at short-term needs rather than long-term investment for the future is critical to enable organizations to adapt quickly to their environments.

Development methodologies according to Avison and Fitzgerald (2003) can also be inflexible, not allowing changes to requirements during development, as well as 'one-dimensional'. Further they argue that a universal development approach may not address all particular organization situations. Recently new technologies (end user computing, fourth generation languages and local area networks) have also brought incremental, fragmented, non-linear development (Baskerville et al., 1992). These new technical innovations will lead to the development of fragmented systems elements in unique ways that contradict the assumptions of traditional methodical thinking.

Therefore new practices and processes for information systems development are required to produce fast changing innovative information systems able to adapt to an ever-changing environment.

3.   REQUIREMENTS OF INNOVATIVE AND ADAPTIVE DEVELOPMENT PROCESSES

According to Riehle (2000) the traditional practice of software development regards the development process as something that has to be planned and controlled in order to reliably achieve the desired result. However, Highsmith (1997) argues that as industries move from predictable, stable environments to unpredictable, nonlinear and fast-changing environments, traditional information systems development practices based on determinism, predictability and stability will be inadequate to foster innovative development. Therefore according to Highsmith (1998) the basic tenants of process improvement, software engineering, and command and control management, are unsuitable.

Traditional rigorous methods according to Baskerville et al. (1992) are based on rational scientific reasoning that assume empirically testable propositions that in turn give a 'truth value'. Highsmith and Cockburn (2002) indicate that traditional methodologies provide what they call 'all-inclusive rules', i.e. strict steps or procedures to be applied under all situations, depending on voluminous written rules that will apply for every situation. Instead software developers need to depend on their creativity to respond to diverse circumstances.

Highsmith (1997) believes that RAD practices, that involve evolutionary development, customer focus groups, JAD sessions, technical reviews, time-boxed project management and continuous software development, are a step in the right direction to achieve innovation and adaptability. This is because they establish innovative collaborative environments and ensure the fast movement and feedback necessary to adapt to change. Riehle (2000) explains that evolutionary prototyping views software development as a shared learning experience based on requirements negotiation and timely feedback from customers, which foster creativity and innovation, as well as accommodating changes in markets.

But according to Truex et al. (1999) development frameworks aiming at rapid and flexible development such as prototyping, end-user development, open systems connectivity and contingent approaches are still insufficient for software emergence, as they still aim to achieve a stable product. They explain that a low-maintenance, stable IT system will battle against the ever-changing environment instead of adapting to it, and that this will inhibit rather than facilitate organizational emergence. Assuming stability will freeze the organizational change instead of achieving continuous change and fluidity (Truex et al, 1999).

Truex et al. (2000) argue that although different development methods apply different techniques, sequences, steps and activities, but despite this seeming diversity they have common assumptions and idealized characteristics. The definition they use for 'method' is the one given by Oxford and Webster's dictionaries; the primary definition given is 'a method is the procedure of obtaining an object', and the secondary definitions describe the method as 'orderly', 'systematic', 'regularity' and 'regimen'.

They suggest instead a shift in the development mode from structured, orderly and methodical to unstructured, emergent and amethodical to fulfill the requirements set by the new development environments. By amethodical they don't mean a descent into chaos and anarchy, but emancipation from structure and regularity. Amethodical systems thinking implies orchestration of the development activities but not a predefined sequence, rationality or universality. Thus, amethodical views of information systems development supports conflict over consensus and forces the developers to attend to the different voices and interests and not seek simply a compromise solution.

The amethodical view also appreciates innovation and 'organizational shake-ups' that lead to adaptation, experimentation and in turn to accidents and opportunism. Truex et al. (2000) outline the differences between methodical and amethodical information systems development, by characterizing the methodical view as controlled, linear, universal and rational, and the amethodical view as random, non-linear, unique (occurs in completely unique and idiographic forms) and capricious.

This paper agrees with these arguments and adopts new assumptions based on social construction and autopoiesis. In the following section we examine the phenomenon of organization emergence which provides the theoretical underpinning of how development processes emerge as response to specific problem situations.

4.   ORGANIZATION EMERGENCE

Organization emergence according to Baskerville et al. (1992) is characterised as the organization being in continual change, following no predefined pattern and never reaching a steady state. The organization may in fact exhibit temporal regularities, but it remains constantly in transition without ever becoming fully formed. Hence the theory of emergence rejects the notion of structures, regularities and stability. In the following we examine the phenomenon of emergence and its properties.

4.1   The Theory Behind Emergence

According to Riehle (2000) high speed and high change of the market make businesses unpredictable and hence unplannable in the traditional sense of control and optimization. Highsmith (1997) argues that high speed and frequent change induces complexity that has more to do with the number of interacting agents, and the speed with which those agents interact, rather than with size and technological complexity. Today's organizations involve many independent agents; customers, vendors, competitors, stockholders interacting with each other, at a speed that implies that linear cause-and-effect rules are no longer sufficient for success.

Highsmith (1998) stresses the need for a new way of thinking about how complex software products come into being. He believes that by studying life itself we can adopt mechanisms for self-production and evolution. He compares information systems to a self-contained, enclosed, living ecosystem. Kelly (1994) explains how in the dawn of the new era systems become so complex and autonomous as to be indistinguishable from living organisms. He provides a reasoning of evolving complex systems instead of engineering them by looking to them as self-sustaining living systems.

Complex Adaptive Systems (CAS) theory (Highsmith, 1997) provides three main concepts to explain the world: agents, environments, and emergent outcomes. Agents compete and cooperate to get work done, but the final result is not the outcome of the work of any particular agent or process. Effectively, the result emerges from the overall competition and cooperation of the agents, therefore system behavior cannot be predicted because simple cause-and-effect reasoning has broken down. Therefore Highsmith (1997) argues that the absence of group behavior rules within organizations is a good phenomenon as there are no more algorithms or procedures that define the results expected from the group, hence the group behavior emerges, rather than being predefined.

4.2   Emergence is not just a notion of Change

Change is a transition over time from one state to another, but emergence is about evolution, which means emergent organizations are always in the process of "remaking" themselves, never reaching a steady state. Therefore adaptation in the 'Internet-time' environment is not just another word for change management or following a plan (Highsmith, 1998).

4.3   Emergence is not just Contingency

Contingency approaches to information systems development are an attempt to introduce flexibility into the systems development process. However, the contingency approach still provides a structure or a framework that guides the developers in choosing the appropriate techniques and tools according to the specific situation. In emergent environments selecting a path from an almost unlimited number of alternate paths will freeze the organization's assumptions about itself and its environment, which is argued to contradict and inhibit emergent behavior. The produced plan will become out of date very quickly and, rather than focusing on the outdated plan, it is important to deal with the changing realities directly (Highsmith and Cockburn, 2002).

4.4 Adaptability through Spontaneous Order not Predictability

Predictability and plan-driven strategies result in rigidity and are thus less able to respond to change, and according to Highsmith (1998) process improvement or change control management activities, based on predicted outcomes, negating our product development efforts. Schwaber (2002) makes clear that team structures and requirements emerge during the course of the project rather than being determined at its outset, therefore it is critical for organizations to adapt quickly to their environments rather than tending to create a fit between the information system and the existing structure. Baskerville et al. (1992) talk about 'seizing the opportunity' to develop the necessary organizational flexibility to cope with the situation and its possible opportunities based on small-scale and short-term planning. But of course the team is guided by preliminary and sketchy visions of requirements and architecture, not planning for long-term outcomes.

Having examined the notion of organization emergence, or as Highsmith (1998) terms it, the 'spontaneous order', we now look at how this might be achieved. In the following section we explore ways to operationalize these concepts to become more creative and adaptive to change.

5. SELF-ORGANIZATION FOR INNOVATIVE AND UNPREDICTABLE ENVIRONMENTS

As mentioned above Highsmith (1997) finds that a growing number of software projects are so complex that the development processes are inherently unplannable, yet successful products can emerge from such environments. He argues that creative group behavior emerges unpredictably from spontaneous self-organization. Vidgen et al. (2002) also note that the success of OSS (Open Source Software) is superficially surprising, in fact they suggest that OSS is a triumph of self-organization, rather than hierarchical regulation. Kelly (1994) also points to self-organization as the essence of innovation and the way to cope with business turbulence. Riehle

(2000) concurs and argues that the underlying assumption of equating developers performing the development process to computers executing software is not appropriate, instead it is suggested that self-organizing teams lead to an environment in which innovation and adaptation thrive so that 'local order' can emerge.

Highsmith (1998) defines self-organization as a property similar to a collective "aha," that moment of creative energy when the solution to some nagging problem emerges. Such self-organization arises when independent individuals cooperate to respond creatively to, and reflect on, a specific problem situation. Truex et al. (1999) argue that self-organization is not deterministic, rather a product of a constant social negotiation, continual change of work culture and decision processes where outcome stages arise from previous history and context. This they refer to as the 'dialectics of organizational autopoiesis'. Hence self-organization refers to a theory of social underpinning derived from the unstable environment in which the information system will be developed. The reasoning of autopoietic, or self-referential social systems, lies in social organizations that are continuously self-making via discourse, that will never reach a steady state (Baskerville et al., 1992).

Riehle (2000) indicates that this distinction implies that different development processes are required. The differences focus upon; how a development methodology views customers, how it fosters creativity and innovation, how it tolerates changes in the market and in requirements. For example RAD practices are characterized, according to Fitzgerald (1998), by active user involvement and small-empowered teams, which in turn supports the reasoning of self-organization. In fact the focus on JAD sessions, incremental prototyping and focus groups, engage the development team members and the stakeholders to collaborate and jointly innovate the product solution. But still RAD follows a rigid agenda and uses a traditional top-down approach, often supported by an integrated CASE tool, which reduces creativity and the ability to respond freely to change. On the other hand OSS (Open Source Software) developers have been identified as self-selected, highly motivated and self-deprecating, which has important implications for stimulating co-operative development in order to realize complex and mission-critical software such as operating systems, email servers etc. with less people, within time, within budget and with high quality (Feller and Fitzgerald, 2000) (Vidgen et al., 2002).

Riehle (2000) stresses that self-organization cannot be commanded but must be nurtured. Kelly (1994) points to the need to instill in organizations guidelines and self-governance and relinquish what he calls 'our total control'. In the following a case study experience is explored, a business-business (B-B) e-marketplace, where self-organizing activities in modeling upfront requirements for the portal have been evolved as a way to respond to the turbulent and unpredictable settings that required high degree of innovation and responsiveness. The findings of the case study pose the question of how to provide mechanisms and dynamics to instil self-organization instead of leaving it to evolve by chance. This research is still ongoing and those questions need to be addressed in our further studies.

6.   EVOLVING SELF-ORGANIZING ACTIVITIES IN REQUIREMENTS MODELING: A CASE STUDY EXPERIENCE

6.1   The Case Situation and Research Methodology

A study of a B-B (business to business) e-marketplace application directed to the pharmaceutical industry in Egypt has been carried out. The primary objectives were to explore practical issues in adopting development methodologies within real-case complex and turbulent settings, identify problems and if required suggest new approaches. A portal application had already been developed before the fieldwork was carried out. The fieldwork of this study was initially part of an evaluation of the project, as it was decided that the application needed to be changed/adjusted after its launch to improve its competitive value. The original analysis and design of the e-marketplace identified the functional building blocks of the pharmaceutical trading supply chain.

According to Baskerville and Wood-Harper (1996) the purpose of 'action research' is to observe and create effective organizational change and they consider it ideal for studying new or changed systems development methodologies. The complex, multivariate settings of the e-marketplace made problem diagnosis efforts as well as intervention difficult and not straightforward. Susman and Evered (1978) state that in terms of generating action and knowledge, action research employs a 'process view' of research and they provide a model of the process as a five phase cyclical process; comprising diagnosing, action planning, action taking, evaluating and specifying learning. These processes were adopted and the phases indicated below. Firstly, the research method elements of the case are highlighted and this is followed by the content and discussion elements.

During 4 months of case study work the researcher identified problems resulting from the developers having become overwhelmed with too much detail during the comprehensive traditional analysis of the e-business case. For example, the order lifecycle was decomposed into order preparation, order placement and order fulfillment, each in turn being extensively analyzed. As a response the researcher produced an overview of the major functional elements of the trading process to replace the detailed flowchart decompositions. For example, the following major functional elements were outlined; fixed price purchasing, special offers, return of expired products, billing and invoicing, online payment and delivery tracking. (Diagnosing Phase)

As a way to evaluate and re-define requirements for the portal we identified the importance of what we termed 'random upfront' requirements modeling and design where global strategic requirements and design features can be developed spontaneously. For us the term 'upfront' does not mean 'initial' as the application will undergo continuous, constant development as indicated by Truex et al. (1999) and therefore there is a need to go back again and again to review the global requirements. Of course there will be a need to carry out a comprehensive analysis and to get to the level of detail needed to build the application, but this comes after the proposed 'spontaneous upfront' analysis exercise. (Action Planning Phase)

As an aid to analysis a modelling tool designed for e-business projects, the 'e-Business Issues Roadmap' (Figure 1) was introduced. This provides a model of the problem situation and by re-visiting it stakeholders identified problems and issues that emerged over time and accordingly self-organized to the needs of the problem situation. Three user-based analysis sessions were undertaken to try out the modelling tool based on brainstorming sessions and as response to that design ideas and/or appropriate actions were triggered to address the specific problems faced. As the objectives were to model upfront strategic issues only a few such sessions were needed. (Action Taking Phase)

The source of data collected for the research was primarily interviews (unstructured at the beginning then semi-structured) conducted mainly with the business consultant and the IT manager, but also with the general manager, the business analyst, members of the IT and technical staff, the legal representative and the users (pharmacists, pharmaceutical companies and distributors). A total of fifteen interviews were conducted. The other sources of data included meetings (a total of seven), observational notes, documents and protocols of the application.

Afterwards two additional meetings were held with the general manager and the IT manager to discuss ways to operationalize some of the design ideas and actions produced from the proposed upfront exercise. Their response was very positive about the business-value of the ideas triggered. (Evaluation Phase) This research work is ongoing and further work is required to improve the reasoning of self-organization within development processes and analysis of benefits, insights and drawbacks associated with its use. (Specifying Learning Phase)

6.2   The Process Characteristics and Outcomes

To identify and analyze innovative requirements a heuristic approach is used as it is more appropriate than a rigorous approach, we initially thought brainstorming sessions would be appropriate for this purpose. Joint Requirements Planning (JRP) as well as Joint Application Development (JAD) sessions (Martin, 1991) have been suggested as a means to set up the initial requirements for a project through brainstorming activities, but we feel they are too narrowly focused. They concentrate on producing detailed requirements and artifacts such as functional decomposition diagrams, dependency or data flow diagrams, etc. They are also highly structured and involve formal meetings that have defined rules of behavior e.g. the use of well-defined agendas, keeping official meeting minutes, the involvement of a qualified facilitator, etc. These we felt unlikely to engender the necessary creativity and found agile modeling sessions to be more appropriate.

Ambler (2001b) suggests that agile modeling sessions be highly iterative, with the agile modeler iterating back and forth; as it is more common to identify a requirement, analyze it, and propose a potential design strategy within minutes, hence the need to iterate quickly between phases. Also the formulation of requirements is likely to happen by asking questions, therefore Ambler (2001a) suggests the use of user stories (Beck, 2000) that are essentially very high-level wish list of stakeholders. Ambler (2001a) characterizes user stories as a reminder to conduct a conversation with project stakeholders and capture high-level requirements, including business rules, constraints, and technical requirements.

However, for internet-based applications we believe that the complexity of such applications, together with the variety of multi-disciplinary issues that interplay, make the use of user stories too vague and loose. Therefore the 'e-Business Issues Roadmap' (Figure 1), mentioned above, was derived (Alaa and Stockman, 2001). This provides a template that categorizes the different underlying issues and concerns that typical internet-based projects face. The classification proposes issues by sector, e.g. business, technological, legal, political, economic and social, and the level they work at, for example, intra-organizational, customers and partners and environmental. Due to the complexity and the large variety of potential issues, the proposed roadmap acts as a 'balance' between structure and freedom by providing an initial set of topics/issues that need to be thought about. It is used to drive and guide the triggering of issues under the various identified categories but still leaves space to the modeler to be creative. The

Roadmap provides a starting point without which some of the many relevant issues might be missed and too much time wasted but it is not restrictive as other issues are readily triggered by discussion and added.

This resulted in the running of the three modeling sessions. Only three were undertaken because a comprehensive analysis was not required, just an up-front set was sought. This also helped to reduce complexity and stimulate creative thinking. As the underlying issues are multi-disciplinary, the participants of the modeling sessions included representatives of the different kinds of members (stakeholders) involved with operationalizing the e-marketplace portal.

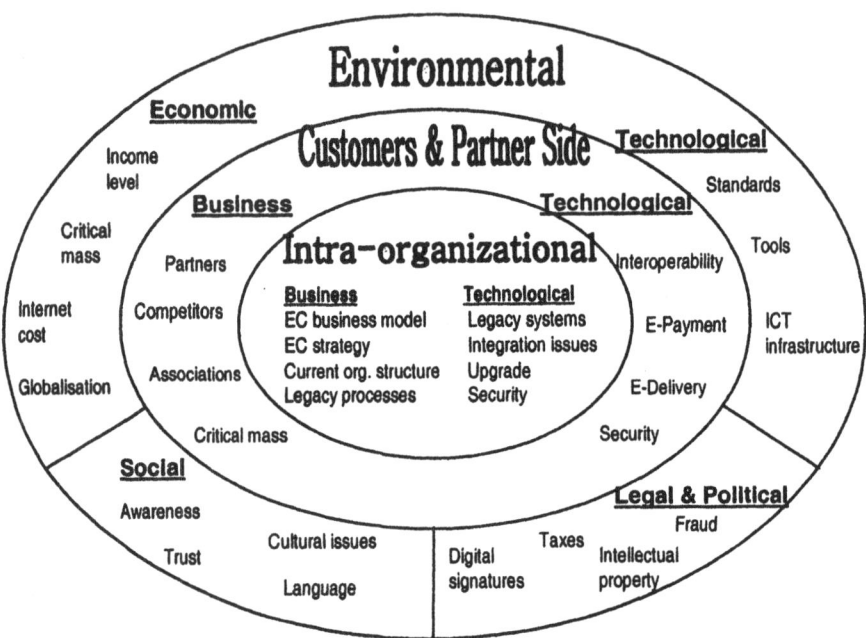

Figure 1. e-Business Issues Roadmap

The first brainstorming session was attended by the business consultant, the business analyst and the researcher where issues were triggered under the different identified categories in the roadmap and consequently the implications of the issue and possible solutions and drawbacks were discussed. One of the issues triggered in this session was that the organizational structure needed to suit the e-marketplace architecture and not vice versa. Another was that a sales pipeline was identified as necessary to fulfill the sales process, as some of the major activities were still carried out offline. Some of the solutions triggered in this session were the 'awareness campaigns', the 'installment program' and the different value-added services, such as the customization of product dispatch as part of the order placement cycle (Figure 2).

There was a need to hold other sessions with representatives of the technical staff as well as senior management to take the different issues perspectives into consideration. So another brainstorming session was held and some of the issues that came out in this session were that the e-trading process is invisible and ambiguous over the traditional one, which it was felt might make the buyers reluctant to use the e-marketplace application. This resulted in a focus on the importance of the confirmation of the order before it is placed. Another idea was to provide virus-shield software as a way to further increase customer trust and satisfaction. These and other ideas are shown in Figure 2.

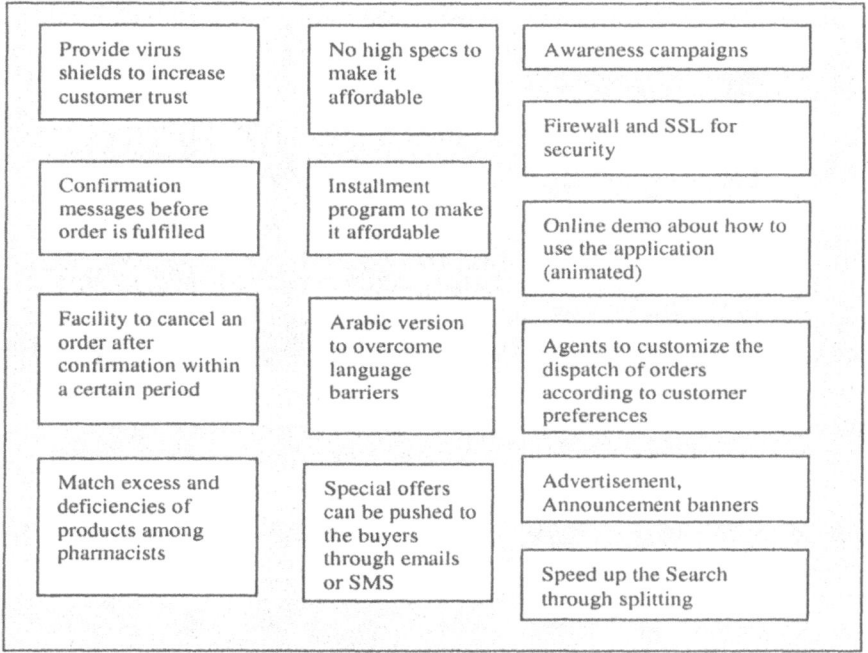

Figure 2. Design Ideas and Actions Triggered

Another session was arranged later with the general manager, the business consultant and the researcher where the attendees started to reflect on and evaluate (in an informal way) these ideas/solutions (Figure 2). The ideas they really liked were the focus on value-added services to further publicize the site and providing virus shield software. Using agent-based software was postponed for the next release, as the company wanted to make some revenue before undertaking further investment. Other ideas were hard to implement, as the current system architecture would not allow this easily, such as the customization of the order preparation function, which would necessitate the redesign of the order module.

This is an example of how the proposed 'spontaneous upfront modeling' has been applied. The resulting design ideas and actions (Figure 2) show a number of new and innovative features for the e-business application, e.g. SMS messaging of special offers, the introduction of an installment program, etc. that were not identified before and did not emerge from the traditional functional decompositions and use cases. The 'e-Business Issues Roadmap' provides a sort of a global picture of the whole situation, and by visiting the proposed exercise in a periodical basis will keep the 'picture' updated. This helps to respond to unpredictable problems and issues that emerge over time, in this way project stakeholders self-organize according to the implications of the problem situation. A more detailed description and derivation of the approach are provided in Alaa and Fitzgerald (2004).

7.   TOWARDS 'THE PRACTICE': THE DEVELOPMENT OF SELF-ORGANIZING ACTIVITIES

The problems encountered in the case by the initial use of rigorous analysis techniques focused attention on the limitations of these traditional techniques. Their systematic, prescriptive and plan-driven nature appeared to reduce creative thinking and the ability to respond to change. The authors proposed approach, which is argued to enhance creativity and spontaneity, is quite different. This leads us not to use the term 'technique' for it, as it relies on people to drive it according to the opportunities and events that arise, rather than following, or being driven by it. We thus call it a 'Practice'. The dictionary definition of the word 'practice' relates to exercise, training, run through, habit or ritual. Applying these meanings to requirements for software development, we identify the proposed 'Practice' as a 'discourse during which software development activities and responses evolve'.

Unlike traditional requirements and software development techniques, the notion of a 'Practice' will support developers in a self-organizing way to create innovative development environments driven by opportunity and events. Therefore the 'Practice' is neither prescriptive, nor anchored in rationality, reductionism, empirical or scientific underpinning. However, some initial rules or guidelines are required to frame the activities, such as the 'e-Business Issues Roadmap' (Figure 1) which sets an initial boundary but leaving space to the development team to react independently to the problem situation. Hence the 'Practice' depends on the individuals and their creativity to find ways to solve problems as they arise, rather than providing them with inclusive rules, i.e. all things they could possibly do under all situations.

The 'Practice' can be compared to playing a non-chance 'game' where the general game rules are known, but the game path evolves in use according to the situation and the responses. To win a game you need to understand the game rules and have the appropriate skills, then through 'practicing' the game you will become more skilful and more likely to win. Whereas the 'technique' can be compared to 'cooking instructions or a recipe', that has to be followed strictly in order to get the meal done. Any flaw in the instructions or the ingredients will spoil the meal and its taste.

However, the concept of the 'Practice' is not introduced to replace the 'Technique' nor to find a kind of a compromise, as we believe that they are complementary. The 'Practice' will address habits of the development team and dynamics such as agile modeling sessions and continuously modeling upfront requirements to help stakeholders respond creatively in a rapid, emergent way. Whereas later the 'Techniques' will be needed to carry out a rigorous detailed analysis and design to get to the level of detail needed to build the application.

The concept of the 'Practice' builds also on the requirements of innovative approaches suggested in the literature (section 3). It supports the amethodical view suggested by Truex et al. (2000). This requires development processes to be unstructured, opportunistic and driven by accident. It also builds on the AM (Agile Modeling) methodology suggested by Ambler (2001 a), in fact it fills in some of the gaps of AM that provides only a set of guidelines such as the importance of active stakeholder participation, applying the right artifacts, creating several models in parallel and others (Ambler, 2001 c) without providing a mechanism to operationalize them.

We conclude that the 'Practice' is an exercise carried out by the development team. Unlike the technique, the 'Practice' is a creative and communicative process. It doesn't follow a set of all-inclusive rules; it rather acts as a 'parable' that guides the participants based on a set of concepts, guidelines, action/interaction medium, previous experiences and best practices, and then the 'development game' evolves in use. This research is still ongoing and further studies will address elements and dynamics of the 'Practice' that will further instill and cultivate self-organization within organizations instead of leaving it to evolve by chance.

8.   CONCLUSIONS

A highly turbulent business environment and rapid technological developments require a high degree of innovation and responsiveness. The proposed 'spontaneous upfront modeling practice' facilitates business innovation and emergence through informal interaction and global strategic reasoning, about the required solution. It helps by providing the guidance of a non-prescriptive, stimulating requirements modeling tool, the 'e-Business Issues Roadmap', to identify issues, resolve contradictions and overlaps within the e-business project settings and as response to that trigger possible solutions and design ideas according to what the specific situation implies. This is a creative and communicative process that may have to be visited a number of times in order to continuously adapt to ever-changing shifting environments. This is an example of how development teams can self-organize according to specific problem situation instead of following a change plan. Self-organization can be instilled and cultivated within organizations as suggested through 'practicing' or something like 'exercising' that will develop/train the skills needed to maneuver, improvise and innovate, instead of leaving it to evolve by chance.

REFERENCES

Alaa G., Fitzgerald G., 2004, A Proposal of an Upfront Requirements Modeling and Design Practice for e-Commerce Projects, Proceedings of the Thirty-Seventh Annual Hawaii International Conference on Systems Sciences, January 2004

Alaa, G. and T. Stockman, 2001. An Investigation into the Nature of E-commerce, Requirements and Needs in Developing Countries, BITWORLD 2001 Proceedings, Cairo, Egypt.

Ambler, S., 2001, Agile Requirements Modeling, The Official Agile Modeling (AM) Site, Copyright 2001-2002, http://www.agilemodeling.com/essays/agileRequirements.htm

Ambler, S., 2001, Rethinking Modeling Sessions, The Official Agile Modeling (AM) Site, Copyright 2001-2002, http://www.agilemodeling.com/essays/modelingSessions.htm

Ambler, S., 2001, The Practices of Agile Modeling, The Official Agile Modeling (AM) Site, Copyright 2001-2002, http://www.agilemodeling.com/practices.htm

Avison D., Fitzgerald G., 2003, Where Now for Development Methodologies?Communications of the ACM, January 2003, Vol. 46, No.1

Baskerville R., Travis J., Truex D., 1992, Systems Without Method: The Impact of New Technologies on Information Systems Development Projects, IFIP Transactions A-8 North-Holland 1992

Baskerville R., Wood-Harper T., 1996, A critical perspective on action research as a method for information systems research, Journal of Information Technology (1996) 11, 235-246

Beck, K., 2000, Extreme Programming Explained: Embrace Change, Addison Wesley

Feller J., Fitzgerald B., 2000, A Framework Analysis of the Open Source Development Paradigm, in W. Orlikowski, P. Weill, S. Ang and H. Krcmar (eds) Proceedings of 21st

Annual International Conference on Information Systems, Brisbane, Australia, December 2000

Fitzgerald B., 1998, A Preliminary Investigation of Rapid Application Development in Practice, Proceedings of 6th International Conference on Information Systems Methodologies, editors Wood-Harper AT, Jayarantna N., Wood J R G, pp. 77-87

Highsmith, J., 1997, Messy, Exciting, and Anxiety-Ridden: Adaptive Software Development, Cutter Information Corp. , http://www.jimhighsmith.com/articles/messy.htm

Highsmith, J., 1998, Order for Free, Life—the Artificial and the Real, Software Development, http://www.jimhighsmith.com/articles/order.html

Highsmith, J., 2000, Retiring Lifecycle Dinosaurs, Software Testing and Quality Engineering, July/August pp 22-28

Highsmith J. and Cockburn A., 2002, Agile Software Development: The Business of Innovation, IEEE Software, September/October 2002

Kelly K., 1994, Out of Control: The New Biology of Machines, Social Systems and the Economic World, London : Fourth Estate

Martin J., 1991, Rapid Application Development, Macmillan Publishing, New York

Riehle D., 2000, A Comparion of the Value Systems of Adaptive Software Development and Extreme Programming: How Methodologies May Learn from Each Other, Proceedings of the First International Conference on Extreme Programming and Flexible Processes in Software Engineering (XP 2000). Page 35-50.

Schwaber K., 2002, The Impact of Agile Processes on Requirements Engineering, International Workshop on Time-Constrained Requirements Engineering 2002

Susman, G. I. and Evered, R. D. ,1978, An Assessment of the Scientific Merits of Action Research, Administrative Science Quarterly, Vol. 23, pp. 582-603.

Truex D., Baskerville R. and Klein H., 1999, Growing Systems in Emergent Organizations, Communications of the ACM, August 1999, Vol. 42 No.8

Truex D., Baskerville R., Travis J., 2000, Amethodical Systems Development: The Deferred Meaning of Systems Development Methods, Accounting Management and Information Technology 10: 53-79

Turban, E., Lee, J., King, D. and Chung H. 2000. Electronic Commerce, a managerial perspective, New Jersey, Prentice Hall.

Vidgen R., D. Avison, B. Wood, T. Wood Harper, 2002, Developing Web Information Systems, Butterworth-Heinemann Information Systems Series, Elsevier Science

ENRICHING VIEWS OF INFORMATION SYSTEMS WITHIN ORGANIZATIONS: A FIELD THEORY

Shaila Miranda and Robert Zmud
Division of MIS, Michael F. Price College of Business, University of Oklahoma, USA

Abstract   Three perspectives have dominated research on IT adoption, implementation, use, and impacts: technological determinism, organizational imperatives, and the emergent perspective. While the last is most realistic in its assumptions, it falls short in two respects. First, structuration theory, which has served as the underpinning for most research based on this perspective, while acknowledging the potentially constraining effects of structure, presumes that all agency is equally unconstrained within a set of structures. Given this presumption, a second shortfall of the emergent perspective is that it fails to shed light on the systematic manner in which agency is constrained.

In this paper, we first consider the advantages of field theory in augmenting our current understanding of IT in organizations. We then identify three specific organizational fields that may be useful in modeling the disparate constraints on entities' agency vis-à-vis IT. The first is the production field, which encompasses organizations' core or primary business processes. The second is the coordination field, comprised of organizations' activities at the interface of various core activities as well as supportive activities and internal relationship management. The third is the field of co-opetition, i.e., the arena of organizations' management of its relationships with external entities.

Keywords: IT Theory, Field Theory

1.   INTRODUCTION

The last 40 years have seen dramatic changes in the manner and extent to which information systems (IS) are applied within organizations, and in the focus and the depth of scholarly research examining management, design, and use of IS within organizations. In practice, IS have moved from technologies used for the sporadic automation of discrete tasks, with minimal influence on organizational performance, to technologies that are embedded deeply within and broadly across organizations. In scholarship, IS research has moved from a science framed through relatively simple, static models and a positivistic epistemology to a science that applies increasingly complex, dynamic models and epistemological diversity. Success in domains of both practice and research is dependent on the availability of models that capture the richness of IS phenomena, their application within organizations (also organizational constellations), and their evolutionary paths within organizations.

The objective of this paper is to develop a field theory of IS that portrays and enables analyses of capabilities, values, agency, and evolution of IS within organizations. This should facilitate more robust explanations of both organizational stability and transformation, as well as the differential agency of actors within the organizational fields in which IS are embedded.

2.   DEFINITIONS OF TECHNOLOGY AND OF IS

A key challenge involves achieving clarity and consensus around the meaning of the central object of study, in our case of IS. However, the concept of technology, alone, is elusive:
-   Technology "is oriented …to the problem, given the end, of choosing the appropriate means" (Weber, 1978, p. 67).
-   It is a "collection of plant, machines, tools and recipes … for the execution of the production task and the rationale underlying their utilization" (Woodward, 1994/1965, p. 4).
-   It is "systematic knowledge transformed into, or made manifest by, tools" (Tornatzky & Fleischer, 1990, p.9).
-   Technology is the product of one-dimensional (non-reflexive) understanding. It "becomes mastered only when from the first and without reservation 'yes' is said unconditionally to it. This means that the practical mastering of technology in its unconditional unfolding already presupposes the metaphysical submission to technology" (Heidegger, 1941, as cited in Zimmerman, 1990, p. 89).

A common theme across the first three definitions is that technologies are *tools* built from knowledge held by organizational actors to accomplish desired ends. IS are distinguished from other types of technology in that the nature of their work involves transmission, processing, and framing of data, information, and knowledge. As with other technologies, IS are not neutral in these processes, but rather embody capabilities, i.e., actionable knowledge, and values as do human agents. Such capabilities and values are activated and manifested only in the *relationships* between IS agents and human agents. Thus, IS assume a reality only in their relationships with other agents within organizational fields. In this perspective of IS, we are consistent with Heidegger's view of technology – in his view the problem of constrained consciousness derives not from modern technologies themselves, but rather from the manner in which we relate to them (Zimmerman, 1990). Our perspective differs from prevailing perspectives not in our definition of IS, but in our view of the agency of IS and of that of human actors in relation to IS.

In the MIS literature, IS are defined in terms of their 'structural features', i.e., "the specific rules and resources, or capabilities, offered by the system" and their spirit, i.e., the "normative frame with regard to behaviors that are appropriate in the context of the technology" (DeSanctis and Poole, 1994, p. 126). An IS, then, represents a functionality bundle that reflects the intentions of its builders, yet is appropriable toward users' intentions, even when objectives of builder and user collide. An IS thus is "both engine and barrier for change; both customizable and rigid; both inside and outside organizational practices. It is product and process." (Star and Ruhleder, 1996, p. 111).

More recent structurational perspectives take exception with such a view, suggesting that IS structures emerge via practice, rather than being embedded a priori within IS or within organizations (Orlikowski, 2000). Such an amendment, while consistent with Heidegger's focus on our relationships with technology as being its defining characteristic, falls short of locating rules and resources within the *relationships* among technological and human agents. Furthermore, it fails to explicitly recognize that some organizational actors have more freedom in their relationships with technology than do others (e.g., Lamb and King, 2003). This is a key difference between structuration and a theoretical perspective representative of field theory (as developed here), which posits that constraints that operate upon agents and their freedom to enact structures is a function of their position within a given field.

In summary, we argue that existing conceptualizations of organizational IS evince three weaknesses. First, they fail to account for interdependencies across organizational IS. Second, they fail to adequately account for the evolution of an IS. Third, they fail to account for the complex sets of relationships that exist among a focal IS, other implemented IS, members of the organization, and the organization itself. In their focus on the IS 'artifact', existing conceptualizations tend to privilege the technical over the social, even though the boundary between "technical" (i.e., social artifacts) and "social" (i.e., an organization and its human actors) is a control-oriented social-construction (Bloomfield and Vurdubakis, 1994). These limitations can be overcome through developing the notion of organizational fields in conceptualizing the IS role within organizations.

3.   TOWARD A THEORY OF IS WITHIN ORGANIZATIONAL FIELDS

Field theory represents a median ground between structuration theory's arena of unconstrained agency and critical theory's arena of completely constrained agency, focusing instead on constraints that arise as a function of agents' positions within a given field. In developing this field theoretic perspective on organizational information systems, we rely primarily on theoretical articulations of fields by Bourdieu, derived from sociological studies of fields such as academics, sports, and aesthetics (e.g., Bourdieu, 1984) and augmented by subsequent research on fields such as gastronomy (Ferguson, 1998). We temper this sociological perspective on field theory with the notions of the field developed by Lewin in his psychological work based on individuals and groups (e.g., de Rivera, 1976) and that of DiMaggio and Powell (1983) on organizational fields. Field theory accounts for changes without causal or triggering agents: instead, location within the field evokes change in an element, which has "particular attributes that make them susceptible to field effects" (Martin, 2003, p. 4). In other words, the impetus for change and constrictions on agency are a function of an agent's position in a field. Field theories are both similar to and different from mechanistic theories, which utilize lower-level, intermediate constructs to explain a phenomenon, and functionalist theories, which rely on explanations of instrumentality. Like mechanistic theories, but unlike functionalist theories, they focus on local action; like functionalist theories but unlike mechanistic theories, they look for global patterns that explain local action (Martin, 2003).

A field is a "configuration of objective relations between positions. These positions are objectively defined, in their existence and in the determinations they impose upon their occupants, agents or institutions, by their present and potential situation in the structure" (Bourdieu and Wacquant, 1992, p. 97). Here, the 'structure' being referenced is based on and reproduces distributions of power and capital, creating relations of domination and subordination as well as homology (equivalence). These distributions of power and capital relate to the gradients of field forces that differentially impinge on the actants within the field. The structure of a field is "self-regulating, self-validating, and self-perpetuating" (Ferguson, 1998, p. 598). A field translates "external economic or political phenomena into its own terms for its own use or, rather, for the use of its occupants" (Ferguson, 1998, p. 597). In other words, the forces within an organizational field are oriented toward the perpetuation of the field, not toward the survival or prosperity of the individual agents that constitute it. Field theory highlights the fact that actants' behaviors are not simply triggered by environmental stimuli, but rather that the effects of stimuli on actants' responses is mitigated by field forces such as habit, habitus, and group behavior (Bourdieu, 1984; Lewin, 1951).

While in Bourdieu's view sociological field forces are an objective reality, in Lewin's (1951) perspective, psychological forces derive from agents' subjective experiences, rather than from objective, physical phenomena. These perspectives on field forces are not contradictory. Rather, psycho-social field forces are the objectification of actants' collective subjective experiences over time, culminating from their individual enactments of those experiences. These enactments constitute actants' "life spaces", i.e., socio-psychological fields that are products of actants' states, history, and context (Lewin, 1951).

Fields are constituted by distinct sets of values or forces and capabilities. The values within sociological fields are the forces that define the field. An important facet of the field is the capital available to each agent, as capital makes games not a matter of mere chance but rather of historically-accumulated capability (Bourdieu, 1983). Value forces are inextricably linked to the distribution of capital within the field, differentially subjugating actants based on their accumulated capital. Recognition of capital is therefore imperative if we are not to view organizational fields as locations of "instantaneous mechanical equilibria between agents who are treated as interchangeable particles" (Bourdieu, 1983, p. 241).

A distinct form of capital underlies every field (Bourdieu, 1983). A sufficient amount of that capital is a condition of membership in the field (Bourdieu and Wacquant, 1992). Capital acquires meaning within a field only by virtue of the values ascribed to it by actants. For example, artistic ability holds little value in fields of economic endeavor, and would therefore convey little position power (Bourdieu, 1983). Position in a field, or relative power, is then a function of actors' accumulated capital relative to that of others and the values subscribed to within the field.

Capital and values represent dualities that are constitutive of fields. Within the context of a defined set of values, capital "confers a power over the field, over the materialized or embodied instruments of production or reproduction whose distribution constitutes the very structure of the field, and over the regularities and the rules which define the ordinary functioning of the field, and thereby over the profits engendered in it" (Bourdieu and Wacquant, 1992, 101). In other words, competencies and prescriptive rules provide the basis for agreement and the resolution of disputes within different fields (Boltanski and Thevenot, 1987). In the context of organizational fields, the capital that human actors have is their portfolio of capabilities, i.e., actionable knowledge (Churchman, 1971). "To be capable of some thing is to have a generally reliable capacity to bring that thing about as a result of intended action" (Dosi, Nelson, and Winter, 2002, p. 2). Thus, a defining characteristic of an organizational field is its *capability requirements*.

The connotation of "power" in field theory is not about agents' ability to directly constrain each other's actions; rather, it "refers to a 'possibility of inducing forces' of a certain magnitude on another" (Lewin, 1951, p. 40). Thus, agents' power is enabled by and exercised only indirectly through field forces. The distribution of capabilities in organizational fields coincides with the distribution of power. "There is a distinction between the execution of high-frequency, repetitive daily business by low-level employees and the decisions of executives about the development and deployment of capabilities" (Dosi et al., 2002). Position in organizational fields thus accrues from one's capability to create rather than to simply execute the creations of others. Note that 'capability' is used in the dual sense of competence as well as freedom to perform. While fields require and permit certain capabilities in their actants, the presence of certain actants may attenuate or heighten the salience of others' capabilities. When actants are deficient in capability-based capital, they may trade based on an alternate form of capital, e.g., time. Consider, for example, specialists deficient in the typing skills required by a new IT: "Specialists improvised ways of dealing with backlogged data entry... working after hours to get caught up" (Orlikowski, 1996, p. 73).

Underlying the notion of the field is recognition that "external factors – economic crises, technological change, political revolutions, or simply the demand of a given group – exercise an effect only through transformations in the structure of the field where these factors obtain. The field *refracts*" (Bourdieu, 1988, p. 544). The asymmetric perception of a threat or an opportunity, based on actants' positions within the field, creates disequilibrium. Differential perception of threat or opportunity within a field yields disparate efforts by actants to develop capabilities to address the threat or leverage the opportunity. Differences in actants' capabilities may subsequently transform the field as new capabilities and/or value systems become salient. Alternatively, actants may collectively ignore external factors, thereby inducing stagnation.

Thus, a field is "also a field of struggles (Bourdieu and Wacquant, 1992, p. 101). 'Struggle' implies both the production of continuity and the production of change. Underlying struggles may be human intentionality:

"Acts of intention occur when the subject's needs lead him to foresee a situation in which he desires to act in a certain way, but knows that the situation will not automatically bring about the desired action. When successful, the subject's act of intention functions to transform the future psychological field" (de Rivera, 1976, p. 46). However, the idea of struggles in the field is not inconsistent with Orlikowski's (1996) notion of organizational transformation being wrought through "ongoing improvisation enacted by organizational actors trying to make sense of and act coherently in the world". Rather, we view such sensemaking as inherently a struggle, entailing dissonance and conflict that culminates in revised action and cognition (Weick, 1995). Nor does the term 'struggle' always entail strategy: "There is a production *of* difference which is in no way the product of a *search for* difference" (Bourdieu and Wacquant, 1992, p. 100, italics in original). Yet, these struggles too often yield transformations of the field (Bourdieu and Wacquant, 1992).

Strategic actions at play within a field are neither unconstrained nor equally constrained for all actants (Bourdieu and Wacquant, 1992). The ability to withstand pressures or susceptibility to those pressures is a direct function of human agents' position in the field. The boundary of the field is established by the marginalization of field forces. The field terminates at a point where its forces cease to constrain agency. While bounded, fields do not operate autonomously. Rather, interdependencies exist among fields, e.g., within literary and political fields (Bourdieu, 1988). These interdependencies develop through actants' simultaneous participation in multiple fields and are necessary for alignment (Garvin, 1995). However, the consequent referencing of multiple logics of action yields problems in resolving disagreements (Boltanski and Thevenot, 1987). Field interdependencies also play out in the "structural subordination" of one field to another (Bourdieu, 1988, p. 546). Such subordination operates differently on actants, based on their respective positions in each of the fields. Positions are not simply inherited, but are also the product of actants' strategic efforts at differentiation and subsequent field transformations. Such differentiation enables actants to configure unique niches within the field so as to minimize their competition with other actants and, perhaps, even provide them with a monopoly regarding their niche. (Bourdieu and Wacquant, 1992, Bourdieu, 1988).

Field theory thus allows for simultaneous accounts of structure and history, reproduction and transformation, and statics and dynamics (Bourdieu and Wacquant, 1992, p. 90). In fact, field theory requires that we integrate these analytical dichotomies: "We cannot grasp the dynamics of a field if not by a synchronic analysis of its structure and, simultaneously, we cannot grasp this structure without a historical, that is, genetic analysis of its constitution and of the tensions that exist between positions in it, as well as between this field and other fields" (Bourdieu and Wacquant, 1992, p. 91).

In sum, the salient contributions of field theory to our reconceptualization of organizational IS are the following. First, we view both technological artifacts and human actors as agents occupying different positions in a field. Fields constitute and are constituted by capital and value forces, and agents' positions are a function of their accumulated capital that is afforded prominence based on the values of the field. Position constrains agency differently making access to capital and deviations from established routines more difficult for some agents than for others. Through intentional agency and the intended and unintended consequences of that agency, agents' positions in fields shift. Whereas unintended consequences are treated as a matter of chance in structuration theory, they are viewed as a function of field dynamics in field theory.

What is the role of IS within such organizational fields? Consistent with actor network theory, we attribute agency to societal artifacts, specifically to IS (Law and Hassard, 1999). As agents within an organizational field, IS possess capital as do human agents. However, the capital possessed by IS appears in an objectified form in contrast to the embodied capital of human actors (Bourdieu, 1983). Embodied capital, i.e., "long-lasting dispositions of the mind and body" (Bourdieu, 1983, p. 243), or "habitus" (Bourdieu, 1988, p. 546) are a product of one's historical environment. Objectified capital – also known as structural capital – is owned by the organization, while embodied capital – the property of human actors – may simply be rented by the organization (Stewart, 1999).

As an actor with capital, an IS can be both instrumental in redistributing capability-based capital within a field, e.g., downgrading experts' capability-based capital and upgrading that of novices (Orlikowski, 1993). It may alter human actants' positions in a field by spotlighting the location of such capital, e.g., call center databases provide a "brag record" for specialists with "more calls in there than anybody else" (Orlikowski, 1996, p. 77). In substituting requirements of technical capabilities for domain capabilities, IT may erode domain capabilities in an organization, e.g., attrition of domain knowledge following implementation of a community knowledge system for biologists (Star and Ruhleder, 1996). Similarly, changes in capability requirements may engender resistance, and consequently, transformation of the field (Gill, 1996; Markus and Pfeffer, 1983).

4. FIELDS OF ORGANIZATIONAL PRACTICE

Fields are instantiated through practice or routines: "The field is a critical mediation between the practices of those who partake of it and the surrounding social and economic conditions" (Bourdieu and Wacquant, 1992, p. 105). Practices or routines are characterized by repetition, by recognizable patterns of action, by multiple participants, and by interdependent actions. They have ostensive and performative aspects, i.e., the idea versus the enactment, and there is a recursive relationship between ostensive and performative. The ostensive can guide enactments, serving "as a template for behavior or a normative goal" (Feldman and Pentland, 2003, p. 106). It provides a basis for accounting or justifying enactments, and supplies a label that is useful in communicating about what is being done. The performative maintains, but can also modify and create the ostensive (Feldman and Pentland, 2003). Modifications may or may not be legitimized and incorporated into the ostensive (e.g., the performance of virtual visits in the ostensible practice of recruiting). When they are, modifications to the cognitions underlying the field or new cognitions emerge: "When substantive actions are taken, words come to life and become meaningful" (Feldman and Pentland, 2003, p. 108).

Thus, in performance, routines get improvised, subsequently modifying underlying cognitions. In the reification of routines via IS, though, reflexivity is eliminated, thereby conflating idea and enactment and precluding improvisation. Hence, as an "embodiment of standards" of practice (Star and Ruhleder, 1996, p. 113), the influence of an IS within a field can be substantial. This is particularly the case with transformations of a field through "recurrent and reciprocal practices over time" (Orlikowski, 1996, p. 66). That transformation in the field can occur without change in practice is of essence to a field-based theory (Martin, 2003). While designed practices within a field may be viewed as constraints, they also allow for "empowering people and unleashing their creative juices" and can be "liberating" (Garvin, 1995, p. 83).

A field's effect on actants is exerted toward the perpetuation of organizational means-end values (Weber, 1978, pp. 118-137). Such values may pertain to (1) material means of production or (2) coordination of diverse activities (Weber, 1978). However, effective organizational functioning also depends on (3) defining, negotiating, and coordinating a plurality of interests among multiple stakeholders within and outside the organization. These three value domains correlate with three core organizational capabilities: capabilities that "accomplish the work of the organization", capabilities to mobilize people or behavioral processes that "infuse and shape the way work is conducted by influencing how individuals and groups behave", and capabilities to mobilize ideas or change processes that "alter the scale, character, and identity of the organization" (Garvin, 1998, p. 41). Based on these notions of organizational capabilities, we characterize the three principal organizational fields as fields of production, coordination, and co-opetition.

Within each of these three fields, IS serve as "steering mechanisms" that supply capabilities and reify value-control (Myers and Young, 1997). Within each field, a force gradient exists that is amplified or muted, by the presence of IS (Martin, 2003), differentially constraining or enabling actants within the field. These force fields are depicted in Figure 1 and are now discussed.

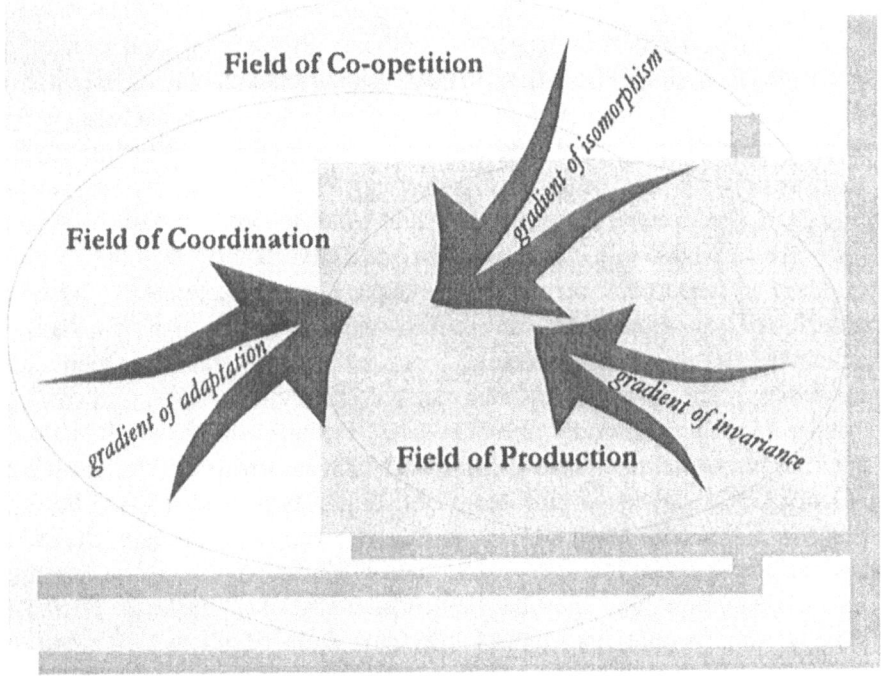

Figure 1. Organizational Force Fields Circumscribing Human and Technological Agency

4.1   Field of Production

The field of production is constituted by organizational units concerned with improving the organization's core competencies, i.e., the core production capabilities that are fundamental to organizations' competitiveness (Prahalad and Hamel, 1990). Capabilities in this field entail "technological and production expertise" (Stalk, Evans, and Shulman, 1992, p. 66), specifically in the primary or core processes along a firm's value chain. In addition to capabilities, as noted earlier, fields are defined by their underlying values. The value orientation of production fields is invariance and efficiency in the production of products and services. This is manifest in the focus of operations management and engineering disciplines on quality control and process planning and scheduling. The gradient of forces toward efficiency and invariance within this field weakens as one moves away from the firm's core value processes. This is visible in notations that experimentation via skunk works takes place at the peripheries of organizations and is precluded at the organization's core (Ciborra, 1993) as is visible with firms' experimentation with eBusiness applications (Willcocks and Plant, 2001). While core capabilities promote such experimental skunkworks at organizations' peripheries, they are not tainted by the peripheral experimentation (Meyer and Utterback, 1993).

Given the field's underlying values of invariance and efficiency, IS implemented within this field are typically oriented toward the automation of tasks within a process (Schein, 1992). Automation can eliminate jobs entirely and the efficiency and reliability of production IS are unquestionable (Gill, 1995a). In addition to themselves assuming invariant agency in production fields, IS agents also foster invariance in task performance by human agents (Gill, 1995b). In such situations, we note the power of IS agents to induce field forces upon human agents in the field.

The agency of production IS can also modify the capability requirements of human agents. Automation has been found to reduce manual and cognitive skill requirements in production tasks (Kelley, 1990; Orlikowski, 1993). However, strategic agency by human agents may enable them to circumvent or resist the forces induced by IS and their consequent capability modifications to the field. Strategic agency may even permit human agents to leverage the IS agency to increase their power, as is evinced in the conversion of blue-collar production jobs into white-collar programming jobs following automation, particularly in larger organizations with more specialized division of labor (Kelley, 1990). Automation has also been found to reduce the ranks of middle managers too, when their "structured and standardized roles facilitate top managers' efforts to substitute IT for middle managers"; in contrast, in organizations where they play unstructured roles, middle managers have been able to leverage IT to "enlarge their ranks" (Pinsonneault and Kraemer, 1993, p. 288). The distribution of power subsequently changes from a seniority basis and negotiated by collective bargaining to vestment with the bureaucracy (Kelley, 1990).

4.2   Field of Coordination

Coordination refers to "a pattern of *decision-making and communication* among a set of *actors* who perform *tasks* in order to achieve *goals*" (Malone, 1987, p. 1319). The value underlying this field is adaptiveness, which enables synchronized action despite organizational and environmental variations. The field of coordination is constituted by organizational units and is concerned with their "ability to integrate, build, and reconfigure internal   and   external   competences   to   address   rapidly   changing environments" (Teece et al., 1997, p. 516). While the value underlying this field is adaptability, capabilities from the production arena, specifically those enabling organizational units to meet and exceed organizational goals, mitigate units' position in this field. Units directly focused on executing the capabilities required to deliver on organizational goals occupy more peripheral positions in the field of coordination and are therefore less constrained by the forces of adaptation characteristic of this field. Adaptation entails managing resource allocations (Crowston, 1997), information flows (Terwiesch, Loch, and Meyer, 2002), and the media through which the first two are accomplished (Adler, 1995).

In coordination fields, IS serve as connectivity agents – transmitting information about interdependent actants and facilitating common interpretations of work tasks and resources, e.g., through negotiating acceptable results, use of resources, managing conflict (Crowston, 1997). This technological agency produces adaptability in fields of coordination by making global information available locally, and thus permitting centralized or decentralized decision-making as required by the situation (Malone, 1997). In this field, communication technologies have been observed to collaborate with human actors in the definition of communication genres: While   human   agents   in   authority   positions   deliberately   instituted communication practices around a communication technology, the passive agency of the technology in its forum management capabilities engendered contradictory agency in the appropriation and constitution of genres by human agents occupying lower positions within the field (Yates, Orlikowski, and Okamura, 1995). In this example, we note the mutation of the structure of field positions following the technology implementation. IS have also evinced strong agency in the field of coordination, relative to human agents, resulting in the de-integration, i.e., modularization, and downsizing of organizations (Brynjolfsson, Malone, Gurbaxani, and Kambil, 1994). IS may also be co-opted by human agents to heighten their position in the field of coordination by improving their ability to enable organizational adaptation or by reifying valued production capabilities, which can then relieve them from forces of adaptation.

4.3   Field of Co-opetition

The field of co-opetition is constituted by organizations' customers, suppliers, competitors, and complementors, i.e., those who increase the value of the organization's products and services. There is a fundamental duality underlying organizations' relationships with these constituencies: "Whereas creating value is an inherently cooperative process, capturing value is inherently competitive" (Brandenburger and Nalebuff, 1996, p. vii). Co-opetition refers to this dual process of cooperating and competing at all times.

The "product" of the co-opetition field is organizational identity. In the process of cooperating and competing, organizations establish their affinities and distinctions in their social space, thus defining their identity (Tajfel and Turner, 1986). Identity may also be viewed as a platform – a collection of "organizational mechanisms and forms" and the "collective, cognitive engine" that facilitate activities organizations' relationships with their co-opetitors (Ciborra, 1996, p. 104). This identity is produced in practices across organizational boundaries, and in response to environmental conditions: "Any identity comes out of diametrically competing energies", through negotiations among those within a competitive space (White, 1992, p. 7). The production of identity is not unconstrained. Rather, the field of co-opetition makes available a portfolio of identities from which an organization can select (Ruef, 2000).

An organization's identity informs its goals and values, thereby constraining and enabling its choices (Albert and Whetten, 1985). Effective identity renewal is key to organizational success, even survival. Too stable an identity compromises preparedness for and responsiveness to environmental changes (Gioia, Schultz, and Corley, 2000). In contrast, 'adaptive instability' is beneficial because it permits responsive to changing environments (Gioia, Schultz, and Corley). However, identity change can be costly to manage and can damage self-esteem and is therefore often resisted (Brown and Starkey, 2000). Identities may also be pluralistic and such pluralism increases an organization's requisite variety and enhances learning and creativity and future strategic value, i.e., strategic agility (Pratt and Foreman, 2000).

In attempting to create niches that are competitive within the inter-organizational field, organizations also strive for a distinctive identity (Bourdieu and Wacquant, 1992). While adaptive instability, pluralism, and distinctiveness are hallmarks of identity renewal, they can also be dysfunctional. The efficacy of pluralistic identities is also determined by how component identities inter-relate (Pratt and Foreman, 2000), and the organization's possession of "meta-identity" that is recognized by all constituents: "Lack of a shared identity creates dissonance and makes collective action and collective sensemaking impossible" (Brown and Starkey, 2000, p. 113). Similarly, identity continuity over time is critical as a lack of a unifying storyline creates ontological insecurity, which impinges on organizations' risk-taking (Giddens, 1991). "When major change is required, a wise way to proceed in motivating people to accept the necessity of change is to demonstrate how at least some elements of the past are valued and will be preserved" (Brown and Starkey, 2000, p. 113). Thus, successful renewal necessitates not only the repositioning of the organization so as to be distinctive, but also the maintenance of a central and continuous story line. In this sense, the tension between compliance and creation is most visible in this field (Ghoshal and Bartlett, 1995).

Renewal is a function of a specific type of organizational competence – "effective identification and linking of technological options and market opportunities, and for identifying the strengths and weaknesses of existing resources relative to the requirements of a new product or process" (Dosi et al., 2002, p. 6). The capability requirement within this field is therefore boundary-spanning. We distinguish boundary-spanning from brokering, which simply entails acquiring and transferring information about opportunities and who can benefit from them. In contrast, boundary-spanning capabilities entail learning, reconfiguration, and transformation (Teece et al., 1997). It requires developing the richness of the information used in formulating strategic intent, budget objectives, and corporate policy, as well as reconciliation of disparate perspectives (Ghoshal and Bartlett, 1995). It necessitates not just the identification of significant actants, but also the elicitation of cooperation from actants with competing interests. It implies awareness, even prescience, of environmental conditions and co-opting of multiple agents and indigenous and external competencies "to handle change by transforming old capabilities into new ones" (Dosi et al., 2002, p. 7) and the ability to integrate and reconcile disparate perspectives over time and across multiple stakeholders. It requires imagination. While individual imaginations enable individuals' strategic positioning of themselves within a field, the space of this field is a corporate imagination. Within this field, organizations "unleash corporate imagination, identify and explore new competitive space, and consolidate control over emerging market opportunities" (Hamel and Prahalad, 1991, p. 82). A corporate imagination necessitates "escaping the tyranny of served markets; searching for innovative product concepts; overturning traditional assumptions about price/performance relationships; and leading customers rather than simply following them" (Hamel and Prahalad, 1991, p. 83).

The forces underlying the field of co-opetition are those of isomorphism (Martin, 2003; DiMaggio and Powell, 1983). While a distinctive identity may be a desirable response to an organization's contingencies, isomorphic forces differentially constrain such distinctiveness, depending on the organization's position in its organizational field. This is visible in research findings on the differential benefits that accrue to early versus late technology adopters within organizational fields. Research in the manufacturing and financial sectors has demonstrated that investments in innovative IT, i.e., the first use of a particular technology within a given industry or the deployment of technology towards development of new products, services or standards, yielded firm value increments that were not available for investments in non-innovative IT (Dos Santos, Peffers, and Mauer, 1993). Banks that first adopted ATMs were able to leverage the technology toward increased market share and sustained income gains, but this advantage was not available to later adopters of ATMs (Dos Santos and Peffers, 1995). Early adopters are then able to customize the innovations they adopt to the specific needs of their organizations, thereby improving their operations; in contrast, late adopters are precluded from such customization and their technological adoptions are instead circumscribed by the implementation patterns of early adopters, thereby providing them with none of the efficiency advantages that accrue to early adopters (Westphal, Gulati, and Shortell, 1997).

While first mover advantages may accrue to early adopters, collaboration among later-adopting competitors may generate network externalities in excess of the gains by initial adopters. Recognition of such network externalities prompted Citibank, the first-mover with ATM technology, to ultimately join the network of banks after initially refusing to do so (Brandenburger and Nalebuff, 1996). Thus, we see evidence of the forces within the co-opetition field constraining the implementation and benefits of IT in the fields of production and coordination. Prowess in the fields of production and coordination facilitate early adoption (Meyer and Utterback, 1993). Again, we see the intersection of field forces in the extent to which agents are constrained or are freed from constraint.

In the field of co-opetition, technology serves as an agent of equal, often greater, potency to/than human agents in the re-production of isomorphic field forces. This is effected via technological standards, e.g., EDI, XML, and by *de facto* industry practices enabled through technology, e.g., SAP within vertical industries. While horizontal standards permit compatibility and interoperability, vertical standards facilitate integration and coordination (Markus, Steinfield, and Wigand, 2003). In the adoption of horizontal or platform standards, human and technological agents reciprocally constrain future choices and inter-operability (Dedrick and West, 2003).

5.   IMPLICATIONS FOR IS INNOVATION

This paper outlined organizational fields, within which IS possesses an agency equal to that of human actors. The articulated field theory, though only partially developed at this point, possesses a robustness absent in earlier conceptualizations of the role of IS in organizations. This is evident in our examination of the limitations of prior IS conceptualizations noted earlier. We now re-examine these limitations, and our proposed field theoretic solution, within the context of IT adoption and implementation.

First, as with human actors, IS may assume agency in one or more organizational fields, thereby generating and accounting for interdependencies among the fields and among IS within and across fields. In fact, an adopted IS is likely to manifest varying levels of functionality across all three fields. For example, a web portal focused on enabling an inter-organizational supply chain simultaneously automates procurement transactions, defines communication genres through which the conversations required for inter-firm coordination occur, and puts organizational identities into play in re-structuring supply chain relationships. Thus, viewing an IS as simultaneously constituted within fields of production, coordination, and co-opetition emphasizes that all IS are multivalent in their organizational influences.

Such an observation has implications for IS adoption and implementation in that the logic driving the introduction of an IS lies within each of these three fields. Innovation champions must therefore understand the facilitating and inhibiting pressures at work within each field. They should ascertain which field has precedence at the introduction and during the innovation's long-run sustenance as an active, morphing agent within the organization. Such an understanding would enable champions to anticipate the constraints on human agency within each field. Additionally, while the human actants involved with an IS implementation initiative are influenced by these three fields, the organizational 'life space' of each individual, e.g., their functional membership, will differentiate the operant force fields. For example, while production invariance might drive a technologist to seek to increase the leveragability of an IS across the entire organization, to adopting managers, forces of production invariance may necessitate a more local effort, specifically increasing the productivity of their work processes.

Second, the complex relationships both formed and affected by an IS within organizations is reflected in recognition that the focal IS becomes just one actor among many, all of whom compete and collude with each other, and among whom contests are played out. Just as the agency of IS can constrain and facilitate human actors' development and use of their capabilities, human actors constrain and facilitate the nature and deployment of an IS' capabilities and each other's use of it. Such complex effects produce and are reproduced by path-dependent, episodic interactions amongst actants, thus shaping the socio-technical relationships that characterize organizations.

Here again, the implications for IS adoption and implementation are profound. Any newly introduced technology agent replaces, substitute for, augments and/or serves as a catalyst for other technology as well as human agents, each of which is influenced by operant force fields and, over time, contributes to the inherent nature of these operant force fields. Those responsible for the implementation of a new technology agent must not only recognize the salience of this network of affected/affecting agents, but must interpret their collective ramifications through filters that reflect the relative and shifting influence of the three force fields.

Finally, understanding a focal IS' evolution over time requires accounting for both (1) the interdependencies that exist among a focal IS and the other actors within an organization and (2) the interactions that parallel these interdependencies. Most important, recognizing fields of production, coordination, and co-opetition will facilitate the surfacing and examination of the focal IS' interactions with other social and artifactual actants as they play out contests across fields, and the development and deployment trajectories of the focal IS based on those interactions. Thus, while an organization's implementation of a new IS might initially be largely influenced by one force field (e.g., coordination), the trajectory of its evolution may be more heavily influenced by another force field (e.g., co-opetition). For example, a web services architecture might initially be viewed as a mechanism for enterprise application integration, but then be seen as a core enabler of inter-organizational process integration. The resultant dynamics of project sponsorship, funding, leadership and membership are likely to shift dramatically as a reflection of the capital and values inherent in the operant force fields at a given point in time and space.

6.   CONCLUSION

This paper provides a basis for studying unexplored or under-explored research domains regarding the introduction, evolution, and organizational impacts of IS. While work by researchers such as Orlikowski (2000) hints at the intersection of organizational fields via IS, there has been little explicit attention to this in IS research. Field theory not only offers a rich palette with which to outline the complexities that arise when technologies are introduced into intersecting fields, but also allows for accounts of endogenous change, i.e., that isn't externally driven but rather emerges via struggles within the organizational space (Bourdieu and Wacquant, 1992) that are both episodic and continuous (Martin, 2003). It thus promises to enable researchers interested in studying IS innovation to account for both planned and emergent interventions that constitute an IS implementation initiative as human agents grapple with and apply the new technology within the complex and constantly evolving network of organizational routines and practices.

These ideas also have methodological implications for IS innovation research. While prior IS innovation research has considered the individual, group, business unit, technology, or practice as a unit of analysis, this paper highlights an alternate unit – the field. Field theorists have offered the following guidance for the empirical study of fields: (1) analysis of positions occupied by actants within the field, (2) the structure of the field, and (3) genesis (history) of actants' embodied (and institutionalized) capital (Bourdieu, 1988, 1983). While ideas regarding the agency of IS in organizational fields have surfaced in IS research, we hope that our formalization and extension of these ideas invites further conceptualization and novel empirical approaches to the study of IS phenomena.

REFERENCES

Adler, P. (1995) "Interdepartmental Interdependence and Coordination: The Case of the Design/Manufacturing Interface." *Organization Science*, 6(2), 147-167.

Albert, S. and D.A. Whetten (1985) "Organizational Identity." In L.L. Cummings and B.M. Staw (eds.), *Research in Organizational Behavior*, 7, 263-295, Greenwich, CT: JAI Press.

Bloomfield, B.P. and T. Vurdubakis (1994) "Re-presenting Technology: IT Consultancy Reports as Textual Reality Constructions." *Sociology*, 28(2), 455-477.

Boltanski, L. and L. Thevenot (1987) "Judgment in a Multi-Natured Universe: The Problem of Agreement in a Complex Society." *Symposium on Political Communication: Foundations and New Approaches*, Paris, May 14-15.

Bourdieu, P. (1980) *The Logic of Practice*. Stanford, CA: Stanford University Press.

Bourdieu, P. (1983) "The Forms of Capital." In J.G. Richardson (ed.) *Handbook of Theory and Research for the Sociology of Education*, NY: Greenwood Press, 241-258.

Bourdieu, P. (1984) *Distinction: A Social Critique of the Judgment of Taste*. Cambridge, MA: Harvard University Press.

Bourdieu, P. (1988) "Flaubert's Point of View." *Critical Inquiry*, 14, 539-562.

Bourdieu, P. and L.J.D. Wacquant (1992) *An Invitation to Reflexive Sociology*. Chicago, IL: The University of Chicago Press.

Brandenburger, A.M. and B.J. Nalebuff (1996) "Co-opetition." New York, NY: Currency Doubleday.

Brown, A.D. and K. Starkey, 2000 "Organizational Identity and Learning: A Psychodynamic Perspective." *Academy of Management Review*, 25(1), 102-120.

Brynjolfsson, E., T.W. Malone, V. Gurbaxani, and A. Kambil (1994) "Does Information Technology Lead to Smaller Firms?" *Management Science*, 40(12), 1628-1644.

Churchman, C.W. (1971) *The Design of Inquiring Systems: Basic Concepts of Systems and Organization.* New York, NY: Basic Books, Inc.

Ciborra, C.U. (1996) "The Platform Organization: Recombining Strategies, Structures, and Surprises." *Organization Science,* 7(2), 103- 118.

Ciborra, C.U. (1993) *Teams, Markets, and Systems: Business Innovation and Information Technology,* Cambridge University Press.

Crowston, K. (1997) "A Coordination Theory Approach to Organizational Process Design." *Organization Science,* 157-175.

Dedrick, J. and J. West (2003) "Why Firms Adopt Platform Standards: A Grounded Theory of Open Source Platforms." MISQ Special Issue Workshop on Standards, WA: Seattle.

DeSanctis, G. and M.S. Poole (1994) "Capturing the Complexity in Advanced Technology Use: Adaptive Structuration Theory." *Organization Science,* 5(2), 121-147.

DiMaggio, P.J. and W.W. Powell (1983) "The Iron Cage Revisited: Institutional Isomorphism and Collective Rationality in Organizational Fields." *American Sociological Review,* 48, 147-160.

Dosi, G., R.R. Nelson, and S.G. Winter (2002) *The Nature and Dynamics of Organizational Capabilities.* Oxford University Press.

Dos Santos, B.L. and K. Peffers (1995) "Rewards to Investors in Innovative Information Technology Applications: First Movers and Early Followers in ATMs." *Organization Science,* 6(3), 241-259.

Dos Santos, B.L., K. Peffers, and D.C. Mauer (1993) "The Impact of Information Technology Investment Announcements on the Market Value of the Firm." *Information Systems Research,* 4(1), 1-23.

Feldman, M.S. and B. Pentland (2003) "Reconceptualizing Organizational Routines as a Source of Flexibility and Change." *Administrative Science Quarterly,* 48(1), 94-118.

Ferguson, P.P. (1998) "A Cultural Field in the Making: Gastronomy in 19th-Century France." *American Journal of Sociology,* 104(3), 597-641.

Garvin, D.A. (1995) "Leveraging Processes for Strategic Advantage." *Harvard Business Review,* 73(5), 77-90.

Garvin, D.A. (1998) "The Processes of Organization and Management." *Sloan Management Review,* 33-50.

Ghoshal, S. and C.A. Bartlett (1995) "Changing the Role of Top Management: Beyond Structure to Processes." *Harvard Business Review,* January-February, 86-96.

Giddens, A. (1991) *Modernity and Self-Identity: Self and Society in the Late Modern Age.* Stanford, CA: Stanford University Press.

Gill, T.G. (1996) "Expert Systems Usage: Task Change and Intrinsic Motivation." *MIS Quarterly,* 20(3), 301-330.

Gill, T.G. (1995a) "High-Tech Hidebound: Case Studies of Information Technologies that Inhibited Organizational Learning." *Accounting, Management, and Information Technology,* 5(1), 41-60.

Gill, T.G. (1995b) "Early Expert Systems: Where are They Now?" *MIS Quarterly,* 19(1), 51-81.

Gioia, D.A., M. Schultz, and K.G. Corley (2000) "Organizational Identity, Image, and Adaptive Instability." *Academy of Management Review,* 25(1), 63-81.

Hamel, G. and C.K. Prahalad (1991) "Corporate Imagination and Expeditionary Marketing." *Harvard Business Review,* 69(4), 81-92.

Kelley, M.R. (1990) "New Process Technology, Job Design, and Work Organization: A Contingency Model." *American Sociological Review,* 55(2), 191-208.

Lamb, R. and R. Kling (2003) "Reconceptualizing Users as Social Actors in Information Systems Research." *MIS Quarterly*, 27(2), 197-236.

Law, J. and J. Hassard (1999) *Actor-Network Theory and After.* Blackwell Publishers.

Lewin, K. (1951) *Field Theory in Social Science*, New York, NY: Harper and Brothers.

Malone, T. W. (1987) "Modeling Coordination in Organizations and Markets." *Management Science*, 33(10), 1317-1331.

Malone, T.W. (1997) "Is Empowerment Just a Fad? Control, Decision-making, and IT." *Sloan Management Review*, 38(2), 23-35.

Markus, M.L., C.W. Steinfield, and R.T Wigand (2003) "The Evolution of Vertical IS Standards: Electronic Interchange Standards in the U.S. Home Mortgage Industry." MISQ Special Issue Workshop on Standards, WA: Seattle.

Markus, M.L. and J. Pfeffer (1983) "Power and the Design and Implementation of Accounting and Control Systems." *Accounting, Organizations, and Society*, 8(2), 205-218.

Martin, J.L. (2003) "What is Field Theory?" *American Journal of Sociology*, 109(1), 1-49.

Meyer, M.H. and J.M. Utterback (1993) "The Product Family and the Dynamics of Core Capability." *Sloan Management Review*, 34(3), 29-47.

Myers, M.D. and L.W. Young (1997) "Hidden Agendas, Power, and Managerial Assumptions in Information Systems Development: An Ethnographic Study." *Information Technology and People*, 10(3), 224-.

Orlikowski, W.J. (2000) "Using Technology and Constituting Structures: A Practice Lens for Studying Technology in Organizations." *Organization Science*, 11(4), 404-428.

Orlikowski, W.J. (1996) "Improvising Organizational Transformation Over Time: A Situated Change Perspective." *Information Systems Research*, 7(1), 63-92.

Orlikowski, W.J. (1993) "CASE Tools as Organizational Change: Investigating Incremental and Radical Changes in Systems Development." *MIS Quarterly*, 17(3), 309-340.

Prahalad, C.K. and G. Hamel (1990) "The Core Competence of the Corporation." *Harvard Business Review*, 68(3), 79-91.

Pinsonneault, A. and K.L Kraemer (1993) "The Impact of Information Technology on Middle Managers." *MIS Quarterly*, 17(3), 271-292.

Pratt, M.G. and P.O. Foreman (2000) "Classifying Managerial Responses to Multiple Organizational Identities." *Academy of Management Review*, 25(1), 18-42.

de Rivera, J. (1976) *Field Theory as Human-Science.* New York, NY: Gardner Press, Inc.

Ruef, M. (2000) "The Emergence of Organizational Forms: A Community Ecology Approach." *The American Journal of Sociology*, 106(3), 658-714.

Schein, E.H. (1992) "The Role of the CEO in the Management of Change: The Case of Information Technology." *In T.A. Kochan and U. Useem (eds.), Transforming Organizations.* Oxford, Oxford University Press, 80-96.

Stalk, G., P. Evans, and L.E. Shulman (1992) "Competing on Capabilities: The New Rules of Corporate Strategy." *Harvard Business Review*, 70(2), 57-69.

Star, S.L. and K. Ruhleder (1996) "Steps Toward an Ecology of Infrastructure: Design and Access for Large Information Spaces." *Information Systems Research*, 7(1), 111-134.

Stewart, T.A. (1999) *Intellectual Capital: The New Wealth of Organizations.* New York, NY: Currency-Doubleday.

Tajel, H. and J.C. Turner (1986) "The Social Identity Thoery of Intergroup Behavior." In S. Worschel and W.G. Austin (eds.) *Psychology of Intergroup Relations*, Chicago, IL: Nelson-Hall.

Teece, D.J., G. Pisano, and A. Shuen (1997) "Dynamic Capabilities and Strategic Management." *Strategic Management Journal*, 18, 509-533.

Terwiesch, C., C.H. Loch, and A. Meyer (2002) "Exchanging Preliminary Information in Concurrent Engineering: Alternative Coordination Strategies." *Organization Science*, 13(4), 402-422.

Tornatzky, L.G. and M. Fleischer (1990) *The Processes of Technological Innovation*. Lexington, MA: Lexington Books.

Weber, M. (1978) *Economy and Society*. Berkeley, CA: University of California Press.

Weick, K.E. (1995) *Sensemaking in Organizations*. Thousand Oaks, CA: Sage Publications.

Westphal, J.D., R. Gulati, and S.M. Shortell (1997) "Customization or Conformity? An Institutional and Network Perspective on the Content and Consequences of TQM Adoption." *Administrative Science Quarterly*, 42, 366-394.

White, H.C. (1992) *Identity and Control: A Structural Theory of Social Action*. Princeton, NJ: Princeton Books.

Willcocks, L.P. and R. Plant (2001) "Pathways to e-Business Leadership: Getting from Bricks to Clicks." *Sloan Management Review*, 42(3), 50-59.

Woodward, J. (1994/1965) *Industrial Organization: Theory and Practice (2nd edition)*. New York, NY: Oxford University Press.

Yates, J., W.J. Orlikowski, and K. Okamura (1995) "Constituting Genre Repertoires: Deliberate and Emergent Patterns of Electronic Media Use." *Academy of Management Proceedings*, 353-357.

Zimmerman, M.E. (1990) *Heidegger's Confrontation with Modernity*, Bloomington, IN: Indiana University Press.

IT INNOVATION THROUGH A WORK SYSTEMS LENS

Steven Alter
School of Business and Management, University of San Francisco, USA

Abstract: This paper presents what I think is an innovation, the use of work system concepts as a lens for understanding IT innovation in organizations. It starts by presenting the ideas at the heart of the work system approach, the work system framework for describing an existing or proposed work system and the work system life cycle model for describing how work systems change over time. To demonstrate how a work system lens can illuminate both practice and theory related to IT innovation in organizations, it returns to the previous IFIP 8.6 Conference (Copenhagen, October 2003) and applies work system ideas to the situations and theories in the papers presented there. It closes by using CASE adoption as a final example about IT innovation within IT groups.

Key words: IT innovation, adoption, work system, system development, implementation, actor network theory, work system framework, work system life cycle model

1. INTRODUCTION

This paper presents what I think is an innovation, the use of work system concepts as a lens for understanding IT innovation in organizations. This approach builds on the term *work system*, which has been used by a number of socio-technical researchers and by some practitioners for over 20 years, but has not been defined carefully or used as a rigorous concept until recently. The term work system appeared in two articles in the first volume of *MIS Quarterly* [Bostrom and Heinen, 1979a, 1979b]. Mumford and Weir [1979, p. 3] spoke of "the design and implementation of a new work system." Davis and Taylor [1979, p. xv] mentioned "attempts at comprehensive work systems design, including the social systems within which the work systems are embedded." For other past uses of the term *work system* see Alter (2003).

This paper addresses a number of points in the Call for Papers, including ontological issues about the meaning of IT innovation, perceptions of IT innovation, and the observation that "successful innovations in IT, unlike those in consumer markets, must work through the dynamics of the organization." The concepts presented apply equally to internal innovation within IT groups and to innovations in business functions supported by IT groups. Regardless of whether the focus is an IT group attempting to implement OOP or a sales organization trying to implement CRM, IT innovation in organizations is about doing work differently.

Understanding IT innovation from a work system viewpoint is consistent with the observation by Brynjolfsson (2003) that "for every dollar of IT hardware capital that a company owns, there are up to $9 of IT-related intangible assets, such as human capital -- the capitalized value of training -- and organizational capital -- the capitalized value of investments in new business-process and other organizational practices. Not only do companies spend far more on these investments than on computers themselves, but investors also attach a larger value to them." If this ratio is even close to correct, a totally IT-centric view of IT innovation is bound to ignore or underplay important aspects of how IT innovations succeed, survive without success, or simply die in organizations.

This paper shows that an IT-centric view of IT innovation is self-limiting and inherently biased. For example, why should an IT professional or anyone else view IT innovation within organizations (as opposed to a new chip inside a computer) as being about IT? Except possibly from the viewpoint of vendors trying to sell specific commercial products, IT innovation is about improving business operations and business results. This paper develops this point by showing that thinking about IT innovation from a work system viewpoint might generate new insights and new ways to make IT innovations more successful.

The paper proceeds as follows: The next section presents the ideas at the heart of the work system approach, the work system framework for describing an existing or proposed work system and the work system life cycle model for describing how work systems change over time. To demonstrate how a work system lens can illuminate both practice and theory related to IT innovation in organizations, it returns to the previous IFIP 8.6 Conference (Copenhagen, October 2003) and applies work system ideas to a variety of situations and theories in papers presented there. It closes by using CASE adoption as a final example about IT innovation within IT groups.

2. THE WORK SYSTEM FRAMEWORK AND WORK SYSTEM LIFE CYCLE MODEL

A work system is a system in which human participants and/or machines perform work using information, technology, and other resources to produce products and/or services for internal or external customers. Typical business organizations contain work systems that procure materials from suppliers, produce products, deliver products to customers, find customers, create financial reports, hire employees, coordinate work across departments, and perform many other functions. It is possible to view an entire organization, firm, or even an industry in work system terms, but that is not the intention here. We view organizations as consisting of multiple work systems rather than as a single, large work system combining many independent or partly dependent business processes and groups of participants. On the other hand, a work system such as a supply chain or other interorganizational system can extend across multiple business enterprises.

Identifying and organizing work system concepts is one part of a long term project attempting to develop "the work system method," a broadly applicable set of ideas that use the concept of work system as the focal point for understanding, analyzing, and improving systems in organizations, whether or not IT is involved. The work system method includes both a static view of a current (or proposed) system in operation and a dynamic view of how a system evolves over time through planned change and unplanned adaptations. These views are summarized in Figures 1 and 2, whose specific terms are explained succinctly in Alter (2002). A subsequent switch from "business process" to "work practices" is explained in Alter (2003).

The static view is summarized as the work system framework (Figure 1), which identifies the basic elements for understanding and evaluating a work system. The work system itself consists the four elements inside the large trapezoid. The other five elements must be included in even a rudimentary understanding of a work system's operation, context, and significance. This framework is prescriptive enough to help in describing the system being studied, identifying problems and opportunities, describing possible changes, and tracing likely impacts as changes propagate across the system. An immediate implication is that business and IT professionals discussing a real world situation should avoid focusing solely on information needs, computerized information, or IT.

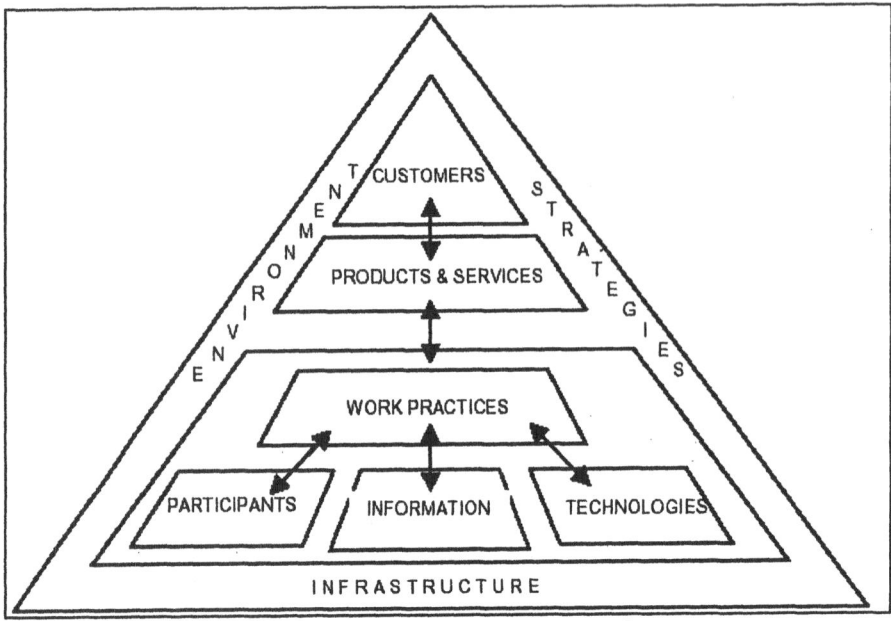

Figure 1. The Work System Framework

The dynamic view is based on the work system life cycle model (Figure 2), which shows how a work system may evolve through multiple iterations of four phases. This life cycle model encompasses both planned and unplanned change, and differs substantially from project models that are often called life cycle models when applied to software projects. Planned change in an established organization typically starts from a work system whose operation or relationship to other systems requires improvements that involve more than small adapations. The initiation phase identifies the problem or opportunity, the general approach, and the necessary resources. The development phase acquires, modifies, or creates hardware, software, and other resources needed before the revised system can be implemented in the organization. The implementation phase ends with acceptance of the new work system, and another operation and maintenance phase begins.

The arrows in the work system life cycle model represent the fact that change may occur at three levels:

1. Incremental adaptations and small changes that tend to be unplanned and localized, and can occur during any of the four phases.
2. Projects that accomplish planned changes in the work system.
3. Evolutionary change that occurs through multiple iterations of planned and unplanned change. IT innovation is frequently a key enabler of a new iteration, although some iterations may be driven by other issues, such as an organizational change or changes in the external environment.

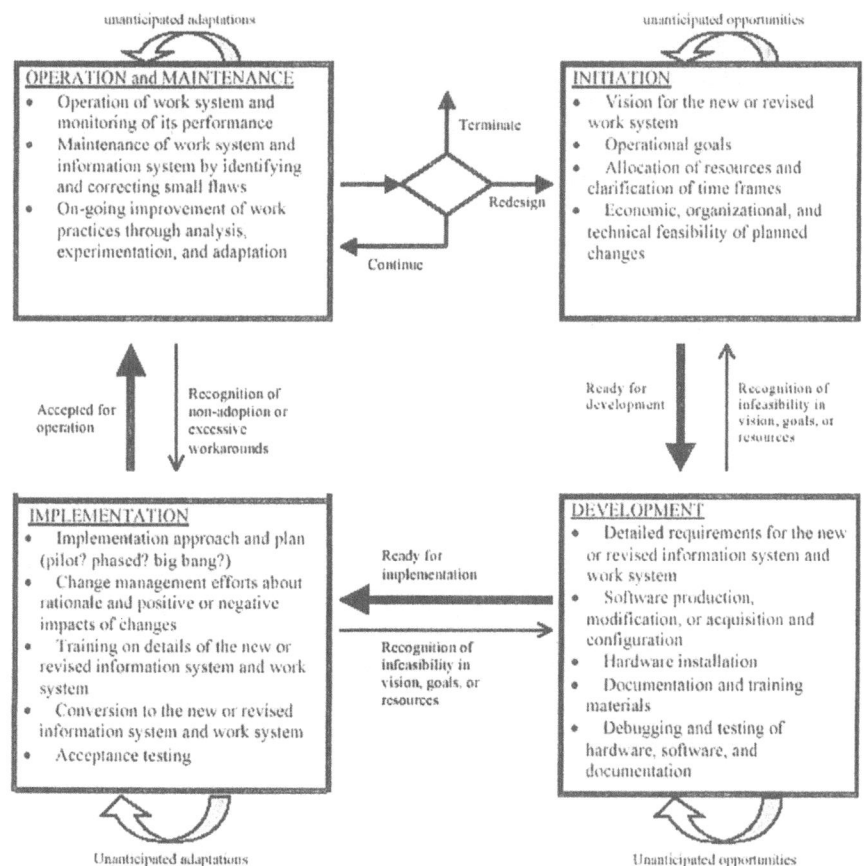

unanticipated adaptations   unanticipated opportunities

OPERATION and MAINTENANCE
• Operation of work system and monitoring of its performance
• Maintenance of work system and information system by identifying and correcting small flaws
• On-going improvement of work practices through analysis, experimentation, and adaptation

Terminate

Redesign

Continue

INITIATION
• Vision for the new or revised work system
• Operational goals
• Allocation of resources and clarification of time frames
• Economic, organizational, and technical feasibility of planned changes

Accepted for operation

Recognition of non-adoption or excessive workarounds

Ready for development

Recognition of infeasibility in vision, goals, or resources

IMPLEMENTATION
• Implementation approach and plan (pilot? phased? big bang?)
• Change management efforts about rationale and positive or negative impacts of changes
• Training on details of the new or revised information system and work system
• Conversion to the new or revised information system and work system
• Acceptance testing

Ready for implementation

Recognition of infeasibility in vision, goals, or resources

DEVELOPMENT
• Detailed requirements for the new or revised information system and work system
• Software production, modification, or acquisition and configuration
• Hardware installation
• Documentation and training materials
• Debugging and testing of hardware, software, and documentation

Unanticipated adaptations   Unanticipated opportunities

Figure 2. The Work System Life Cycle Model

Combining work system concepts with general problem solving ideas yields a systems analysis method that avoids separating social and technical analysis (in contrast, see Mumford and Weir (1979, p. 29) or Taylor and Felten (1993, p. 5)), is more prescriptive than soft systems methodology (Checkland, 1999), and is less detail-oriented than tools for specifying procedures, database structures, or computer programs. The static and dynamic views can be used together in a systems analysis method that treats the information system as part of the work system until a final step when it distinguishes between work system changes that do and do not involve the information system. A unique aspect of this method (in comparison with other systems analysis methods for information systems) is its explicit use of principles that describe a typical system's desired behavior and results. After the system and problem or opportunity are described, the principles are used to examine the current system, to help identify potential improvements, to explore whether potential improvements in one area might cause problems elsewhere, and to sanity-check the recommendation.

The concept of work system is a general case of systems operating in organizations. Special cases of work systems include information systems, projects, value chains, supply chains, and totally automated work systems. These and other special cases should inherit most of the properties of work systems, which include success factors, risk factors, and generalizations related to efficiency and effectiveness. (Alter, 2002; 2003; 2004)

The purpose of most information systems is to support one or more work systems. Although information systems and the work systems they support were often quite separable decades ago when most business computing was still card-based and batch-oriented, today many important information systems overlap significantly with the work systems they serve. For example, in a payroll system or ecommerce web site that sells and downloads software, most or all of the work system is an information system. In work systems related to producing, transforming, or delivering physical things, information systems may play a less dominant role even if they are essential for a work system's effective operation and success. And in extreme cases such as highly automated manufacturing, the information system and work system overlap so much that the manufacturing is largely controlled by the information system. Turn off the information system and this type of manufacturing grinds to a halt.

3.   VIEWING LAST YEAR'S IFIP 8.6 PAPERS THROUGH A WORK SYSTEM LENS

One way to demonstrate the potential usefulness of the work system lens is to show its broad applicability to the situations described in papers from last year's IFIP 8.6 conference, Diffusion and Adoption of Networked Information Technologies, in Copenhagen, Oct. 6-8, 2003. Although the following discussion shows that the work system approach contains ideas that help any business or IT professional identify key issues related to IT innovation, I make no claim that similar ideas have never been stated. Consider several excerpts from the papers:

"It is necessary that the members of the human organization express a great amount of care to incorporate the new system into their daily work life. To reach a full appropriation of the system, the involved actors should engage actively to cope with the involved uncertainties and not rely on a passive detached process of acknowledgement. (Ciborra, 1996)" (cited by Nilsson et al 2003)

"It seems that we -- as a professional community of system developers -- tend to treat the systems as separate units from the work activities, stressing the development of computer-based artefact much more than the development of work (Forsman and Nurminem, 1994). However, as soon as we change our scope from the computer artefact to the activity itself, the information system can be seen as a means of structuring and developing a social system. (Nurminem, 1988)" (cited by Heikkilä et at, 2003, p. 2)

The uniqueness of the work system approach lies in its attempt to combine and amplify selected ideas from sociotechnical analysis, implementation studies, organization behavior, and project management to make them readily usable by typical business and IT professionals trying to describe and understand systems in organizations. In terms of IT innovation, the work system approach says that IT implementation or IT diffusion is not the problem. The problem is improving specific work systems. Accordingly, one should start the analysis with the work system, not the IT innovation.

Next we present direct or indirect evidence of the applicability of work system ideas to situations in 9 of the 11 papers presented at the Copenhagen conference. The other two papers concerned stakeholder theory and the possibility of buyers shaping IT markets.

3.1   IT innovations in healthcare

Applying a work system lens to compare accounts of three situations related to healthcare reveals striking similarities. The situations were the of a large Swedish hospital's radiology department (Nilsson, Grisot, and Mathiassen, 2003), elderly care in Trondheim, Norway (Lines, Viborg Andersen, and Monteiro, 2003) and elderly care in a Swedish locality (Hedström, 2003).

An initial step in analyzing any work system is to use the work system framework (Figure 1) to summarize the work practices, participants, information, and technology. In the radiology department, the work system was about entering orders for radiology services. The existing work system used paper-based orders, with the conveniences and shortcomings of paper. Switching to an electronic ordering system would impose more structure and would absorb more physician time in entering all orders, but ideally would improve consistency and reliability in recordkeeping. In the two elderly care situations, the information system provided administrative information and supported caregivers. In both cases, the work system seemed nebulous. If there was an intention to support care giving, it wasn't clear exactly how the information system was supposed to help. Similarly, even the administrative information to be recorded was unclear from the accounts.

Applying the work system life cycle model (Figure 2) to the situations leads one to look for the initiation, development, and implementation phases. In all three cases, the initiation phase seemed distant from most participants. In one elderly care situation, the decision was attributed to "politicians," implying that little or no analysis of work systems took place. In all three situations, the development phase involved acquiring software produced elsewhere, but it was unclear whether that software was configured to suit local situations. All three accounts described attempts to implement the information system before competing an analysis of how it would be integrated into the work practices. There was little or no effort to identify how the information system would help the work system participants or to reconcile divergent stakeholder interests prior to implementation.

Based on the above, just using Figure 1 and Figure 2 at various times as these situations unfolded would have immediately identified significant problems that should have been resolved before implementation. At minimum, attention to the most basic work system concepts might have been a reminder about simple principles that anyone can apply, but that seem not to have been applied by the main protagonists in these cases:

– *Work practices:* Don't assume that an IT innovation will fix a work system if you can't or won't define that work system.
– *Participants:* Don't assume that all work system participants will find a particular IT innovation valuable. Recognize that incentives matter.
– *Information:* Don't assume that IT innovations necessarily provide better information.
– *Technology:* Don't assume that IT innovations always help. IT innovations may absorb effort rather than reducing it.
– *Customers:* Recognize that different work system customers and other stakeholders have different objectives and priorities.
– *Initiation phase:* Make sure stakeholders including work system participants or their representatives are involved. Obtain agreement about the need to change the work system and about intentions to enforce changes or allow change to occur voluntarily.
– *Development phase:* Make sure the hardware and software fit the situation.
– *Implementation phase:* Before implementing make sure all participants will either benefit directly or will have a strong reason to play their roles.

3.2 IT innovations related to interorganizational systems or eGovernment

As with the papers about healthcare, the theme of inadequate work system definition was found in these three papers. For example, a particular interorganizational information system, "was built to support the execution of work-tasks performed according to the process descriptions, but failed to take into account the fact that in the field the processes for executing different work-tasks, possibly because of long-term relationships between organization A and some of its suppliers, varies by suppliers." (Heikkilä et al, 2003) The issue about process descriptions not matching work practices has been described by ethnographic researchers for decades. Any description of a real world system is only an approximation, but settling for the idealized business process, an approximation that is likely to be inaccurate, seems a recipe for creating implementation problems when work system participants discover the mismatch between the IT innovation and their reality.

A study of the diffusion of ICTs in small to medium enterprises (SMEs) in Italian industrial districts found that most firms had adopted relatively simple technologies such as email, ISDN, and websites, but had not adopted complex technologies that require deeper work system analysis, such as ERP, EDI, or ecommerce. Unlike simple technologies, complex technologies require substantial work system analysis before they can be implemented and used effectively. Despite general reluctance, "technology providers and SMEs themselves have started to pay greater attention to internal process reconfiguration in order to deal with the actual competitive challenges in the market: globalisation, quality standards, time-to-market, variety and differentiation of the production." (Muzzi and Kautz, 2003) Contrary to the largely voluntary technology adoption in the three healthcare examples, competitive pressure is forcing Italian SMEs to adopt IT innovations they may not prefer. A related example in the United States is Wal-Mart's directive to its top 100 suppliers to attach RFID tags to every box and pallet shipped to Wal-Mart by January 2005. Smaller suppliers have until 2006. "Flout Wal-Mart's orders and 'you potentially get thrown off the shelf of the largest retailer in the world.'" (Boyle, 2003) In work system terms, adopting RFID is an attempt by Wal-Mart to improve inventory management work systems. The improved work systems include new work practices enabled by RFID, new roles, abilities, and incentives of work system participants, new information encoded on the tags, new technology for reading the tags, new infrastructure supporting the entire effort, and a business environment in which large retailers are trying to force compliance with desired technical and physical standards. Future IT innovations by Italian SMEs need to conform to work practices, information requirements, and technology standards imposed downward by retailers.

Finally, Sørgaard (2003) notes that ambitions for e-government "are often unrealistic, and that political goals seem to dominate over effective, stepwise approaches to coordination. On a pragmatic level, there is a need to focus on simpler, process-oriented mechanisms for coordination ..." In other words, whether or not e-government seems an attractive banner to march under, the real progress will occur through analyzing and improving specific work systems that span organizations.

3.3   An IT innovation to support virtual workplaces

A study of the diffusion of the virtual workspace product Lotus QuickPlace in a large European financial institution found that different adoption situations "differed to such a degree that the it is problematic to denote it as one single innovation. (Billeskov Bøving and Bødker, 2003) Thus, the sale of Lotus QuickPlace, an IT innovation from a vendor's viewpoint, might be viewed more realistically as an opportunity to improve the performance of specific work systems that differ substantially. Each implementation effort might be viewed as a separate IT innovation.

"The decision to introduce the QP technology" ... was taken without thorough studies of needs and possibilities." It was a quick and dirty, web based approach that needed "no integration with existing security infrastructures" and "could be implemented very quickly from an IT operations point of view." Analysis after 10 months showed a large percentage of dead (unused) documents. Initial use of QP for a virtual workplace occurred 37 times, but sustained use occurred in around a third of the cases. In many instances, use of QP was intertwined with competing or supplementary media such as email or telephone, illustrating that IT innovation involves much more than diffusion of a technical artifact. "The DOI [diffusion of innovation] framework -- when applied to the diffusion of a collaborative technology -- overlooks, however, a second innovation process ... [that] does not come from a central source, but is based on local situated actions. It is the innovations produced when groups of people agree on using the technology to support novel protocols (in the words of CSCW) or genres (in the words of genre theory)." (Billeskov Bøving and Bødker, 2003) The second innovation process is work system innovation, partly represented in the work system life cycle model as the small adaptations denoted by the inwardly directed arrows going back into each phase.

3.4 Paper about application service providers

To an outsourcer, an application service provider (ASP) provides IT services that would be difficult or impractical to provide using internal resources. From the ASPs viewpoint, its job is to provide those services through its particular work systems. As defined by Johansson (2003), "an ASP enterprise [is] a third party firm that deploys, manages, and remotely hosts software applications through centrally located data centers on a pay-as-you-use basis." A detailed look at an ASP would find work systems for deploying and managing software, hosting software applications, performing client billing and performing other necessary functions. Many of those work systems are largely automated due to the nature of the services an ASP provides. A large part of the IT innovation of creating an ASP is creating and managing those work systems, each of which evolves through the same phases and adaptations that characterize the evolution of any work system.

Table 1. Reasons for Project Success and Failure in Projects cited by Pries-Heje (2003), organized by phase of the work system life cycle

	In successful projects	*In unsuccessful projects*
Initiation phase		- No one responsible for whole diffusion process - No one with in-depth knowledge
Development phase	- User expectations harmonized early in the project - System and workflow matched each other	- Written documentation unfit for use - Totally new graphical interface - Rushing to meet over optimistic estimates, developers produced a system filled with defects
Implementation phase	- Implementation planned long in advance - Training developed and given to target users - Managers prepared for dialogue with associates - Effective communication with users - Pilot users involved early - Top management backing - Quality meetings	- Training months before the system was launched - Training rushed and presented when many users were on holiday. - User expectations not met
Operation and Maintenance phase	- Effective support strategy - Technology intuitive to use - Technology transparent to users - Clear advantages and time savings for the users (work system participants)	- Inadequate resources for the system - Not possible to print screen content for lack of color printers

3.5   Paper about stakeholders roles in successful and unsuccessful projects

As part of the background for a discussion of stakeholder roles, Pries-Heje, (2003) compared five successful versus unsuccessful diffusion and implementation projects. Table 1 lists the reasons for success and failure, categorized by phase of the WSLC. In some instances, it was unclear whether a particular factor such as "top management backing" pertained mostly to a particular phase or applied across all the phases.

4.   SEEING ORGANIZATIONAL INNOVATION THROUGH A WORK SYSTEM LENS, WHETHER OR NOT INFORMATION TECHNOLOGY IS INVOLVED

The previous section demonstrated that Figures 1 and 2 could be applied readily to the situations discussed in papers in the 2003 IFIP 8.6 working conference. This section explores ramifications of using a work system lens to understand IT innovation. It covers topics including different views of IT innovation, the possibility of measuring the size of an IT innovation, and a comparison with other frameworks for exploring IT innovation.

4.1   IT Innovation: Thing or Process? About IT or about a Work System?

Table 2 compares four views of IT innovation based on two distinctions: (1) IT innovation can be viewed as a process or as a thing and (2) IT innovation can be about IT or about a work system. The four views are named the vendor view, the diffusion view, the work system view, and the organizational change view.

The views in Table 2 have different associations. The vendor view initially focuses on product engineering and later focuses on sales and marketing. The diffusion view focuses on communication, awareness, beliefs concerning early or late adoption, and individual choice. The work system view is about organizational performance. The organizational change view is about how organizations change over time.

All of these views are linked. For example, a vendor's success depends on diffusion of products in a marketplace. For IT products used in organizations, success depends on whether they are incorporated into work systems (or into infrastructure that is hidden from work system participants.)

Similarly, in an organizational context (when usage is not voluntary and independent), the diffusion view merges into the work system as product awareness leads to adoption and usage. Links in the opposite direction are also worth noting because many work system innovations would never be imagined or realized without the vendor efforts to create products and raise awareness, and without diffusion phenomena that spread awareness and demonstrate potential applicability. Because we are focusing on IT innovation in organizations, we will pursue the work system and organizational change views.

Table 2. Comparing four views of IT innovation

	Innovation as a thing	Innovation as a process
The innovation is about IT	Vendor View: The innovation is a change in hardware and/or software. Example: a new computer chip that uses less energy	Diffusion view: The process is the diffusion of usage of the hardware and/or software Example: Process by which teenagers adopt cellphone cameras.
The innovation is about a work system	Work system view: The innovation is a change in a work system Example: Improvement in inventory tracking through use of RFID tags.	Organizational change view: The innovation is a process of changing a work system Example: Process of changing from an old inventory system to a new inventory system.

From a work system viewpoint, an IT innovation is a change in a work system that is unique enough to be called an innovation in the setting. Although IT innovations always involve IT in some way, they can actually start in different parts of a work system. For example:

- *Starting from technology*: New technology (such as RFID tags) or better use of existing technology (such as fuller use of Word templates) makes it possible to change work practices.
- *Starting from information:* Intention to use different information or provide information in a different form or level of detail (e.g., new graphics or new ETL (extract, text, load) requirements in a data warehouse) leads to innovative use of existing or new technology.
- *Starting from participants*: Enrolling tech-savvy participants or providing training on technology in a work system leads to new possibilities for doing work differently (e.g., train CAD users to use more capabilities of software that they have used at a 20% level for the last three years).
- *Starting from work practices*: Change the business process or change aspects of sense making and articulation work that make it possible to use technology more effectively for better results (e.g., change from assembly line logic to case manager logic).

- *Starting from the products and services*: Improve a work system's products and services by incorporating digitized information or even new IT hardware that provides additional value for customers (e.g., present bills electronically, provide automatic analysis of bills).

The assumption that IT innovation need not start with IT might seem unusual, or possibly self-contradictory, but this broader view could motivate deeper understandings of important phenomena. For example, better understanding of complementary efforts, assets, and investments related to IT diffusion and assimilation might help IT vendors provide more value for their customers through better, more adaptable products and better help in configuring and using those products to make more of a difference for users.

4.2   IT Innovation as Technology Upgrade or Work System Change?

Previous comments about successful and unsuccessful projects mentioned by Pries-Hege (2003) indicate that the phases in the work system life cycle model map to real situations involving IT innovation. Comments about negotiation and accommodation in that paper and other conference papers are consistent with inclusion of Figure 2's inwardly curved arrows that represent unplanned incremental changes and adaptations. To look at life cycle issues further, consider the distinction between innovations that are technology upgrades and innovations that are changes in work practices.

In a technology upgrade, the users often continue doing the same work in basically the same way, but with better tools. Such technology upgrades are less disruptive than changes in work practices because the upgrades can be accomplished as a technical task. The simplest changes of this type might be called "plug-ins" because the change is accomplished by substituting a tool (e.g., a cellphone or computer chip) with greater capabilities for another tool having lesser capabilities. Upgrades that require parameter-drive tailoring to the specifics of the situation (e.g., upgrade to a new personal computer) might be called "plug-in and configure." In both cases, an upgrade process almost transparent to the user might provide significant benefits in cost-effectiveness plus access to infrequently used features or capabilities.

In a work system change involving more than a technology upgrade, the innovation is visible to the users and may require substantial change by users. Alternative processes for innovations of this type include "acquire, configure, and implement" and "design, build, and implement." ERP and CRM projects exemplify the first type of process; building a new software application for internal company use exemplifies the second. The advantages and disadvantages of acquiring commercial application software versus building customized software are well know and will not be discussed here.

Although pure versions of those processes differ substantially, success in each case depends on up-front analysis that improves fit to reality, assures technical quality prior to implementation, and facilitates implementation. Both the technical upgrade processes and the work practice change processes fit directly into the work system life cycle model.

4.3   Is it Possible to Measure the Size of an IT Innovation?

The foregoing qualitative distinction between technology upgrades and work practice changes did not try to quantify the extent of changes related to an IT innovation in an organization. Although an innovation's success is often measured as changes in costs, productivity, or output, performance measures such as these do not describe the size of the changes an IT innovation enables. A work system approach might help in quantifying the size of an organizational IT innovation because larger organizational changes generate higher levels of inherent risk, resistance, and controversy.

It is not obvious how to compute the size of an IT innovation. The degree of change in the technology itself says little about the extent of effects felt by work system participants. The planned monetary expenditure and the number of people whose work practices will change are partial characterizations of the amount of change, but each has shortcomings. Brynjolfsson (2003) notes that the vast majority of expenditures on IT innovations are for complementary processes, assets, and knowhow, rather than for technology. Many of those expenditures are hidden in operational budgets not designed to highlight efforts absorbed by changing work practices. Estimates based on the number of people whose work practices change may be misleading because that number may not reflect the extent or significance of the change. In addition, implementation efforts required for similar changes in different organizations differ based on factors including the quality of leadership, the organization's recent history, organizational inertia, and personal aversion to change in those organizations.

The work system framework leads to preliminary ideas that can only be sketched here, but that indicate possible directions for developing usable measures based on the following 3 assumptions:

1. Innovation is more than change. Innovation is change in a direction that is unfamiliar in the setting. Thus, an ERP implementation in a company might be an innovation even today after hundreds of companies have taken that path. If innovation were limited to first time inventions or contemporaneous second tries, recognizing an innovation would require global knowledge of all comparable situations.

2. Innovation is gauged by the degree of newness in the changes introduced across the work system, not just within a particular element such as the technology. The earlier example of RFID tags shows that these changes occur throughout the work system:
   - new IT and non-IT technology (such as the physical layout required by the limitations of RFID transmission),
   - new information made available
   - change in work practices to use the information and accommodate the technology
   - change in the knowledge and capabilities of the participants
   - change in the products and services produced by the work system.
3. The size of a work system is related to the number of participant hours devoted to the work system within the relevant time frame. This gives equal weighting to participant hours for all participants, but recognizes that some work system participants may contribute to many different work systems during any week (e.g., during a week a manager may plan for the future, help subordinates do their work, and serve on external task forces). Note that the participant hours are hours performing the work in the operational system, not hours devoted to creating that system.

Ignoring typical concerns about scaling, assume that all of the changes and newness measures are expressed in a numerical scale from 0 to n. The size of an IT innovation can be quantified roughly as follows:

Size of innovation =
    {WS size} X {WP newness} X {sum of other newness}, where
- WS Size = work system size, measured as the number of participant hours used in the work system's operation within the relevant time frame
- WP newness = the newness of the work practices
- Sum of other newness = sum of the newness within four other work system elements, technology, information, participants, products and services.

Although not a definitive in a mathematical sense, the formula represents a direction for comparing the size of IT innovations. It implies:
- IT innovations are not just about introduction of new IT.
- Major IT innovations tend to affect every aspect of a work system.
- A major change in IT infrastructure hidden from the users (e.g., new middleware) is a small IT innovation in an organizational sense.
- An IT innovation that is visible to users could still be a small innovation in an organizational sense if there is little change in work practices, such as when new laptops perform existing functions or when new application software is configured to replicate existing work practices.

– Introduction of new application infrastructures such as ERP affect many work systems. Therefore the scale of change for an ERP introduction may be orders of magnitude greater than the scale of change for localized software innovations or new communication technologies such as instant messaging whose usage is voluntary.

– IT innovations introduced without reference to an existing or planned work system have a higher probability of being ignored or having little impact. (e.g., previously mentioned examples related to healthcare systems and Lotus QuickPlace)

4.4   Does the Work System Lens Fit with Other Frameworks for Exploring IT Innovation?

The various papers presented at the 2003 IFIP 8.6 meeting mentioned a number of methodologies and theories including actor network theory, diffusion of innovation (DOI), soft system methodology (SSM), and several models of stages of implementation. Comparing some of these with the work system lens helps in recognizing its generality and limitations.

4.4.1   Comparison with Actor Network Theory

Actor network theory (ANT) models the process of organizational change as a set of interactions among different actors.

"ANT develops from the idea that entities take their form and acquire their attributes as a result of their relations with other entities (Law, 1999). In this scheme entities have no inherent qualities as being large or small, human or nonhuman, etc., but rather, as Law points out, such divisions or distinctions are understood as effects or outcomes. They achieve their form as a consequence of the relations in which they are located." (Nilsson et al, 2003)

The non-human actors can include technology and infrastructure:

"Modern technology will, at least as seen from each organization or each government, appear as an independent actor (Monteiro, 2000)" ...... "Ciborra (2000) argues that since an information infrastructure is so deeply sunk into social practices and shaped by factors not in control by one company, it makes sense to view an information infrastructure as an actor and to describe the relationship between a company and the infrastructure as more symmetrical." (cited by Sørgaard, 2003)

If a work system's technology can be viewed as an actor and the infrastructure that supports it can be viewed as an actor, it is possible that an entire work system and its specific elements such as work practices, information, and products and services can be viewed as actors in the actor-network. Further, the principle of alignment says that these elements change over time (through the planned and unplanned change in Figure 2) in order to remain aligned internally and externally.

Accordingly, it is possible to say that the work system framework identifies classes of actors that should be considered in the actor-network, and further, that the arrows within the work system framework (Figure 1) indicate principle vectors along which those specific actors should align. From this perspective, the work system framework might be seen as a template for identifying relevant actors for an ANT analysis related to an IT innovation. Similarly, the work system life cycle model could be useful in describing part of the path along which the actor network will travel as it negotiates and configures the role of an IT innovation or rejects it. Thus, ANT might provide a vocabulary and analytical approach that supports the use of work system concepts. In return, work system concepts might identify likely classes of actors in an ANT analysis of an IT innovation.

4.4.2   Comparison with Soft System Methodology (SSM)

SSM is a method for defining and analyzing system-related problems and opportunities. Checkland (1999, pp. A3-A15) notes that SSM has evolved over three decades. An iterative seven-stage methodology (p. 163) emerged during the 1970s only to be revised as four main activities (p. A15):
1. "Finding out about the problem situation, including culturally/ politically;
2. Formulating some relevant purposeful activity models;
3. Debating the situation, using models, seeking from that debate both (a) changes which would improve the situation and are regarded as both desirable and (culturally) feasible and (b) the accommodations between conflicting interests which will enable action-to-improve to be taken;
4. Taking action in the situation to bring about improvement."

SSM calls for thinking about the "root definitions" of systems in terms of a set of elements somewhat similar to those of the work system framework (Figure 1). The corresponding elements in SSM have the acronym CATWOE, which stands for customers, actors, transformation process, Weltanschauung (world view), owner, and environmental constraints.

The work system approach tries to be more prescriptive than SSM through the use of pre-defined frameworks for describing a system (Figure 1) and how it evolves over time (Figure 2). Not discussed in this paper, the work system approach also calls for explicit use of work system principles and a large number of system concepts in a semi-structured systems analysis approach. (Alter, 2002). The system concepts are being compiled and organized as "Sysperanto," an ontology of the IS field that is currently under development. (Alter, 2004)

4.4.3   Comparison with Innovation or Implementation Models

The IFIP 8.6 papers from 2003 also mention the diffusion of innovation model (DOI) plus several other innovation or implementation-related models that can be compared with the work system life cycle model (WSLC).

Diffusion of innovation (DOI) theory is centered on the diffusion of technical innovation. The spotlight is on the technology, with the implicit assumption that technology users have the right to adopt it or not. This assumption is incorrect in many situations in which users must adopt the technology in order to play their roles as work system participants. The work system approach avoids a technology-focused view of change by assuming that an IT innovation must be assimilated into a work system to survive forces of rejection and attain significant impact. Accordingly, diffusion-related phenomena such as path dependence and windows of opportunity (Tyre and Orlikowski, 1993) can be interpreted in terms of the details of the development and implementation phases of the WSLC. Similarly, assimilation gaps (Fichman and Kemerer, 1999) and marketing "chasms" (Moore, 1991) can be interpreted as a summation of occurrences and non-occurrences across work systems that might adopt a particular technology.

Nilsson et al (2003) mention Swanson's (2001) "concept of innovation processes as a story having four phases: comprehension, adoption, implementation, and assimilation." The phases of the WSLC are directly comparable with Swanson's four phases. Initiation corresponds with comprehension, but development involves much more than adoption. Adoption sounds as though the crux of the matter is acceptance or rejection of a technical option. In contrast, the development phase might involve acquiring and configuring hardware and software, but might also involve building custom software. Implementation is the third phase of both, but operation and maintenance extends further in time than assimilation. The WSLC assumes that operation and maintenance continues with current technologies until another iteration of the cycle begins.

Figure 1 in Heikkilä et al (2003) is "an IS implementation process" that proceeds through four phases, planning, procurement, implementation, and use. Because this focuses on IS implementation rather than phases of innovation processes, it is closer to the WSLC model than Swanson's model (above). An important difference, however, is that the IS implementation model assumes that the topic is an information system that is being implemented and then "used" in the final phase. Although the WSLC can be used to describe an information system project (because an information system is a special case of work system), the WSLC is designed to describe the evolution of any work system. Furthermore, as information systems are increasingly integrated with the work systems they support, the idea of information system "use," rather like use of a hammer or lawn mower, is increasingly less realistic.

5.   A FINAL EXAMPLE: IT INNOVATION IN IT GROUPS

Instead of closing by summarizing the foregoing discussion, we will look at one more example, the disappointing adoption of an IT innovation called CASE (computer aided software engineering) technology. It once seemed that CASE would be "the new big thing," and that it would revolutionize system development in IT groups. The promise and hype were great, but the extent of adoption has been disappointing, as demonstrated by a lengthy "assimilation gap" between acquisition of CASE and use for 25% of new applications. (Fichman and Kemerer 1999) Based on data through 1993, after 54 months only 24% of CASE acquisitions in their survey had resulted in that level of use. Around the same time, Orlikowski (1993) compared two CASE implementations and concluded that "the adoption and use of CASE tools should be conceptualized as a form of organizational change and that such a perspective allows us to anticipate, explain, and evaluate different experiences and consequences following the introduction of CASE tools in organizations." Noting that "even with its many benefits, most organizations have found it difficult to implement CASE," Sumner and Ryan (1994) tried to find an explanation by identifying critical success factors in information systems analysis and design and determining whether CASE tools support these critical success factors." They concluded that CASE tools support technical analysis, the identification of "the processes and data which are needed for correct task accomplishment by the work system [that is being improved]," but that "social analysis is not well-supported by existing CASE tools." Furthermore, looking at the work within the IS organization, they concluded, "for CASE to be effective, an organization may need to view information systems development as a work system."

The basic questions that might have been asked to assess CASE as an IT innovation are the same as the questions that should be asked when thinking about improving any work system, whether or not IT is involved and whether or not the setting is an IT group or another area of the business:

- What is the work system, how well does it operate, and how well should it operate?

- How good are the work system's products and services, and how good should they be?
- Within the work system, what are the strengths and shortcomings of current work practices, technology, information, and staffing?
- What are the desired changes in work practices, products and services, information, and technology?
- What will the work system participants have to learn and how will they be motivated to change?

Thinking about CASE (or any other technology) as an IT innovation to be diffused, installed, or adopted does not address these practical issues with enough force and direction. Following an IT innovation from the original inspiration to the reality of better performance requires something like a work system approach.

REFERENCES

Alter, S. (2002) "The Work System Method for Understanding Information Systems and Information Systems Research", *Communications of the AIS* 9(6), pp. 90-104

Alter, S. (2003) "18 Reasons why IT-Reliant Work Systems Should Replace the IT Artifact as the Core Subject Matter of the IS Field," *Communications of the AIS*, 12(23), pp. 365-394

Alter, S. (2004) "Sysperanto – A Theory-Based Ontology of the IS Field," Under review.

Billeskov Bøving, K. and Bødker, K. (2003) "Where is the Innovation? The Adoption of Virtual Workspaces," IFIP 8.6 Working Conference on Diffusion and Adoption of Networked Information Technologies, Copenhagen, Denmark, Oct. 6-8.

Bostrom, R.P. and J.S. Heinen (1977a), "MIS Problems and Failures: A Socio-Technical Perspective. PART I: The Causes." *MIS Quarterly*, 1(3), December 1997, pp. 17-32.

Bostrom, R. P. and J. S. Heinen (1977b), "MIS Problems and Failures: A Socio-Technical Perspective. PART II: The Application of Socio-Technical Theory." *MIS Quarterly*, 1(4), December 1997, pp. 11-28.

Boyle, M. (2003) "Wal-Mart keeps the change," *Fortune*, Nov. 10, 2003, p. 46.

Brynjolffson, E. (2003) "The IT Productivity Gap," *Optimize*, Issue 21, July 2003, accessed at http://www.optimizemag.com/issue/021/roi.htm on Oct. 20, 2003

Checkland, P. (1999) *Systems Thinking, Systems Practice*, Chichester, UK: John Wiley.

Ciborra, C., (1996), "Introduction: what does groupware mean for for the organization hosting it?" in *Groupware and Teamwork. Invisible aid or technical hindrance?* edited by C.Ciborra, J.Wiley and Sons.

Davis, L.E. and J.C. Taylor *eds.* (1979) *Design of Jobs*, 2nd ed., Santa Monica, CA: Goodyear Publishing Company.

Fichman, R.G. and Kemerer, C.F., (1999) "The Illusory Diffusion of Innovation: An Examination of Assimilation Gaps," *Information Systems Research*, (10:3), Sept. 1999, 255-275.

Forsman U., & Nurminen M.I. (1994). "Reversed Quality Life Cycle Model," in *Human Factors in Organizational Design and Management - IV*, Bradley G.E., and Hendrick H.W., *eds.*, Elsevier Science B.V., 393-398.

Hedström, K, (2003) The Socio-Political Construction of CareSys: How Interest and Values Influence Computerization," IFIP 8.6 Working Conference on Diffusion and Adoption of Networked Information Technologies, Copenhagen, Denmark, Oct. 6-8.

Heikkilä, J., Vahtera, H, and Reijonen, P., (2003) "Taking Organizational Implementation Seriously: The Case of IOS Implementation," IFIP 8.6 Working Conference on Diffusion and Adoption of Networked Information Technologies, Copenhagen, Denmark, Oct. 6-8.

Johansson, B. (2003) "Exploring Application Service Provision: Adoption of the ASP Concept for Provision of ICTs," IFIP 8.6 Working Conference on Diffusion and Adoption of Networked Information Technologies, Copenhagen, Denmark, Oct. 6-8.

Law, J. (1999) "After ANT: complexity, naming, and topology," in *Actor Network Theory and After*, ed. By J. Law and J. Hassard, Blackwell Publishers.

Lines, K., Viborg Andersen, K. and Monteiro, E., (2003) "MIS and the Dynamics of Legitimacy in Health Care," IFIP 8.6 Working Conference on Diffusion and Adoption of Networked Information Technologies, Copenhagen, Denmark, Oct. 6-8.

Moore, G.A. (1991) *Crossing the Chasm*, HarperCollins.

Mumford, E. (2000) "Socio-technical Design: An Unfulfilled Promise?" *Proceedings of IFIP W.G.8.2 Working Conference 2000, IS2000: The Social and Organizational Perspective on Research and Practice in Information Systems*, Aalberg, Denmark, June 2000.

Muzzi, C. and Kautz, K., (2003) "Information and Communication Technologies Diffusion in Industrial Districts: An Interpretive Approach," IFIP 8.6 Working Conference on Diffusion and Adoption of Networked Information Technologies, Copenhagen, Denmark, Oct. 6-8.

Nilsson, A, Grisot, M. and Mathiassen, L. (2003) "Imposed Configurations by Networked Technology," IFIP 8.6 Working Conference on Diffusion and Adoption of Networked Information Technologies, Copenhagen, Denmark, Oct. 6-8.

Nurminen M.I., (1988). "People or Computers: Three Ways of Looking at Information Systems", Studentlitteratur, 1988

Orlikowski, W.J. (1993) "CASE Tools as Organizational Change: Investigating Incremental and Radical Changes in Systems Development," *MIS Quarterly*, Sept. 1993, pp. 309-340.

Pries-Heje, J. (2003) "Modeling the Primary Stakeholders in IT Diffusion and Adoption," IFIP 8.6 Working Conference on Diffusion and Adoption of Networked Information Technologies, Copenhagen, Denmark, Oct. 6-8.

Sørgaard, P. (2003) "Management and Co-ordination of eGovernment," IFIP 8.6 Working Conference on Diffusion and Adoption of Networked Information Technologies, Copenhagen, Denmark, Oct. 6-8.

Sumner, M. and Ryan, T. (1994) "The impact of CASE: Can it achieve critical success factors?" *Journal of Systems Management*, 45(6), pg. 16, 6 pages.

Swanson, E.G. (2001) "Telling an Innovation Story," in *Diffusing Software Product and Process Innovations*, eds., M.A. Ardis and B.L. Marcolin, Kluwer Academic Publishers

Taylor, J.C. and Felten, D. F. (1993) *Performance by Design: Sociotechnical Systems in North America*, Englewood Cliffs, NJ: Prentice-Hall, 1993

Tyre, M.J. and Orlikowski, W.J. (1993) "Windows of Opportunity: Temporal Patterns of Technological Adaptation in Organizations," *Organizational Science*, 5(1), 1994, pp. 98-118.

SUCCESS AND FAILURE REVISITED IN THE IMPLEMENTATION OF NEW TECHNOLOGY: SOME REFLECTIONS ON THE CAPELLA PROJECT

Tom McMaster[1] and David Wastell[2]
[1]Information Systems Institute, University of Salford, UK; [2]Department of Computation, UMIST, Manchester, UK

Abstract   Reports about success and failure in IS implementation initiatives are plentiful, but the use of such labels is too often uncritical, pejorative and simplistic. We reflect on a two-year implementation project that had elements of both failure and success, using this to revisit and examine more closely the meanings of the terms, and to consider other key dimensions of the innovation process, such as the pivotal role of organizational culture and crises in decisively shaping outcomes.

1.   INTRODUCTION

In mid-July 2003, 250 British Airways check-in and customer service staff staged a two-day wildcat strike at London's Heathrow Airport in protest at the company's unilateral decision to implement a new swipe-card time-recording system. Although the proposal for such a technological innovation undoubtedly looked extremely attractive on paper, the upshot was that around 80,000 passengers suffered massive inconvenience by the action, and it cost the company in the region of £50 million, not to mention the incalculable damage to its image and passenger relations according to subsequent and extensive media coverage. By the end of that month the plan had been shelved if not entirely abandoned, thus demonstrating once again that a long tradition of innovation failure – especially when it comes to the implementation of new technology – is a phenomenon that appears no less a problem today, even in the largest and most recognisable of branded companies, than it has been in the past with others.

It is generally considered that the failure to implement innovative technical systems, for whatever reason, is synonymous with the term 'system failure' (see for example Lyytinen and Hirschheim, 1987). We suggest that the designation of such judgemental and politically loaded terms as failure and success merely reflects a single view. In the above case, and perhaps more often than not, it tends to reflect a managerial perspective that may be prone to 'pro-innovation bias', that is, the assumption that the innovation is intrinsically good and desirable, while subjugating or denying the views of those who see it as an imposition and a threat. Nevertheless, alternative views may be at least as valid and compelling as those promoted by managers. The strikers at BA might well consider the system to have been a complete success, since its trajectory seems to have ensured that so-called 'spanish practices'[1] among themselves and other staff are likely to remain unchallenged for the foreseeable future. It would seem that pinning the label of success or failure to a project sometimes depends on which side one finds oneself politically aligned.

The purpose of this paper is not merely to highlight the fact that those with different perspectives will interpret the implementation of the same technical innovation in very different ways, important though this is. We wish equally to draw attention to the troublesome nature of the over-simplified one-dimensional concept that is frequently signified by the labels 'failure' and 'success' whenever they are used to describe the introduction of new technical systems into organisational arenas. Too often, these terms are employed to describe the object of implementation efforts, when closer examination would instead reveal far less certainty, much wider ambiguity, and immensely greater complexity surrounding the implementation of new technology, where the product of these efforts is more likely to be a whole range of benefits and disbenefits. Some of these are no doubt predictable, but others may not be. We feel therefore that a more critical approach toward these terms and concepts is more likely to provide richer explanations about why there are often very different perceptions about the same processes, the elusive nature of success and failure, and the peculiar fact that what is often considered failure today, may not be considered so tomorrow (Bignell & Fortune, 1987, Larsen & Myers, 2000).

---

[1] A term in common currency in press reports of the time to denote unofficial rule 'bending' – in this case, leaving the job early and having colleagues 'clock-out' for them – as an example of one of the practices that the system was designed to combat.

Here we recount an attempt to introduce a CASE tool along with a related methodology into the software development department of a large public service organisation that in one sense seems to have failed – that is, the technology failed to infuse or become incorporated into the organisation's working practices, and was finally abandoned. Despite this however, at the end of the 2-year implementation exercise the organization found and felt that it was significantly better off in a number of respects than it was at the start of the project directly as a result of the processes involved, and in ways that were not, and could not have been foreseen. Was this then a failure? Such stories concerning the ephemeral and ambiguous nature of success and failure are not common, but we are sure that while rarely reported, nevertheless the phenomenon itself is far more widespread than it seems.

The work presented here represents the authors' reflections about an EU funded project known as the CAPELLA project, in which they played important roles. The paper is organised in the following manner; in the next section we discuss some issues related to success and failure and its meanings in the context of IS based innovation that may be found in the literature. In section 3, we comment upon the research approach adopted in the course of the CAPELLA project, which we describe in more detail in section 4. In the final section, we provide a discussion along with some tentative conclusions.

2. BACKGROUND

Various works discuss failure both in terms that are generally applicable, and in terms that are specific to the IS and IT research communities. We include the seminal works of Bignell & Fortune (1984), and Lyytinen & Hirschheim (1987). The former describe failure as being an assessment of the outcome of an activity based upon the values of the person making the judgement, and we find that failure is not a single thing, but is usually multi-causal, with multiple effects. While it remains an elusive and indistinct concept, failure is nonetheless about the evaluation of a performance set against an expectation, and there is a requirement for it to be historically contextualised, since, as suggested above, often what might have once been perceived as a failure, may not be seen as such on a different occasion.

Lyytinen and Hirschheim (1987), who are specifically concerned with the term as it applies to the IS research communities describe four 'classes' of failure, namely, *correspondence failure*, *process failure*, *interaction failure* and *expectation failure*. Correspondence failure means that design objectives have not been met. Process failure relates to the development process where budget, time or other resource allocations have overrun to the point where any benefits expected from the proposed system have now been negated, or where a workable system has not been achieved. Interaction failure is the argument that the low-level use of the system can be interpreted as failure, and expectation failure is simply that the system has failed to meet the expectations of at least one stakeholder group. These authors additionally take the view that the first three classes of failure may be subsumed by expectation failure, since this is effectively a superset of the others.

Sauer (1993) likewise addresses the IS community, providing as he does a number of case studies. His definition differs somewhat from Lyytinen and Hirschheim insofar as he limits his definition of failure only to those systems that are neither under development, nor in operation, eliminating some at least of Lyytinen & Hirschheim's category of interaction failure. Since the system may be serving at least one person's interest, it therefore cannot be described as a failure. He proposes a 'triangle of dependence' model where the information system depends on the project organisation, the organisation depends on its supporters, and the supporters in turn depend on the information system. The dynamics involved in this tripartite dependency coalition are necessary for understanding system failures, which in short are due to disruption or dissonance in the loop.

Flowers (1996) also provides a number of case studies in IS failure, and his definition of failure generally corresponds to those offered by Lyytinen and Hirschheim, however he proposes a factor-based causal model ('critical failure factors'). These include factors relating to the *organisational context*, such as hostile culture and poor reporting structures; *project management*, including over-commitment and political pressures; and *conduct of the project*, including factors related to various phases of the project that he envisages as a traditional waterfall sequence of activities. Flowers also suggests that most failures result in the players 'burying the facts' of the case, no doubt to avoid embarrassment and fault attribution.

In addition to the above, various individual studies of specific system failures have been reported in the IS research literature. These include Robinson (1994) on the failure of the London Ambulance Service, Kautz and McMaster (1994) on the failure to adopt SSADM in a local government IT department. Wastell (1996) on a similarly abortive attempt to introduce SSADM into the private sector, and McMaster et al (1997) on the failure to adopt a technological innovation into a car-park environment in a UK Institute of Higher Education, to name a few among many others.

Similarly, studies of IS 'success' are not in short supply. These include Teng, et al. (1998), Bussen and Myers (1997), Seddon (1997), and Carr and Johansson (1995) to provide some examples. In addition, the IS literature –

especially that with a focus on diffusion studies – is not short on examples proposing 'critical success factors' to guide would-be implementers through the hazards of their strange esoteric activities. However, almost as rare as hens' teeth are studies that treat failure / success as the ambiguous and ephemeral entities that we take them to be here.

We conclude this section by observing that studies of successful projects, and studies of unsuccessful projects are relatively plentiful in the IS research literature, while studies where the focus is not on the project but on the nature of success and failure per se, are far less so, and it is upon this that we focus our attention. Before moving to a description of our case, we offer some commentary on the research approach.

3. RESEARCH APPROACH

The CAPELLA project was a two-year software process improvement (SPI) experiment funded by the European Commission's Esprit initiative (known as ESSI – the European Systems and Software Initiative). Under this scheme, funding was available to software development organisations in order to promote 'best practices'. The organisation concerned was the software development section of the Information Technology Services Division (ITSD) of Salford City Council, a medium-sized UK local government council.

The authors / researchers represented the Information Systems Research Institute (ISRI) at the University of Salford, which, along with the Norwegian Computer Centre (NR) in Oslo, acted as advisors / consultants to the ITSD. The consultancy roles included the provision of:
– advice on CASE implementation
– advice on research design and methodology
– design of the evaluation framework (metrics)
– management of the evaluation phase of the experiment
– data creation, analysis and modelling
– assistance in the dissemination activities, and
– quality review of the various deliverables

Since the researchers were intimately involved in the execution of the experiment alongside managers in the development organisation, then the term 'action research' is most appropriate to describe the approach employed. More specifically, the approach corresponds closely to what Reason and Bradbury (2001, pp. xxv-xxvi) call 'second-person' action research practices, namely face-to-face interpersonal dialogue including the development of communities of enquiry in the learning organisation that focuses on matters of mutual concern. Since the object of our research was the 'organisation', then action research, according to Galliers' taxonomy (1992), is a highly appropriate approach. However while this might describe the approach to the CAPELLA project described further in the following section, it does not necessarily apply to this paper that more accurately represents a reflective historical view of some aspects of the project that was formally concluded in April 1999.

The fact that our case study is an historical reflection has both advantages and disadvantages. Advantages include distancing and reflecting over time. Disadvantages are that personal recollections may be flawed and that accounts of the processes may therefore be factually inaccurate. We are not unduly concerned about these; detailed reports and accounts of the project and its evolution have been meticulously kept by those local government officers involved in the project, and additional accounts have been published elsewhere (Wastell et al., 1999a, Wastell et al., 1999b). In any case, we are not so much concerned here with simply trying to recount bald facts. Instead, we are interested to see what we can learn from reflecting on the processes and conduct of the experiment that form part of our own experiences. Historical reflections, as Gummesson (1988) puts it, are about building "hermeneutic bridges" between the past and the present, to enable an increased understanding of the current and future state of the organization. We now proceed to a description of the CAPELLA project.

4.   THE CAPELLA PROJECT

The CAPELLA project (CAse tools for Process Enhancement in LocaL Authorities) was a two-year Software Process Improvement initiative undertaken between 1997 and 1999, and funded to the sum of £250,000 under the European Commission's ESSI programme (ESSI Project 23832) as a Process Improvement Experiment (PIE). The principal organisation was the ITSD, with the ISRI at the University of Salford, and the Norwegian Computing Centre in Oslo providing consultancy services. The project had two main aims,

To evaluate and understand the implications of implementing a CASE tool and associated methods, and

To form a wider understanding of the issues and impacts of technological change within an organization.

Put simply, CAPELLA aimed to introduce a CASE tool and new working methods into the department, with the expectation that through this process a number of important benefits might accrue, including:

- Improved software quality
- Increased developer productivity
- Increased customer involvement
- Reduced development / maintenance costs
- Increased developer job satisfaction
- Improved customer satisfaction

4.1 The Project plan

The project plan consisted of three discrete phases; the first of these was concerned with the acquisition, installation and initial training in the use of the CASE tool, selecting those who would make up the initial 'centre of excellence' – that is those who were intended to acquire the technical expertise to subsequently advise others in the next phase of the project and beyond. This was to be initiated through the development of a small pilot project. During this period, a base-line project was also identified as a focus for the main experiment and roll-out of the methodology and tool use. In the first instance, this was to be a 48 person-year project for the Housing Department.

The second phase would be concerned with constructing a development methodology that would accompany the use of the CASE tool. It also concerned the development and capture of various metrics – for example the measurement of (developer) productivity, developer job satisfaction, and attitudinal issues such as customer satisfaction. The third phase was to be the main part of the experiment – the roll-out of the method and tool to wider developer use. This phase also included the requirement to reflect and report on the issues, problems and processes insofar as these had an impact upon the ITSD, and as it affected its customer base and customer perceptions of quality and other matters.

4.2 Problems encountered

The project officially began on March 1st 1997, but was almost immediately beset by difficulties starting with the loss of the baseline housing project. Over the course of the following year there were a number of additional unforeseen problems that had negative effects on the project including very high staff turnover, low morale, Y2K demands, and other matters. The baseline project was lost early in the process. This had been a 48 person-year integrated housing system featuring rent accounting, property repairs, property allocation and contract management. This was a hard blow, given that the European funding was based upon a proposal that assumed this as a 'given' and an essential basis for underpinning the experiment. A much smaller single regeneration budget (SRB) system for the Finance Department was found to replace this, though it never quite met the opportunities promised by the housing system.

A snapshot of the organisation in 1997/98 showed that there was a 27% rate in staff turnover. This included the ITSD's CAPELLA project leader, a senior member of staff, as well as two of those trained in the CASE tool that formed the centre of excellence, and others. One ironic aspect of this was that having received training in the tool, these developers found they were now more marketable with their newfound skills, and therefore able to command greater remuneration elsewhere than they could in local government.

Low morale was evident in the ITSD. This took several forms, but included a degree of cynicism towards 'just another management fad' as the project was perceived to be by some. Y2K demands were made on the ITSD that involved all staff during 1998. This came down through the larger bureaucracy (from the upper echelons of the Council) and despite the fact that personnel were supposedly ring-fence protected by European funding, they were nevertheless diverted into the pressing demands and preparations for the imminent new millennium. Other problems included the realisation that inadequate standards and methods existed within the ITSD, which had an immature planning and management culture, and that collating metrics was far more problematic than had been hitherto envisaged.

4.3   Some outcomes

Work on the implementation process progressed throughout the period March 1997 to April 1999 when the project was formally 'concluded'. The original project proposal was for eighteen months, but some of the difficulties necessitated two requests for short extensions. To put it succinctly and simply, the CASE tool was never fully deployed, not only in terms of the numbers of users – this always remained small – but also in terms of its capabilities and functionality, and it was finally abandoned after some limited use. The developers were unable to incorporate the tool's potential into their working practices for a variety of reasons, not least because of the fact that during this period there was a rapid decline in in-house bespoke systems in favour of widespread packaged software procurement. Today no development work outside of package configuration and the maintenance of legacy systems takes place.

Despite the failure to adopt the technology, there were also some very positive outcomes; spin-offs from the implementation effort that have bequeathed long-term benefits to the department. These included:
– The establishment of a working 'standards and methods' group

- Design and implementation of a project planning infrastructure
- Establishment of a peer review process for all present and future projects
- Staff retention measures to combat severe haemorrhaging of skilled staff
- An evolving 'metrics culture'
- Continued and ongoing collaboration with universities
- Establishment of a research and development team

In the face of these contrary results, how should the outcome of this two-year implementation effort be described? What other lessons can be learned from the case?

5. DISCUSSION

There are some similarities between this story and one told by Ginn (1994) where he describes the efforts to implement a computerised time-scheduling system into a US aerospace factory. Its purpose was to try to address the difficulties associated with moving aircraft parts in a timely manner during the construction process. Cranes to move aircraft parts need to be booked sometimes many months ahead with a fine degree of accuracy in terms of timing so as not to burden the company with undue overheads. Although the software system failed to become 'accepted, routinized and infused' (Saga & Zmud, 1994) and was in fact finally abandoned, nevertheless the company found that efficiency in that section was improved by around 15% as a result of the efforts to implement it.

We are not able to quantify the benefits to the ITSD in quite the same way. However all of the players are agreed that the benefits listed above have been realised. In addition, the ITSD continues to work with university-based researchers, and has expanded its collaborative efforts further to develop a methodology for handling process reengineering projects, known as SPRINT (Wastell et al., 2001) among other initiatives. Despite this positive outcome, the CASE technology itself was not adopted.

There would appear to be four main implications of the case for our theoretical understanding of technological innovation and its problematics, particularly in the area of software engineering. First, the case shows again (if further demonstration were needed) that methodological innovation is a perilous endeavor, in the sense that the introduction of new methods and tools runs a high risk of encountering resistance and rejection. IT-enabled transformation is generally fraught with the risk of miscarriage and failure, with the mortality rate for re-engineering efforts running as high as 70% (MacIntosh, 2003; Al-Mashari and Zairi, 1999) and software process innovation is no different. There is a significant body of research on the implementation of new software tools (Hardgrave et al, 2003) which evidences substantial resistance and frequent rejection (Orlikowski, 1993; Iivari, 1996). Much greater resistance might be expected for complete methodologies given the scale of re-structuring involved (Hardgrave, 2003). Methodologies are more embracing than individual tools, requiring substantial organizational change and the introduction of new processes and practices, rather than the more localized change entailed by the adoption of an individual tool. The rejection of CASE in ITSD confirms this general picture.

The second general lesson relates to the notion of "failure", specifically to the relative nature of this pejorative concept. Although the developers in the ITSD failed to adopt the CASE tool, is it fair to describe the project as a failure? We think not, as the technical artefact merely reflects a single component in a raft of activities and aims that made up the CAPELLA project, some of which were usefully deployed, others not so. The labels success and failure are attributed too often to initiatives in a naïve and simplistic way. While reports dealing with IS outcomes are relatively commonplace, few deal with these labels in a way that does full justice to the underlying complexity of situations and ideas. One exception is Larsen and Myers (2000), who provide an example of a BPR project in financial services firm in New Zealand. These authors address directly the problematic nature of success and failure, relating how shortly after completion of the project, the accounts of its outcome really depended on who was telling the story. They also suggest that such descriptions are more political declarations than statements of fact, and that success is a "moving target". Another example is found in Irani et al (2001), wherein the initial failure to implement an ERP system in a manufacturing company ultimately came to be seen more sanguinely as paving the way for a successful in-house development. It is clear that a more critical approach is required to the use of the terms success and failure, and that more circumspection is called for in the appellations given to attempts at technological innovation.

The third implication relates to the importance of organizational culture in shaping the trajectory of methodological innovation. Popular models of technological change, embodied in diffusonist formulations such as the Technology Adoption Model (Davis et al, 1989), tend to privilege attributes of the technology in accounts of outcomes. The present work in contrast underlines the importance of organizational culture. IT-enabled change is highly problematic in any organizational domain. The cultural circumstance of large public sector bureaucracies would seem to offer a particularly challenging milieu. They are hardly seen in the public imagination as the standard bearers of change and modernization. Several authors have contended that the failure rate for technologically-based innovation in public organizations is likely to be high (Kock et al, 1996) due to a range of factors:

greater regulation restricting the range of potential responses, rigid hierarchies and multiple stakeholders for many processes; overlapping initiatives and a rapidly changing policy environment; lack of resources and a staff culture emphasizing the need for consultation and consensual change (MacIntosh, 1997).

Internally, local councils tend to be deeply hierarchical, bureaucratic entities, often steeped in traditional methods and processes that mean changes to practice can be immensely difficult to achieve. Often positions of authority are held by those who have worked their way up through the hierarchy, having joined the organisation as juniors at a relatively young age. Bad practices are thus often preserved and perpetuated as part of the embedded institutionalised culture. We do not suggest that was a problem specific to the ITSD. Nonetheless it was clear that a significant constituency of developers saw the attempts to introduce the CAPELLA project as just another 'management fad' and perhaps therefore not worth making too much of an effort over. Another feature of local government is that there are few other organisations that are quite so explicitly political in nature. All councils are composed of elected members who are typically aligned with specific political parties. Within legal constraints and through controlling committees, they then attempt to implement policies through the departmental structures of the organisation that best reflect the political aspirations of the parties they serve. Due to the relatively transient status of elected members and the often poor understanding they have of the nuts-and-bolts detail of departmental service delivery, senior officers (full-time employees) often wield considerable influence and power. The present initiative was certainly supported politically, with a number of elected members and Chief Officers attending the CAPELLA launch. However whether this support cascaded down much beyond the upper hierarchy of the ITSD is not certain. It is possible that the enrolment of politicians by management had been seen as a means of trumpeting their own accomplishments in the eyes of those they strove to impress. Winning European funding by such a small strategic business unit is not common; many large councils often have special units, usually part of their Chief Executive's Departments, set up to exploit European funding opportunities. Making the launch of CAPELLA such a high profile affair also sends out the message, 'we know we have problems, but look, we're serious and we're doing something about it'.

Externally, local government is influenced by the vagaries of central government to a degree far in excess of most other organisations. New legislation and advice means that they have to respond, sometimes at relatively short notice to accommodate expediency and law, such as was the case for example in the introduction of the Community Tax (known as the

Poll Tax) in 1990. Councils were obliged to build or buy new computer-based systems to deal with this tax. Within two years the then Conservative government under Margaret Thatcher had to abandon this when in open defiance of government policy, many people chose to go to jail rather than pay this highly unpopular tax. There are no published details of the total national cost of the exercise, but it almost certainly ran to hundreds of millions of pounds to UK taxpayers. The 'Council Tax' immediately replaced the Poll Tax, so that yet another major round of software acquisition was demanded of UK councils, and there are many other on-going legislative requirements that do not attract the same level of public interest. The ITSD are among those that prefer to try and build their own in-house systems such as Council Tax, rather than purchase from commercial vendors. It is easy to imagine therefore that software developers in local government are not keen to have additional and 'unnecessary initiatives' such as the CAPELLA Project thrust upon them, as it must have appeared to some.

This brings us to our final point. Drawing on Actor Network Theory, we take a general, paradigmatic view of technological change as a dynamic process involving a shifting coalition of "actants", including inanimate elements such as the technology itself. Successful transformation reflects the progressive enrollment of these elements into a stable network that ultimately comes to embody the transformed organizational order. Critical in this process is the first stage, of *alignment*. This refers to the initial drawing in of "actants" (human and non-human) through *problematisation* (i.e. establishing that some problem exists and redefining the world accordingly, with the innovation as the key to the solution) and *interressement*, i.e. attempting to impose on others the new roles and identities implicit in the redefinition. If "successful", interressement leads to the progressive *enrolment* of actants in a stronger and stronger network of alliances supporting and promulgating the innovation. Finally there is *congealment,* whereby the innovation consolidates itself as an accepted fact of life, a "black box", as part of the transformed order.

Problematisation is the decisive first step in this transformation. "Success" depended on enrolling key constituencies within ITSD to see the CASE tool as a solution to the key problems perceived as facing the department. Wastell et al (2003) contend that the most important stimulus for organizational change is the perception of a serious threat to the organization. Without such a sense of crisis, inertia will tend to prevail and any mobilization for change will quickly peter out. The work here would seem to confirm this. There was no such sense of crisis strongly aligned with the messianic appearance of CASE. A successful bid for funding had been made, around the need of the organization to improve software quality and productivity. Whilst this sufficed to unite three groups of actants (the researchers, senior ITSD management and ESSI) around the rhetorical need for CASE, other key groups (i.e. the software developers who would use the tool) were not enrolled within this problematisation. Amongst these staff groups, CASE certainly figured in some minority problematisations, e.g. the need to advance individual careers, but these would tend to produce organizational fragmentation rather providing a strong force for collective transformation. It is clear that staff groups saw the key threat coming from other sources, Y2K in particular, and the CASE tool came to be seen as marginal within this problematisation. The study thus confirms the contested, dynamic nature of the innovation process and pivotal role of problematisation. The importance of some sense of crisis is attested in which the innovation offers a route to salvation; this is critical in shaping the trajectory and ultimate fate of the innovation. CASE came to suffer the same, unlamented fate as other artefacts, such as the Penny-farthing bicycle, that fail to address relevant problems as defined by influential stakeholder groups (Pinch and Bijker, 1984).

REFERENCES

Al-Mashari, M. and Zairi, M. (1999), 'BPR implementation process: an analysis of key success and failure factors', in Business Process Management Journal, Vol. 5, pp. 87-112.

Bignell, V. and Fortune, J. (1984), *Understanding systems failures,* Manchester University press in association with the Open University, Manchester.

Irani, Z., Sharif, A.M. and Love, P.E.D. (2001). 'Transforming failure into success through organizational learning: an analysis of a manufacturing system', in European Journal of Information Systems, Vol. 10, pp. 55-66.

Bussen, W., and Myers, M., D. (1997), 'Executive information systems failure: a New Zealand case study', in Journal of Information Technology, Vol. 12 (2), pp. 145-153.

Davis, F.D., Bagozzi, R. and Warshaw, P. (1989). User acceptance of technology: a comparison of two theoretical models. *Management Science*, 35, 982-1003.

Flowers, S. (1996) *Software failure: management failure,* John Wiley & Sons, Chichester.

Galliers, R., D. (1992), 'Choosing information systems research approaches' in Galliers, R., D. (Editor), *Information Systems Research: issues, methods and practical guidelines,* Blackwell Scientific Publications, London, pp. 144-162.

Ginn, M. L. (1994), 'The transisionist as expert consultant: A case study of the installation of a real-time scheduling system in an aerospace factory', In *Diffusion, transfer and implementation of information technology* (Ed, Levine, L.) Elsevier Science BV, Amsterdam, pp. 179-198.

Gummesson, E., (1988), *Qualitative Methods in Management Research*, Studentlitteratur, Sweden, Chapter 4.

Hardgrave, B.C., Davis, F. and Riemenschneider, C.K. (2003). 'Investigating determinants of software developers' intentions to follow methodologies', in Journal of Management Information Systems, Vol. 20, pp. 123-151.

Iivari, J. (1996). 'Why are CASE tools not used?', in Communications of the ACM, Vol. 39, pp. 94-103.

Kautz, K. and T. McMaster (1994), 'The failure to introduce systems development methods: a factor-based analysis', in Levine, L. (Ed), *Diffusion, transfer and implementation of information technology*, Elsevier Science BV: pp. 275-287.

Kock, N.F., McQueen, R.J. and Baker, M. (1996). 'Re-engineering a public organisation: an analysis of a case of successful failure', International Journal of Public Sector Management, Vol. 9, pp. 42-50.

Larsen, M., A., and Myers, M., D. (2000), 'When success turns to failure: a package-driven business process re-engineering project in the financial services industry' in the Journal of Strategic Information Systems, Vol. 8, pp. 395-417

Lyytinen, K., and Hirschheim, R., (1987), 'Information Systems Failures – a survey and classification of the empirical literature', *Oxford Surveys in Information Technology*, Vol. 4, pp. 257-309, Oxford University Press.

McMaster, T, Vidgen, R. and Wastell, D. (1997). Technology transfer: diffusion or translation. In: McMaster et al (eds.), *Facilitating technology transfer through partnership: learning from practice and research*, Ambleside, Chapman & Hall, 64-75.

McMaster, T., Jones, M., C., & Wood-Harper, A., T. (1997), 'Designing stakeholder expectations in the implementation of new technology: Can we ever learn our lessons?', in M. Kyng and L. Mathiassen (Eds), *Computers and design in context*, Cambridge Ma, MIT Press: pp. 239-265.

MacIntosh, R. (2003). 'BPR: alive and well in the public sector', in International Journal of Operations and Production Management, Vol. 23, pp. 327-344.

Reason, P. and Bradbury, H. (Eds.) (2001) *Handbook of Action Research: Participative Inquiry and Practice*, Sage Publications, London.

Saga, V. L. and Zmud, R. W. (1994), 'The nature and determinants of IT acceptance, routinization and infusion', In *Diffusion, transfer and implementation of information technology* (Ed, Levine, L.), Elsevier Science BV, Amsterdam, pp. 67-86.

Sauer, C. (1993), *Why information systems fail: a case study approach*, Henley-on-Thames, Alfred Waller Limited.

Teng, J., T., C. (1998), 'Profiling successful reengineering projects' in Communications of the ACM, Vol. 41 (6), pp. 96-102.

Wastell, D.G, (1996), 'The fetish of technique: methodology as a social defence', in Information Systems Journal, Vol 6 (1), pp. 25-40.

Wastell, D., G., Kautz, K., Kawalek, P., and McMaster, T. (1999a), 'Using CASE to Enhance Service Performance in Local Government: The CAPELLA Project', in M. Oivu and P. Kuvaja (eds.), *International Conference on Product Focused Software Process Improvement* (PROFES'99), VTT, Finland.

Wastell, D., G., Kautz, K., Kawalek, P., and McMaster, T. (1999b), 'Software Process Improvement Using CASE: Lessons from the Front-line', European Software Process Improvement Conference (EUROSPI'99), Pori, Finland

Wastell, D., Kawalek, P. And Newman, M. (2003) Plus ça change: defensive translations and resistance to IT-enabled change in local government. *Proceedings of ECIS 2003, Naples, Italy.*

Wastell, D.G., Kawalek, P. and Willetts, M. (2001). Designing alignment and improvising change: experiences in the public sector using the SPRINT methodology. In: Smithson et al. (Eds.) *Global Cooperation in the New Millennium, Proceedings of ECIS 2001*, Bled, Slovenia.

IT SYSTEMS TO SUPPORT INNOVATION
An empirical analysis of IT Systems to Support Innovative Product Development in Analog Devices B.V.

Brian Donnellan
Analog Devices B.V

Abstract: The provision of systems to support and promote innovation has become a significant concern for New Product Development (NPD) processes as they try to cope with the rapid rate of technology development, change of customer's needs, and shortened product life cycles. Companies such as Analog Devices Inc. (ADI) see the creation of an environment that encourages knowledge to be created, stored, shared and applied for the benefit of the organization and its customers as a key strategic activity. Despite the fact that such initiatives have been widely reported in the business press, the role of IT systems to support innovation in new product development is not well understood. This paper describes one such in initiative – as it was executed in practice in Analog Device's NPD organization in Ireland. The work is presented in the context of current research in Knowledge Management Systems (KMS). The structure of the paper is set out in Table 1.

Key words: Innovation, New Product Development , Knowledge Management, Knowledge Management Systems

Table 1. Paper Structure

Section	Contents
1: Introduction	The background for the study in terms of the challenges faced by new product development organizations as they attempt to apply IT systems to improve their competitive advantage in the market place. Introduction to the innovation initiative in ADI.
2: Research Method	Rationale for choosing action research (AR) methods and description of AR cycles.
3: IT Strategies to support Innovation	Review and analysis of current conceptual

Section	Contents
	models of systems to support innovation. Positioning of ADI IT systems in framework illustrating region of operation of each application.
4: IT Systems Framework to support innovation in NPD	Proposed framework synthesized from conceptual models and ADI's stage-gate NPD process. Description of three IT systems in use in ADI.
5: Conclusions	Conclusions and future research.

1.   INTRODUCTION

At a recent address to the Irish Management Institute, Michael Porter pointed out that Ireland is entering a new economic era and that the key element in its competitive agenda is the "strengthening of its innovative capacity" (Porter 2003). Porter's analysis concluded that Ireland's traditional competitive advantages are eroding because (i) competing locations have caught up in terms of business-friendly regulations and tax structure and (ii) rising cost levels are making Ireland's traditional position as a low cost location untenable. He proposed that Ireland needs to develop new strengths to emerge as an innovative economy.

It is clear that at the firm level there is a need for systems that provide an infrastructure to facilitate knowledge creation, storage, distribution and application. Such systems are designed to increase revenue and profits for an organization by

a) Improving the sharing of knowledge and best practice across the organization
b) Providing a faster solution development to technical problems and hence reduce TTM
c) Accelerating innovation rates by bringing diverse views and experience to bear on an issue
d) Breaking down geographic/organization barriers
e) Improving efficiency by learning from others

Both the academics and practitioners who have been involved in developing systems to support knowledge creation have tended to interpret such strategies as bifurcated into (i) those focusing on explicit knowledge (so-called "codification" or "cognitive" strategies) and (ii) those focusing on tacit knowledge (so-called "personalization" or "community" strategies) (Hansen, Nohria et al. 1999), (Swan, Newell et al. 1999).

There are significant innovation-related IT initiatives taking place in ADI, but without an overall guiding framework which could 'make sense'

((Weick 1979), (Weick 1995)) of the various activities. In this study, we present a framework for IT in NPD that we have derived, based on an understanding of the types of knowledge used in the design phase of NPD processes and a synthesis of current research on systems to support knowledge creation. In particular, we argue that a balanced approach needs to be taken when developing such systems. In the past, there has been a tendency to focus solely on codification or personalization strategies. We argue however, that while on one hand, some forms of codified knowledge lend themselves to a repository-based approach, however tacit knowledge is best managed by promoting human interaction.

2.   RESEARCH METHOD

A research method which has proven useful when research needs to be closely aligned with practice is that of action research (AR). Typically, an AR project is a highly participative model where researchers and practitioners focus on a real business project or problem as a starting point. Thus, all the associated risk and unpredictability of a real organisational situation is factored in from the outset.

The site for this research was a U.S. multi-national firm, Analog Devices Inc. (ADI). ADI is a world leader in the design, manufacture, and marketing of integrated circuits (ICs) used in signal processing applications. Founded in 1965, ADI employs approximately 8,500 people worldwide. Innovation has long been an integral feature of the landscape at ADI. Indeed, ADI's Chairman of the Board, Ray Stata, published some of the early research in the area (Stata 1989).

(Lewin 1947) originally described the action research cycle as having four basic steps: diagnosing, planning, acting and evaluating. Lewin saw the process as a "spiral of steps, each of which is composed of a circle of planning action and fact-finding about the result of the action" (p.206). The action research model being applied in this research is similar to that described in (Susman and Evered 1978) and sees the research process as a five phase cyclical process containing the following discrete steps: diagnosis, action planning, action taking, evaluation and learning.

The AR method recognises that a research project should result in two outcomes, namely an *action* outcome and a *research* outcome. Taking each in turn: firstly the action outcome is the practical learning in the research situation. Thus, a very important aspect of the research is the extent to which the organisation benefits in addressing its original problem. This serves to ensure the research output is relevant and consumable to practice. Secondly the research outcome is very much concerned with the implications for the advancement of theoretical knowledge resulting from the project. In this study there were two action research cycles. The first cycle of the action research project produced a new business process called "knowledge" embedded in a new framework for ADI's core business processes in the new product development organization. The second cycle of the action research project involved the deployment of two knowledge management systems to support the knowledge core process and the re-engineering of a the peer review process to make it more effective as a forum for sharing knowledge across product development teams. The KM systems were called "EnCore" and "docK". The research saw the process as a "spiral of steps, each of which is composed of a circle of planning action and fact-finding about the result of the action" (Lewin p.206) and involved a five phase cyclical process containing the following discrete steps: diagnosis, action planning, action taking, evaluation and learning (Susman and Evered 1978).

3.   IT STRATEGIES TO SUPPORT KNOWLEDGE CREATION

There are two generic strategies described in the literature. These approaches have been characterized as – "codification" and "personalization" (Hansen, Nohria et al. 1999), (Alavi and Leidner 1999). The essential difference between the two paradigms is whether you are motivated by a goal to encapsulate knowledge in a form that makes it suitable for re-use in another context (codification) or are motivated by a goal to transmit knowledge by making it easy to locate the relevant experts (personalization). These two approaches will be elaborated on in this section.

3.1   Codification Approach

The goal of the codification approach is to provide a high-quality, reliable, means of re-using codified knowledge though the use of electronic repositories. It is a "people-to-documents" approach. Knowledge is extracted from the person who developed it, made independent of that person and re-used. The approach allows for these "knowledge objects" to be searched for by many people and the codified knowledge retrieved without having to contact the person who originally developed it. Examples of this approach being implemented in the semiconductor industry are described in (Keating and Bricaud 1998) and (Chang, Cooke et al. 1999).

Recent contributions to the theoretical aspects of codified knowledge reuse have come from (Hansen, Nohria et al. 1999), (Swan, Newell et al.

1999), (Dixon 2000) and (Markus 2001). The contributions of Hansen et al. and Swan et al. has been to identify the features of codified knowledge management systems that differentiated them from systems dedicated to supporting the transfer of tacit knowledge (Hansen, Nohria et al. 1999) (Swan, Newell et al. 1999). Swan characterized such systems as applying a "cognitive" approach to knowledge management systems. Hansen characterized such systems as having a "codification" approach. Both authors identify the primary function of such systems as the capture and codification of knowledge. The key enabler of the system was identified by the authors as being information technology. Both authors identify the weaknesses of the codification approach as over-reliance on IT, with not enough attention being paid to human factors in KM.

The contributions of Dixon (Dixon 2000) and Markus (Markus 2001) were in the development of a typology of knowledge transfer and reuse situations. Dixon identifies five types of knowledge transfer: serial transfer, near transfer, far transfer, strategic transfer and expert transfer. She illustrates the types with five case studies. However, her focus is not solely on codified knowledge but on "best practices" i.e. knowledge about how to do some things better. Markus, on the other hand, restricted her focus to codified knowledge. She identifies four types of knowledge reuse situations: shared work producers, who produce knowledge that they later reuse: shared work practitioners, who reuse each other's knowledge contributions; expertise-seeking novices; and secondary knowledge producers. She also identifies the factors that affect the quality of repositories. For companies involved in NPD the goal of a codification approach is to make it easy for individuals in a new product development community to access a repository of previously design products so that the knowledge captured in the repository may reused in a new product.

3.2 Personalization Approach

The goal of the personalization approach is to produce highly customized solutions to unique problems by promoting person-to-person interaction. It is a "people-to-people" rather than "people-to-documents" approach. It focuses on creating opportunities for dialogue between individuals, rather than directing people to knowledge objects in a database. The approach assumes that people arrive at insights by firstly finding out who is knowledgeable on a topic and then going back and forth on problems they need to solve. Some firms use such knowledge maps or skills profiles to connect individuals with other individuals having relevant project knowledge. An example of this approach is described in (Carrozza 2000).

There have been relatively recent developments in IS technology that have facilitated the growth of the "personalization" approach. Traditionally, the systems approaches that have been associated with knowledge management in organizations have been: dedicated knowledge-based systems, document management system, database systems, and data warehouse technologies. With the advent of web technology and markup languages such as XML, new capabilities in IS technology have emerged. The key development is that it is becoming possible to integrate knowledge acquisition into the organization's existing business processes, rather than providing, in retrospect, a means of finding knowledge in existing unstructured data (Attipoe 1999). The approach proves particularly appropriate for companies using the so-called "stage-gate" new product development process where milestones such as product reviews can be captured and made available for the broader product development community. The gates are, in fact, "knowledge events" where teams are required to externalize their knowledge - thereby rendering that knowledge available to be shared and applied elsewhere in the organization.

A central aspect of this approach is the use of meta-data to add structure to product development documentation so that it is more easily located or "harvested" (Gutl, Andrews et al. 1998) (Bellaver and Lusa 2002). In order to provide a degree of added structure and simplify productive access to information from the Web, conventions such as the "Dublin Core" are used in conjunction with standard data formats such as XML, (Rabarijanaona, Dieng et al. 1999), (Finkelstein and Aiken 1999), (Abiteboul, Buneman et al. 2000). For companies involved in NPD the goal of a personalization approach is to make it easy for individuals in a new product development community to make documents relating to their work available to the community so that others can be aware of their contributions to a particular topic. Members of the development community may then take advantage of this facility by contacting the authors directly and seeking their help to get a deeper understanding of the issues in question.

4.   ADI'S IT FRAMEWORK FOR NPD

The NPD process has a requirement for IT that takes account of the different knowledge types inherent in a stage-gate process. The framework being developed in ADI has several applications that target different phases of the NPD flow (see Figure 1). This section contains a description of three of those applications.

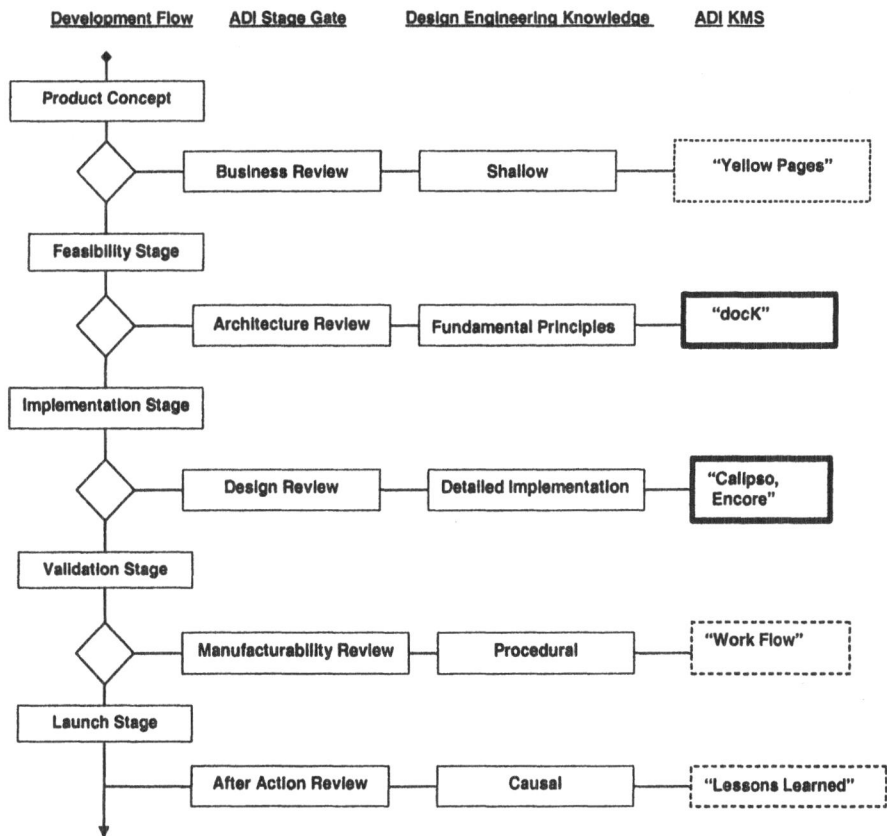

Figure 1. NPD Stage-Gate Process, Design Engineering Knowledge and KMS

4.1 Meta-Knowledge – "docK"

"Conventional explanations view learning as a process by which a learner internalizes the knowledge, whether "discovered," "transmitted" from others, or "experienced in interaction" with others." (p.47) (Lave and Wenger 1991). However, before one can initiate such a process, whether through discovery or interaction, there must be a mechanism by which people can easily find out what knowledge is being created in the organization and by whom. The knowledge being sought is, in fact, knowledge about knowledge or "meta-knowledge" (Swanstrom 1999), (Kehal 2002).

Meta-knowledge attempts to provide answers to questions such as "Where can I get information about a particular technical topic? How can I find out more about this topic? Is there work in progress in this organization on this topic?" The dock application tackles these challenges by making it easy for members of the technical staff to publish and locate technical reviews, notes, articles etc. - items which previously may have required several emails and phone calls to track down ("dock" stands for digital on-line cache of Know-how).

This is achieved by (a) the use of sophisticated resource discovery tools, and (b) the development of rich varieties of resource description.

(a) Resource discovery tools have been characterized as falling into two categories – search engines (SEs) and digital libraries (DLs). The first generation of SEs and DLs defined the basic structures of indices, directories and libraries. The second generation put the first generation tools to work in an operational setting. The third generation emphasized popularity measures such as links, usage and time as well as the use of parallel computing power and advanced search techniques (Hanani and Frank 2001). Through the use of meta-knowledge, the documents become more like databases where search, retrieval and reuse of text elements (explicit knowledge) are promoted while also giving the reader the opportunity to contact the source of the knowledge so that they may have a dialogue to enable tacit knowledge transfer (Braa and Sandahl 2000).

(b) Metadata is used by docK to provide a richer resource description for information on the WWW. *Meta* is used to mean a level above a target of discussion or study. Metadata is data about data and is often used in the context of data that refer to digital resources available across a network. Metadata is a form of document representation that is linked directly to the resource, and so allows direct access to the resource. Internet search engines use metadata in the indexing processes that they employ to index internet resources. Metadata needs to be able to describe remote locations and document versions. It also needs to accommodate the lack of stability of the Internet, redundant data, different perspectives on the granularity of the Internet, and the variable locations on a variety of different networks. There are a number of metadata formats in existence to provide bibliographic control over networked resources. The Dublin Metadata Core Element Set is one of the prime contenders for general acceptance – and is the format implemented in docK (Kunze, Lagoze et al. 1998).

4.2   Knowledge Catalog – "Calipso"

A "Catalog", in this context, is an application that generates a list of previously designed products in the product development community. The catalog would enable product development staff to quickly find out if products were previously designed that were similar to those currently under development.

Calipso is a catalog of functional circuit blocks developed by ADI's development staff. The entries are created and owned by the product development staff. Each entry in the catalog represents is a potentially reusable circuit design. Catalog entries, depending on their utility, are potential candidates for inclusion in a repository. The problems that were identified in the new product development process that were to be addressed by "Calipso" were:

a) (a) a lack of awareness of what previously designed circuit blocks had been created in ADI and might be available for reuse in future projects

b) (b) a mechanism by which product development staff could easily make their products more easily "discovered" by members of the product development organization outside of their own organization unit.

4.3   Knowledge Repository – "EnCore"

A "Repository", in this context, provides a store of previously design products that could be reused throughout the corporation. Each of the repository's elements has an extensive support kit associated with it i.e. thorough documentation, contextual information about previous usage, data formats compatible with existing NPD systems, validation data, interface information, etc.

EnCore is a structured repository for formal knowledge containing previously used circuits that were internally developed and externally procured circuits that may also be re-used in future products. Its purpose corresponds, generally, to what Hansen termed a "codification" strategy where the value of the repository lies in "connecting people with reusable codified knowledge" (Hansen, Nohria et al. 1999) or to what Swan termed a "cognitive" strategy where the primary function of the repository is to codify and capture knowledge so that it can be recycled (Swan, Newell et al. 1999).

The goal of EnCore is to provide a library of robust and supported reusable circuit designs available for download, obtained from both internal and external sources. The repository contains previously designed products packaged in a format suitable to delivery as intellectual property to either internal groups or external groups (or both). They are close to the explicit dimension on the diagram because they represent an attempt to codify the knowledge associated with a product i.e. a people-to-documents approach.

The motivation for both Calipso and EnCore was based on a belief that reuse of design intellectual property (IP) helps keep development costs down while also helping to reduce time to market. The key difference between Calipso and EnCore is that to qualify for entry into the EnCore repository a circuit block needs to satisfy rigorous standards with respect to reusability, supporting documentation, usage potential etc. Calipso's catalog points to a very broad set of circuits that do not necessarily conform to these standards.

The fact that a catalog exists containing these entries gives members of the engineering community an opportunity to contact the originators of the entries and share knowledge about the element in question.

5.   CONCLUSIONS AND FUTURE RESEARCH

A framework has been developed that is based on a stage-gate NPD process and current thinking on IT to support innovation. The framework is being evolved and elaborated in an industrial setting in ADI. The approach being pursued is based on an understanding of the types of knowledge important to NPD and the range of applicability of that knowledge across organizational units. The work builds on earlier work by (Hansen, Nohria et al. 1999), (Swan, Newell et al. 2000), (Dixon 2000) and (Markus 2001).

Figure 1 summarizes the framework by showing the stages in ADI's NPD stage-gate process and the corresponding IT systems being developed to support the appropriate product development stages. The development of the framework is at an early stage so there are gaps in the IT support of some stages of the NPD process.

Where those gaps exist, possible IT solutions for the stage are proposed e.g. a so-called "Yellow Pages" application which would be useful at the conceptual stage of the project. These are applications that provide a centralized database of user knowledge profiles. They offer users multiple ways to find user profiles. Participation is usually voluntary (i.e. no automatic profile creation). Users can create and maintain their profile's visibility and access. An example is described in (Carrozza 2000).

Possible solutions for the launch stage include so-called "Organizational Memory Information Systems (OMIS)" that could capture the results of After-Action-Reviews. An "OMIS" is an organizational memory information system and in this case would be targeted at the results of After Action Reviews (AARs). The defining processes of an OMIS application are acquisition, retention, maintenance, and retrieval (Stein and Zwass 1995). The development and application of these additional components of the overall framework will be the subject of future research in this area.

The IT framework to support new product development that has been described has a strong focus on engineering knowledge. It is recognized that there is other forms of knowledge that is critical to NPD success that is not depicted in the framework at present e.g. customer and marketing knowledge.

This research project exposed some of the issues surrounding the implementation of a codified knowledge reuse system in a product development environment. The lessons learned from the implementation will enrich our understanding of the parameters that need to be accounted for in a comprehensive model of codified knowledge re-use. The key success factors included , upper management support, the "Productization" role of intermediaries between the developers and the end-users, stable domains and architectures, infrastructure support (common systems, robust networks), quality and availability of potential repository entries, cultural acceptance of reuse in the organization, standardized interfaces and formats, and the demonstration of early reuse success stories.

ACKNOWLEDGEMENTS

The applications described in this paper are being developed by the SLI TMIS groups in ADI. The author wishes to thank Sreenivasa Rao, Xavier Haurie, Tanya Mollenauer, Beth Rhinelander and Colin Lyden for their insights and helpful elucidations.

REFERENCES

Abiteboul, S., P. Buneman, et al. (2000). Data on the Web: from Relations to Semistructured Data and XML. San Francisco, Morgan Kaufmann Publishers.

Alavi, M. and D. E. Leidner (1999). "Knowledge Management Systems: Issues, Challenges. and Benefits." Communications of the Association for Information Systems 1(7): 1-37.

Attipoe, A. (1999). "Knowledge Structuring for Corporate Memory." Markup Languages: Theory and Practice 1(4): 27-37.

Bellaver, R. and J. Lusa (2002). Knowledge management strategy and technology. Norwood, MA, Artech House.

Braa, K. and T. Sandahl (2000). Documents: from paperwork to network. Planet Internet. Braa K., Sorensen C. and Dahlbom B. Lund, Sweden., Studentlitteratur.

Carrozza, T. (2000). From Hyperlinks to Human Links at Hewlett-Packard. Knowledge Management Review. 3: 28-33.

Chang, H., L. Cooke, et al. (1999). Surviving the SOC Revolution : A Guide to Platform-Based Design. Norwell, Mass., Kluwer Academic Publishers.

Dixon, N. M. (2000). Common Knowledge : How Companies Thrive by Sharing What They Know. Boston, Harvard Business School Press.

Finkelstein, C. and P. Aiken (1999). XML and Corporate Portals. Building Corporate Portals using XML, McGraw-Hill Companies Inc.

Gutl, C., K. Andrews, et al. (1998). Future Information Harvesting and Processing on the Web. European Telematics Conference: Advancing the Information Society, Barcelona.

Hanani, U. and A. Frank (2001). "The Parallel Evolution of Search Engines and Digital Libraries: their Convergence to the Mega-Portal." IEEE Software.

Hansen, M., N. Nohria, et al. (1999). "What's your strategy for managing knowledge?" Harvard Business Review March-April(1999): 106-116.

Keating, M. and P. Bricaud (1998). Re-Use Methodology Manual for System-On-A-Chip Design. Norwell, Mass., USA, Kluwer Academic Publishers.

Kehal, M. (2002). "Searching For An Effective Knowledge Management Framework." Journal Of Knowledge Management Practice(February, 2000).

Kunze, J., C. Lagoze, et al. (1998). The Dublin Core Metadata Element Set, http://purl.org/metadata/dublin_core.

Lave, J. and E. Wenger (1991). Situated learning: Legitimate peripheral participation. Cambridge, Cambridge University Press.

Lewin, K. (1947). "Frontiers in Group Dynamics." Human Relations 1(1).

Markus, M. L. (2001). "Toward a Theory of Knowledge Reuse: Types of Knowledge Reuse Situations and Factors in Reuse Success." Journal Of Management Information Systems 18(1): 57-93.

Porter, M. (2003). What is Strategy ? Irish Management Centre Seminar on Innovation, Dublin, October 9th, 2003.

Rabarijanaona, A., R. Dieng, et al. (1999). Building a XML-based Corporate Memory. EKAW, Banff, Canada.

Stata, R. (1989). "Organisational Learning - The Key to Management Innovation." Sloan Management Review, Spring 30(3): 63-74.

Stein, E. and V. Zwass (1995). "Actualizing Organizational Memory with Information Technology." Information Systems Research 6(2): 85-117.

Susman, G. I. and R. D. Evered (1978). An Assessment of the Scientific Merits of Action Research. Administrative Science Quarterly. 23: 582-603.

Swan, J., S. Newell, et al. (2000). Limits of IT-Driven Knowledge Management Initiatives for Interactive Innovation Processes: Towards a Community-Based Approach. 33rd International Conference on System Sciences, Hawaii.

Swan, J., S. Newell, et al. (1999). "Knowledge management and innovation: networks and networking." Journal of Knowledge Management 3(3): 262-275.

Swanstrom, E. (1999). MetaKnowledge and MetaKnowledgebases. The Knowledge Management Handbook. J. Liebowitz. London, CRC Press.

Weick, K. E. (1979). "Cognitive Processes in Organisations." Research in Organisational Behavior 1: 41-74.

Weick, K. E. (1995). Sensemaking In Organizations. Thousand Oaks, CA, Sage Publications.

PART II

INNOVATING SYSTEMS DEVELOPMENT AND PROCESS

ASSESSING IMPROVEMENTS OF SOFTWARE METRICS PRACTICES

Helle Damborg Frederiksen[1] and Lars Mathiassen[2]

[1]*Department of Computer Science, Aalborg University, Denmark;* [2]*Center for Digital Commerce, eCommerce Institute, J. Mack Robinson College of Business, Georgia State University, USA*

Abstract: There is an extensive literature on design and implementation of software metrics programs, but our knowledge on how to improve and sustain metrics programs that are already implemented is limited. This paper reports from an action research effort to improve metrics practices in a large software organization. We adopt a comprehensive organizational change model to compare and contrast metrics practices before and after the intervention. Our analysis views the metrics program in its organizational context emphasizing management practices, organizational structures, people, technical systems, and underlying metrics rationale. The paper provides insights into tactics for improving software metrics practices, it provides a specific approach to assess such improvements, and it leads to suggestions for how to make software metrics improvements sustainable.

Key words: Software metrics, assessment, organizational change

1.   INTRODUCTION

Software engineering has matured as a discipline through the past decades (Wasserman, 1996) and the idea of measuring the process and its outcome has come to play an important role. One prominent example of this is Albright's invention of function points as a measure of the size of software (Jones, 1996). Much research has been dedicated to designing software metrics for specific purposes, e.g. object oriented programming (Chidamber & Kemerer, 1994). Another line of research has been concerned with the implementation of metrics programs, e.g. (Berry & Jeffery, 2000;

Hall & Fenton, 1997; Iversen & Kautz, 2001). However, recent studies suggest that the benefits from metrics programs are not as great as expected (Dekkers, 1999). It is therefore important to increase our understanding of how metrics practices can be improved (Mendonça & Basili, 2000). The little research that has been conducted on that issue focuses on adopting data mining to increase the benefit of data (Mendonça & Basili, 2000) and on validating the theoretical robustness of measures (Kitchenham et al., 1995).

The purpose of our research is to study how we can improve and sustain software metrics programs emphasizing the organizational dimensions of the involved change. Our research is exploratory in nature and based on a longitudinal action research effort (Avison et al., 1999; Kock & Lau, 2001; Mathiassen, 2002; McKay & Marshall, 2001) in KMD, a large Danish software organization providing IT-services for municipal institutions. The initial activity of the research effort was a diagnosis of the existing metrics program (Frederiksen & Mathiassen, 2002; Frederiksen & Rose, 2003). From this diagnosis improvement efforts were subsequently identified and performed. This paper describes and evaluates the impacts of the improvement efforts at KMD based on Applegate's model for analyzing organizational changes (Applegate, 1994) combined with Kirsch's studies of approaches to control software practices (Kirsch, 1996; Kirsch, 1997). This comprehensive approach allows us to systematically compare and contrast metrics practices before and after the intervention.

The argument proceeds as follows. Section 2 reviews the theoretical background and Section 3 presents the research approach in detail. Section 4 provides a description of the old metrics practices, the change process, and the new metrics practices. Section 5 analyses metrics practices in KMD before and after the intervention. Section 6 presents the key findings, relates them to the literature on software metrics programs, and discusses the implications for both practice and research. Section 7 concludes the argument.

2.   SOFTWARE METRICS

There is a rich literature on software metrics. Section 2.1 gives an overview of what software metrics are and recommendations on how to implement them. Failing to effectively support software management a metrics program will easily degenerate into a bureaucratic procedure, where data are collected, stored, and reported without having practical consequences (Iversen & Kautz, 2001; Iversen & Mathiassen, 2003; Niessink & Vliet, 1999). Hence, we look in Section 2.2 at different ways in which metrics programs can serve as control mechanisms and in Section 2.3

at the system of organizational mechanisms that are needed to support their effective operation.

2.1 Using and improving software metrics

Our current knowledge on how to define and implement metrics is fairly well developed (Fenton & Neil, 1999). However, implementing a metrics program is difficult, complex, and likely to fail. It has been estimated that as many as 78% of metrics programs fail (Dekkers, 1999) and recent research has consequently addressed success factors, e.g. (Hall & Fenton, 1997; Iversen & Mathiassen, 2003). Niessink and Vliet (1999) stress focusing on the actual use of a metrics program to make sure that it becomes of value to the organization. This is also the focal point in Iversen and Kautz's (2001) advice for software metrics implementation: Use improvement knowledge, use organizational knowledge, establish a project, establish incentive structures, start by determining goals, start simple, publish objectives and data widely, facilitate debate, and use the data. Mendonça and Basili have developed an approach for improving existing metrics, data collection mechanisms, and the use of data within a software organization (Mendonça & Basili, 2000). They combine a top-down approach using Goal-Question-Metric (GQM) and a bottom-up approach based on data mining techniques. Kitchenham et al. (1995) emphasize that validation is critical to the success of a software metrics program. They suggest a framework that can help validate measures and assess their appropriateness in a given situation. The focus in these contributions to improve existing programs is on developing additional managerial support based on available measures and on improving the conceptual validity of the adopted measures. However, there are no efforts directed towards assessing the actual use of a program as a basis for increasing its value for the software organization.

2.2 Metrics as control mechanisms

Control plays a critical role in managing software processes and projects effectively (Kirsch, 1997) and in particular in ensuring progress in ongoing development efforts. Kirsch suggests viewing control as a relationship between a controller and a controllee rather than between a supervisor and a subordinate. The controller can be a supervisor, a project manager, a customer, or a consultant. Kirsch lists four modes of control. The two formal modes are behavior and outcome control and the two informal modes are clan and individual control:

- **Behavior:** Rules and procedures are articulated, e.g. following a methodology in a software development project. Compliance to rules and procedures is rewarded and assumed to lead to the desired outcome.
- **Outcome:** Outcomes or goals are articulated, e.g. setting a deadline for the project's delivery of software to the customer. Reward or sanction is based on fulfillment of goals.
- **Clan:** Common values and beliefs are disseminated and shared by the group (or clan). Identifying, sharing and reinforcing acceptable behavior, e.g. documentation styles for source code, lay the foundation for control.
- **Self:** This mode of control is like clan control but on the individual level. The values can be implicit to the individual.

A controller adopts a portfolio of these modes to exercise control (Kirsch, 1997). The portfolio depends on task characteristics, role expectations, and project-related knowledge and skills. The purpose of control is to regulate behavior: "...monitoring serves as an information system and makes behaviors observable, whereas evaluation and reward constitute the way in which actions are regulated" (Kirsch, 1996). Indicators from a well functioning metrics program play an important role in this.

2.3   Metrics programs as organizational systems

The literature focuses on the internals of metrics programs and less on external factors (Niessink & Vliet, 1999). To remedy this adopt the well-established framework by Applegate (1994) that offers a comprehensive understanding of the key mechanisms required to sustain an organizational effort, see Figure 1.

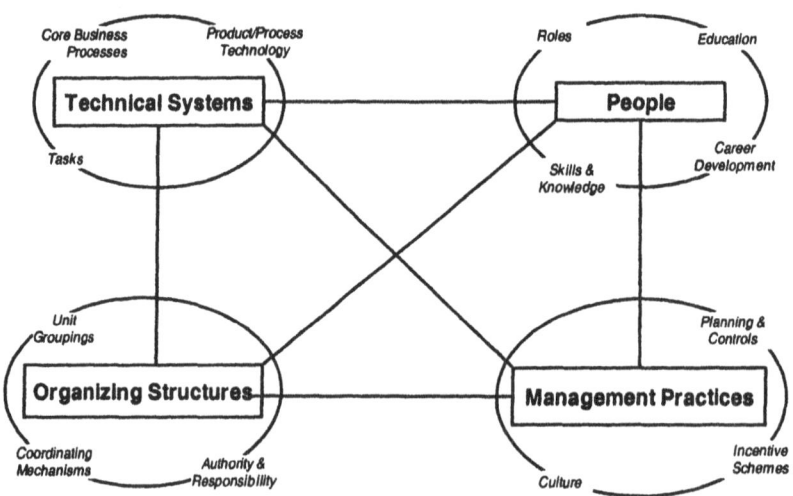

Figure 1. Applegate's model

The technical systems comprise of (a) core business processes, (b) product/process technology, and (c) tasks. Examples of core business processes could be supplying indicators for managers and benchmarking projects. Examples of product/process technologies could be tools for collecting data and for presenting indicators. Examples of tasks could be prompting project managers for data and analyzing data.

The people dimension addresses (a) roles, (b) education, (c) skills and knowledge, and (d) career development. For example, to implement a metrics program you must define and assign individuals to serve in the necessary roles, e.g. function point counter. Skills and knowledge on how to calculate indicators is required, courses need to be followed, and career development for function point counters should be established.

Organizing structures refer to (a) unit groupings, (b) coordinating mechanisms, and (c) authority and responsibility. The unit groupings of a metrics program include forming a metrics staff. Coordination mechanisms for collecting and presenting data should be defined. The authority and responsibility of the metrics staff should be defined, e.g. is the manager or the metrics staff responsible for prompting project managers for due data.

The main issues of concern for management practices are (a) planning and control, (b) culture, and (c) incentive schemes. For example, how are data from the metrics program used for estimating new projects? Are project indicators discussed in public or are they private to managers? Are data used to reward or sanction project managers?

We suggest to use Kirsch's model as an additional dimension in Applegate's model to assess the underlying rationale for a given software metrics program.

3. RESEARCH APPROACH

Our research addresses the following question: How can changes in metrics practices be assessed and made sustainable? In Section 3.1 we describe and argue for the adopted action research approach based on the IDEAL model (McFeeley, 1996). The resulting process and its organization are presented in Section 3.2. Section 3.3 describes the approach to data collection.

3.1 Action research

Galliers (1992) defines action research as applied research where there is an attempt to obtain results of practical value to groups with whom the researcher is allied, while at the same time adding to theoretical knowledge.

The major advantage of action research is the close integration of research and practice. The main critique is that it may be regarded as little more than consultancy (McKay & Marshall, 2001). McKay and Marshall argue that part of the issue concerns the way in which we currently conceptualize action research and they suggest to explicitly include a problem solving interest cycle and a research interest cycle (p. 57).

3.2   Research organization

Our research addresses improvement of a software metrics program and we have therefore based it on the IDEAL model of software process improvement (McFeeley, 1996). This model presents a pragmatic, operational way to improve software practices and it can be seen as a specialized version of Susman and Evered's (1978) classical action research cycle: diagnosing, action planning, action taking, evaluating, and specifying learning. The IDEAL model consists of very similar phases as shown in Figure 2. The purpose of the first phase is to initiate (I) the change process. The purpose of the second phase is to diagnose (D) the organization's current practices. In the third phase the purpose is to establish (E) the change process. The purpose of the fourth phase, acting (A), is to perform the actual changes. The purpose of the last phase, leveraging (L), is to reflect on the successes and failures of the intervention. The knowledge gained from this may be used in the next cycles of the IDEAL model and as a basis for research.

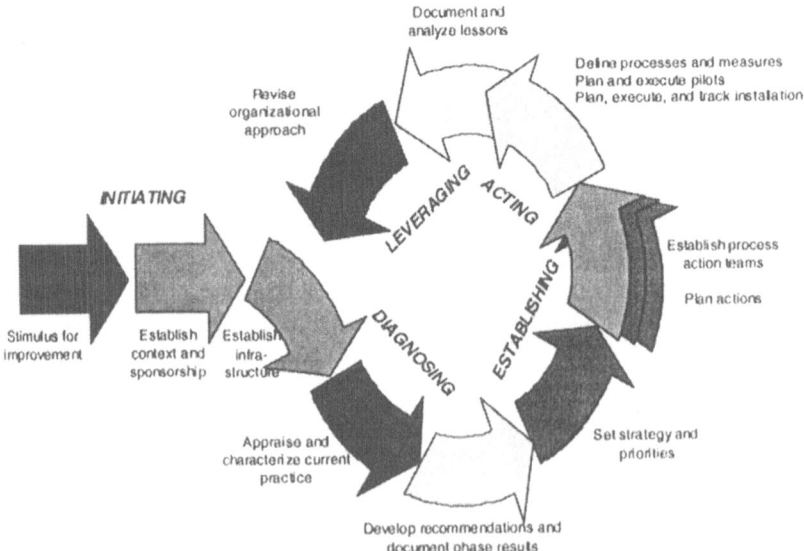

Figure 2. The IDEAL model (McFeeley, 1996)

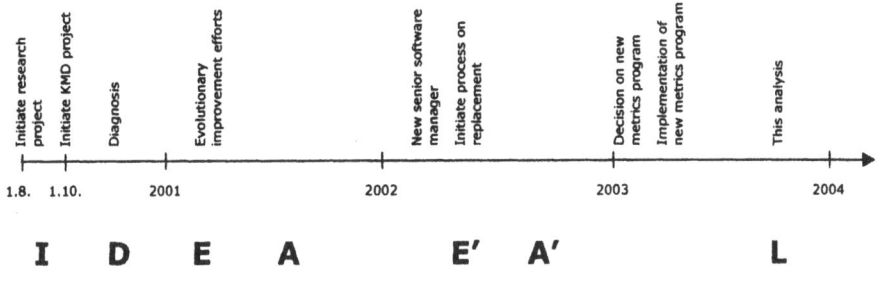

Figure 3. Timeline

Our research collaboration involved all aspects of the IDEAL model, see Figure 3. Key events are presented above the timeline and the corresponding phases of the IDEAL model are represented below the timeline. The research collaboration was initiated August 1st 2000. The practical part, the KMD-improvement-project, was initiated October 1st 2000 (labeled I). A major activity was the diagnosis of the current metrics practices at KMD (labeled D). From the diagnosis a set of evolutionary improvements efforts were established in early 2001 (labeled E). After establishing these efforts the actual improvements were enacted (labeled A). During this phase a new responsible software director was appointed which led to a new and more radical improvement effort (labeled E' and A'). The last phase of the collaborative process was to learn from the improvement effort (labeled L). The paper presents key insights from this assessment of the impact of the improvement effort using the frameworks of Applegate (1994) and Kirsch (1997).

To manage and control the research effort it was helpful to think of it as two interrelated subprojects: 1) the KMD-improvement-project, and 2) the academic-research-project as suggested by McKay and Marshall (2001, p. 57). The KMD-improvement-project was organized as a conventional project with a project staff, a project manager and a steering committee. The researchers were participating actively in the steering committee. The academic-research-project interacted closely with the steering committee and it was driven in close collaboration between the manager of the improvement project and a university researcher (the two authors).

3.3 Data collection

The initial diagnosis (D in Figure 3) helped understand the old metrics practices at KMD. The adopted diagnostic approach (Checkland, 1981; Andersen, 1990; Checkland & Scholes, 1990; Pedersen, 1996) and our experience with it are reported in (Frederiksen & Mathiassen, 2003).

Another research team used the same data in parallel to analyze the old metrics practices based on structuration theory (Frederiksen & Rose, 2003). This analysis was used to challenge and complement the findings of the first diagnosis to counteract the natural bias that is caused by performing action research in your own organization (Coghlan & Brannick, 2001).

Data have been collected through semi-structured interviews with project managers, senior managers, metrics staff, and quality assurance staff. Each interview lasted for 3 hours and was documented as minutes of meetings that were subsequently corrected and approved by the interviewee. We also collected plans, decision reports, and e-mail correspondence related to metrics, minutes of meetings from the metrics staff, and documents from the organization's process improvement initiative, consultant's reports and the company intranet. Finally, one of the authors has worked as a metrics specialist at KMD for 8 years and was also manager for the improvement initiative. Throughout this process she has kept a personal log with experiences from the use of the old metrics program and design and implementation of the new program. A complete record of all emails sent to her in connection with the project was available for study. This deep engagement with the research situation meant a daily familiarity with the company.

4.   IMPROVING METRICS AT KMD

The more than 2400 employees at KMD, including almost 700 software developers, are organized in a conventional hierarchical structure and they work at different sites. Each application team is responsible for correcting faults, supporting their users and for upgrading the applications with additional functionality and up-to-date technology. Software development is organized in projects, with project members drawn from different departments.

4.1   The old metrics program

KMD has a longstanding tradition of collecting simple measures, e.g. the number of faults in an application and other operational parameters as customer satisfaction, which provide visibility into problem areas. In 1996 the CEO decided to implement an elaborate metrics program to increase productivity and quality in software development. In order to identify weaknesses, KMD should be benchmarked against other software development companies. Hence, an external contractor, Compass, supplied a model (Figure 4) and a software tool to collect data suited for benchmarking.

The basic elements of the model are projects (A1-A7) and applications (B1-B5). Furthermore, data on general overhead activities (C, D and E), personnel, development software, hardware, and systems software are included.

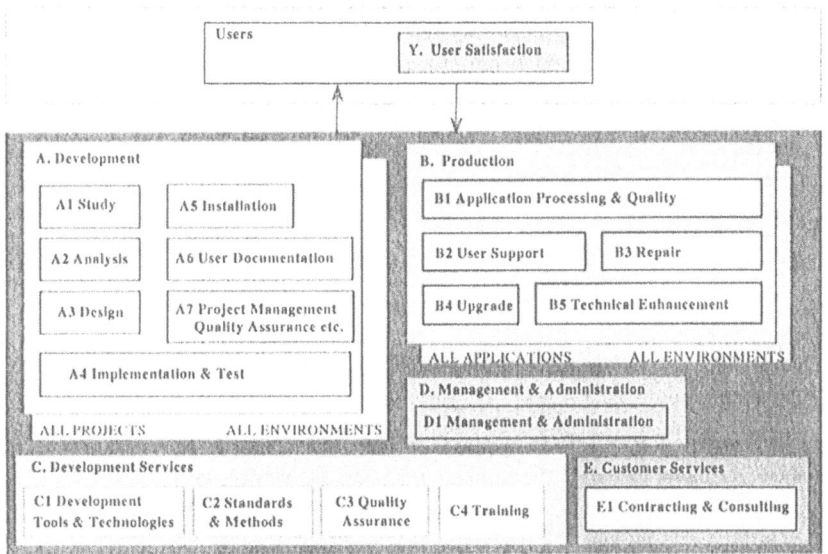

Figure 4. Basic model of old metrics program

The supply of data (measures) to the metrics program is illustrated in Figure 5. The project and team managers supplied measures on their projects and applications to the metrics program every three months. Units external to the software unit supplied measures on personnel, and expenses for consultants and software. Some data were subtracted from the ERP-system (SAP) and fed into the Compass software. Data were processed, validated and packaged at KMD before they were sent to the contractor. Compass sent results and graphs to KMD each quarter. The results were used for benchmarking KMD against a reference group at Compass. Furthermore, the results were used for internal benchmarking among the software units. Compass analyzed annually all available data and delivered a report with recommendations for managerial actions. Compass and the software process improvement organization presented the report to top management. The results informed the management's decisions on general improvement efforts to the software process. Afterwards the results were available to project and team managers, however, the results did only to a very limited extent inform their decisions.

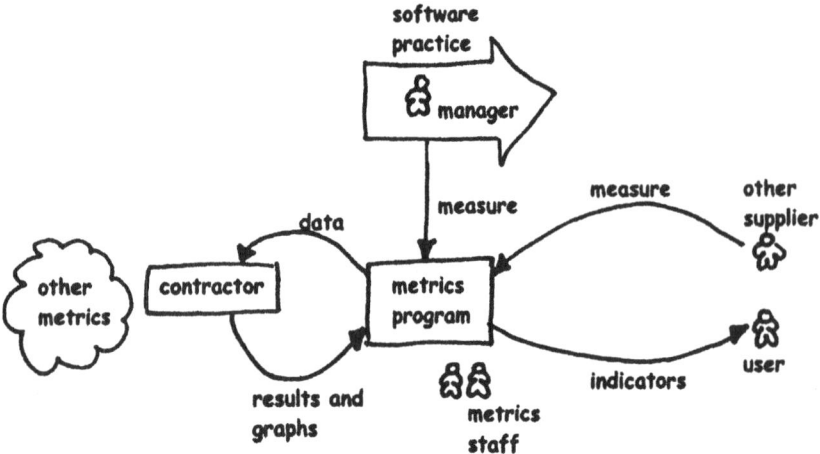

Figure 5. The old measurement process at KMD

Despite five years of intensive work with the metrics program it was difficult to point to areas at KMD where the program had made a significant impact. The metrics program was questioned by different stakeholders and the general situation at KMD made it clear that the current metrics program was far too expensive.

4.2   The improvement process

That situation led to the establishment of the KMD-improvement-project and the related action research as described in Section 3 (I in Figure 3).

4.2.1   Results from diagnosis

The initial activity was to diagnose the current metrics practices (D in Figure 3) involving all relevant stakeholders at KMD (Frederiksen & Mathiassen, 2002; Frederiksen & Rose, 2003). In summary, the diagnosis revealed:
– The data suppliers did not understand the definitions of metrics and measures.
– The metrics program contained a lot of data that are potentially useful for several purposes.
– The primary users were software improvement agents identifying high-level improvements.
– Most software engineers and managers did not use the metrics program. Very few managers saw a need for it.

- A proper interpretation of data required insights into the metrics program, the submitted measures, and the actual software practices. Few people were capable of that.
- There were few managerial responses based on information from the metrics program.
- The project indicators focused on resources and deadlines and ignored the quality of the delivered software.

4.2.2   From evolutionary to radical improvements

The diagnosis led to four suggested evolutionary adjustments of metrics practices. The steering committee acknowledged the diagnosis and chose to initiate two of the efforts (E and A in Figure 3). The responsible software director was, however, replaced March 1st 2002. The latest report from Compass was soon after presented to the group of senior managers. The presentation led to a discussion of the relevance and cost of the Compass-measurements. As a result an effort to replace the old metrics program with a new program was initiated by the newly appointed software director (E' and A' in Figure 3).

The program was designed during the summer and fall of 2002. The annual cost of the new program was estimated to less than half of the cost of the old program. The software director and several other software managers were heavily involved. One of the authors was project manager and facilitated and documented the process and its results. In January 2003 the CEO accepted the new metrics program, the implementation was initiated, and the new metrics group was staffed.

4.3   The new metrics program

The new model includes four key areas as illustrated in Figure 6. Each area includes 1-3 simple indicators. The core of the metrics program is the area Development. This area includes project indicators for single projects such as the precision in the project estimates for delivery date, resources, and functionality. An indicator of project productivity is calculated from these data and published for the total set of projects at KMD. The area Maintenance includes indicators on quality and user satisfaction on applications and an indicator on the effort spent in the software unit on maintenance compared to the income for providing these services, i.e. the size of applications is measured in income and not in function points. The remaining areas, Overhead and Method & Tools, each include a single indicator on the effort spent on management, administration and technology support compared to the effort on software development.

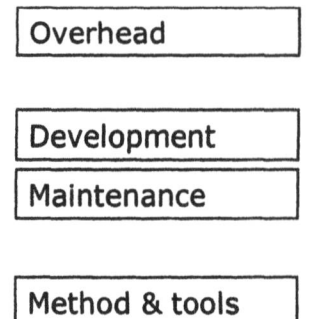

Figure 6. Basic model of new metrics program

The indicator on productivity is benchmarked against a reference group at The International Software Benchmarking Standards Group (ISBSG), a not for profit organization supported by industry sponsors and software metrics associations of many countries. The ISBSG offers a free, flexible project benchmarking service, which provides a facility to benchmark against a repository of software projects.

The new metrics staff consists of a manager (one of the authors) and three metrics consultants. The metrics staff is all full time personnel and each consultant covers a geographical location of KMD. Sharing experiences and knowledge are considered crucial and the staff builds its own simple tools and templates in Word and Excel. The main task of the consultants is to facilitate measures of local software projects. The manager decides which projects to measures based on the consultant's suggestion. Consultants help project managers create the basis for measuring a project and they also assist project managers estimating deadlines, resources, and functionality based on historical data. The consultant keeps track of the different versions of estimates and observes the conditions of the project in general. After a project is completed the project manager and the metrics consultant make a report, which illustrates the differences between estimates and realities. The report also illustrates changes in estimates. The project manager is responsible for sharing the report with relevant stakeholders in the unit. The consultant submits the data to ISBSG. The benchmark report is shared with the project manager and important lessons learned are shared within the software unit and with the manager in charge of software process improvement.

5.   ASSESSING IMPROVEMENTS AT KMD

The assessment of the new metrics program was initiated in June 2003 (L in Figure 3) based on Applegate's model (1994) of key organizational dimensions (section 5.1 to 5.4) in combination with Kirsch's model (1997) of the underlying control rationale (section 5.5). This comprehensive approach allowed us to assess the implied changes and possible improvements to sustain the new program by systematically comparing and contrasting old and new metrics practices.

5.1   Technical systems

Core business processes: The core of the old program was to benchmark KMD to suggest improvements to the software operation. The benchmark included comparisons on projects (e.g. productivity and time-to-market), on applications (e.g. productivity and quality in maintenance), and general indicators (e.g. time for management and administration and cost for software tools). The new program includes benchmark of KMD on only one indicator (project productivity). This benchmark satisfies top management needs based on data from ISBSG. Indicators are now used to set the goals of the software units. There are two major changes in the process. First, the purpose of the metrics program has shifted from identification of improvements towards management support. Second, the focus has shifted from external benchmarking towards learning from specific projects about details on the software operation, e.g. staffing of projects.

Product/process technology: The core technology of the old program was the model in Figure 4 and the related tools and templates supplied by Compass. The metrics staff developed local tools and checklists to support the measurement tasks. Data were extracted from the ERP system, modified and filled into the Compass tools. The new model (Figure 6) has no details on applications and general activities but more details on projects including major changes. These metrics are defined in close cooperation with software managers. The indicators are inspired by the Compass model, but the metrics staff is expected to define other relevant metrics, and the set of data is expected to change and grow over time. The replacement of the model meant that tools, templates, and checklists were replaced, too. Excel is used to record details of function point counts. Data are extracted from the ERP system and primarily used without modification. When needed data are adjusted in the ERP system. There are three major changes in technology. First, the model has shifted from generic Compass metrics towards specific KMD metrics. Second, the measurement process is now simpler, cheaper and more flexible. Third, the process is more tightly linked to the ERP system.

Tasks: The tasks of the metrics staff in the old program were interpretation and adaptation of Compass' model and data collection. The metrics staff was responsible for coordinating data collection, assisting in 70 validation of data and presenting the results. The metrics staff was also available for more detailed presentations and debates of the indicators. The model and routines of the new program are developed and maintained by the metrics staff. The decision on which projects to measure is taken by software management guided by metrics staff and potential relevance for estimation of future projects. The border of the responsibilities between the metrics staff and the data suppliers has changed. Now the metrics staff collects data and prompts project managers and other relevant people for validation. Furthermore, the metrics consultant presents project indicators to the project manager. The metrics staff is still available for more detailed presentations and debates of indicators. The major change is that the metrics staff is more involved in projects and process execution has shifted from being outsourced assisted by internal experts towards being in-house facilitated by internal experts and with considerable degree of management participation.

5.2   People

Roles: A set of roles was explicitly defined in the old program. A Compass-controller coordinated data collection in each software unit and supported the project and application managers in data collection. Furthermore, the Compass-controller supplied data on general activities like time spent on management and administration and the number of employees in the software unit. The software unit also had a local function point counter. The collaboration of the local Compass-controller and function point counter was crucial to obtain valid data. A specialist from the methods department was responsible for the metrics program, and another specialist was responsible for the function point method and for using data for estimation. In the new metrics staff there are only two roles - the metrics consultant and the team leader. The metrics staff subtracts data from the ERP system and asks the local controllers to validate them. The formal roles of the project and software managers haven't changed. The objective of improving the program was to adjust actual behaviors to fit managerial needs. The new metrics program is still used to support software process improvement. However, this role is now distributed with more emphasis on learning within and from projects. The major change is that there are fewer roles and hence fewer interfaces between roles.

Education: Metrics specialists were educated by consultants from Compass in the old program. The software and projects managers were introduced to the Compass-measurements. Later they were given a detailed introduction as preparation for measuring their projects and applications. Each time indicators were presented some kind of introduction was given. The new program was introduced through several sessions with senior managers. Education of the metrics staff was based on available literature and there was more focus on organizational aspects. The metrics consultants now count function points, but no training was needed because they used to be function point counters. The specific details of indicators were introduced through one-on-one presentations of results from the unit in question. These presentations were followed by a joint discussion between senior managers. Project managers are introduced to the new metrics program when their project is to be measured. The major change is that managers on all levels learn more by being part of the process. This is possible since there are fewer people involved and the metrics program is simpler.

Skills & Knowledge: A specialist from the methods department with skills and knowledge on measurements managed the old and now the new program. In the old program the Compass-controllers were appointed based on their insight into administrative routines. They were less skilled on software processes and project management. The Compass-controllers were supposed to help managers with interpretation but it rarely happened. The new metrics consultants are all very skilled on software processes. They are experienced software developers and project managers who have worked at KMD for several years. The major change is the move from Compass related skills to KMD specific skills. Managers are required to understand and share KMD's perception of software processes, and the metrics staff actively facilitates their interpretation of indicators.

Career Development: Working with the metrics program is not part of a formal career at KMD. The work has, however, changed to become more prestigious. The metrics consultants now have a central and acknowledged position in the local management groups. The software director used to prompt the manger of the metrics program for occasional presentation of indicators in the group of senior managers. The present software director treats the metrics manager as a personal assistant on these matters. Overall the position of the metrics staff has become more visible and central.

5.3   Organizing structures

Unit Groupings: The organizing structures of the software units have not been affected. An important player then and now is the forum where the senior managers meet with each other and the manager of the methods department. The old metrics staff was organized in two groups chaired by methods specialists. The function point group included all function point counters, and the Compass-group included all Compass-controllers, the Compass-coordinator and the specialist on Compass. All members worked part time on the metrics and had different managers. The new metrics consultants are organized in one coherent team with one manager, and they work full time on the metrics program. The major change is that the metrics staff now works full time and is organized as a smaller and more coherent team.

Coordinating Mechanisms: The primary coordinating mechanism of the two groups was mutual orientation between the chairmen and coordination between the Compass-controller and function point counter of each unit. The dissemination of results followed the conventional hierarchy of the software units. Some indicators were in addition presented in a monthly report composed by the financial department. Project indicators were presented to the project manager. However, it was not expected that the project manager passed the indicators on. The new metrics staff meets on a regular basis to coordinate and discuss the program. The staff now publishes a set of overall indicators, e.g. the company productivity. The metrics consultant presents indicators from a single project to the project manager, who then is responsible for reporting them through the management system along with other information on the project. The project manager qualifies the indicators and presents them to the next management level. The aggregated indicators are presented in a management information report along with other information. The major change is that project managers own their project indicators, and that the aggregated indicators are reported as an integral part of management information in the software units.

Authority & Responsibility: The old program was introduced by the present CEO and sponsored by the methods department. However, the managerial responses were sparse. The old metrics program was perceived as yet another administrative burden on project managers. The new metrics program is introduced and sponsored by the software director. The director is excited about the program, and he contributed significantly to its design and adoption. The metrics consultants are perceived as representatives of the program and the software director. Hence they have more authority. The major change is that the responsible software director has given the metrics staff authority to measure software processes and act as experts and facilitators on measurements.

5.4   Management practices

Planning & Controls: In the old program Compass' definition of a project was used to benchmark projects. The data collected included various characteristics of a project and effort spent in the different phases. Project managers at KMD are solely responsible for planning and controlling their projects. The project indicators were, however, calculated post mortem by Compass and they ignored changes in projects. The purpose was measuring overall performance more than giving the project manager data for control.

The new metrics consultants assist the project managers in estimation and ensuring that their plans are suitable for calculating indicators. Project data are supplemented with the metric consultant's observations on software quality. The new project indicators are still used for monitoring overall performance, but the main purpose is to support and learn from projects. The metrics consultants are involved in the projects and they calculate the indicators at several occasions to assist the project manager in high-level tracking and control. The ISBSG benchmarking is flexible which means that the program can include or cope with new aspects of software projects. The major change in metrics practices is that the staff is more involved in projects and the metrics program is more flexible in the way projects are measured.

Culture: Despite skepticism towards comparisons with other companies senior managers have increasingly appreciated the usefulness of indicators and demanded that software performance should be benchmarked. Despite unfortunate cases of public blaming of project managers based on indicators, the project managers remained very open about their Compass indicators. The project managers have now taken ownership of their data and teamed up with the metrics consultants to provide valid data and to subsequently use data to improve management practices, e.g. related to estimation, staffing, and monitoring. The major change is that the perceived role of the metrics consultant has changed from supervisor and controller to colleague and team player.

Incentive Schemes: The software managers were not motivated to actively participate in, contribute to, or use the old metrics program. No incentive schemes were set up to facilitate their involvement. The overall incentive schemes at KMD have traditionally given highest priority to meeting deadlines. The new program includes indicators that focus on effort and functionality. However, meeting deadlines is still the primary objective. The strong interest of the software director motivates other managers to become involved in the new metrics program. The indicators are used actively to set goals for the units, and eventually the indicators contribute to the calculation of each manager's bonus. The major change is that the indicators are used in setting and tracking the goals of the software units and software managers on all levels take an interest in the metrics program.

5.5   Metrics rationale

Looking at the metrics rationale underlying these emerging practices we can identify changes in mode of control (Kirsch, 1997).

Outcome: The indicators in the old program were calculated after completing the project. This suggests outcome control. However, no goals were set for outcome, and the indicators from the program generated no managerial response. The response was limited to questioning data quality and discussing possible managerial actions. In the new program goals are set for indicators. The indicators are as a minimum calculated after completing the project, but in many cases they are also calculated during the project. Changes in a project lead to new or modified goals. The software director requires explanations and actions from the software managers when indicators deviate from goals.

Behavior: The old project measures were not an integral part of the project manager's planning and control activities. The metrics program existed in isolation, and the project manager submitted data only when required to do so. There was no behavior control. The new measures are an integral part of project management. The project manager and the consultant work closely together to estimate and plan the project. When a change occurs they adjust the goals and the consultant assist in provision and interpretation of indicators. KMD's software process includes this practice and the project managers accept it. This implies elaborate behavior control.

Clan: The old metrics staff did not share common values and beliefs and some of the Compass-controllers expressed concerns about the value of the metrics program. The software process improvement group was initially optimistic about the program, but they resigned, as they experienced how measures were not used. Very few project managers expressed that they valued measurements. All members of the new metrics staff believe in the value of the program and its intended use. The project managers see the value of measurements to help identify and address the problems they face. However, there are still project managers who have no experience with the metrics program, and therefore don't have enough insight to acknowledge the potential of the metrics program.

Self: This mode of control was not intended or perceived as essential in either the old metrics program beyond what is implied by the other modes of control. In the new program there is an explicit focus on each program and unit manager and the performance of the dedicated metrics consultants. Project managers own data from their project and their performance assessment includes data from the metrics program.

KMD has, in summary, moved from non-specific outcome control towards focused outcome control complemented with behavior control and elements of clan and self control. The key controllers have shifted from the metrics staff (controlling improvement efforts) to software managers (controlling the software process).

6.   DISCUSSION

The above systematic assessment of metrics practices and the following discussion constitutes the key learning from our research (L in Figure 3). Our guiding research question is: How can changes in metrics practices be assessed and made sustainable? We address this question by examining the following more specific issues: Does the new metrics program solve the identified problems? Does the new metrics program represent a sustainable improvement? What are the lessons from the systematic assessment?

Does the new metrics program solve the identified problems? The identified problems with the old metrics problem (Section 4.2.1) were the reason for and the basis for designing the new program. The problems have been addressed as follows. First, the data suppliers did not understand the definitions of metrics and measures. Now, the metrics staff collects the data and supports project managers in measuring their projects. This high level of interaction facilitates a shared understanding of the intended meaning of indicators. Second, the primary use of the metrics was by software improvement agents for high-level software process improvement. Most software engineers and managers did not use the metrics program, and few managers saw a need for it. The metrics program has been redesigned to contain indicators, which are in demand by managers, and the use of indicators for estimating projects has been emphasized. At the same time the metrics program includes data wanted by the metrics staff for low-level software process improvement. Third, a proper interpretation of data required insights into the program, the submitted measures, and the actual software practices. Few people were capable of that, hence, results were often presented in quite general terms and the readers were left to interpret them on their own. Now, the metrics program is simpler with specific metrics tailored for KMD, and the metrics staff facilitates interpretation of indicators. Fourth, the old project indicators focused on resources and deadlines and ignored the quality of the delivered software. The new project measures include data on software quality. Lastly, there were few managerial responses based on information from the program. Now, indicators are used for setting goals for managers.

Does the new metrics program represent a sustainable improvement? The new program constitutes an improvement because it addresses as argued above the identified problems with software metrics at KMD. The program is therefore perceived as being better. At the same time it is considerably cheaper, 3 million kroner per year compared to 7 million kroner per year (section 4.2.2). The new program has also adopted several of the recommendations in the literature. The program now primarily supports managers by supplying indicators of projects and applications (Grady,

1992). At the same time it still supports identification of project specific improvements (Hall & Fenton, 1997). The program generally emphasizes use of indicators over collection of comprehensive sets of data (Niessink & Vliet, 1999). Indicators are used in setting and tracking the goals of software units and software managers on all levels have initially taken an interest in the metrics program. Moreover, data collection is more automated through extensive integration with the ERP system (Hall & Fenton, 1997). The metrics staff now works full time and is organized as a smaller and more coherent team corresponding to Hall & Fenton's (1997) recommendation of a dedicated metrics staff. There has been a considerable degree of management participation in the design and implementation of new metrics program. As a result managers on all levels have learned about metrics, more managers are committed to the metrics program and act as champions, and the metrics program is better suited to managerial needs (Hall & Fenton, 1997). Metrics "can only be useful to those controllers who truly understand the meaning and implications of the data" (Kirsch, 1996, p. 4) and having the metrics staff work closely with managers supports this.

The systematic comparison between the old and the new metrics program revealed a number of key changes as presented in Section 5. However, it also revealed a number of issues that needs to be addressed to further improve and sustain the metrics program. The analysis showed that all the organizational aspects presented by Applegate (1994) had been addressed, but some needs more attention. First, and most importantly, the incentive schemes for the project managers are implicit and long range. Project managers are motivated by the demand of indicators from their managers, and the anticipation of the usefulness of project indicators for estimating future projects. The first measurements have shown that the motivation increases when a project manager experiences how indicators can be used, e.g. to support staffing. Their motivation could, however, be stronger if there were more explicit and short range incentive schemes. It is also at this point unclear how the new program affects the relation between project and software managers. Discussions are needed to clarify what the roles are, and how project and software managers can inspire each other to generate effective responses based on indicators from the program. It still remains to be seen what the sustainable level and quality of managerial responses will be in the new set-up. The tasks, roles, and technology for distilling and spreading best practices based on project measurements are also not well defined. Project managers are expected to learn from their own projects, and the metrics consultants should distill general patterns across projects. To achieve any effects it is critical that the project managers and metrics consultants are observant, analytical and proactive. The insights and impacts from this learning process are of interest to senior managers and the responsible for software process improvement. However, it is not well defined how results are identified, presented and diffused.

What are the lessons from the systematic assessment? The comprehensive organizational analysis can in this way help develop a more sustainable program. The design and implementation of the new program was guided by the identified problems (see Section 4.2.1) and recommendations in the literature. Yet the analysis based on Applegate's framework (1994) offered a comprehensive organizational understanding of the program and its operation and it revealed important areas, which need further attention. Adding Kirsch's concepts of control mechanisms (1997) helped us understand and design the underlying control rationale of the metrics program. This analysis stressed in a very direct way the importance of making active use of indicators (Niessink & Vliet, 1999, Iversen & Kautz, 2001) and it enriched our understanding of different approaches to control software practices. The adopted approach to systematic assessment is easy and inexpensive to use and the process can be completed within a very limited time frame. Our experience shows that the effort is worthwhile. Finally, it is important to note, that the approach focuses on first-order improvements, i.e. on assessing improvements of the metrics program itself. Eventually, a useful metrics program must demonstrate second-order improvements as well, i.e. improvements in software process performance, to legitimize the cost of running the program and to ultimately assess the ROI of the metrics improvement effort.

7.   CONCLUSION

Many software organizations fail to get satisfactory benefit from their investments in metrics programs (Dekkers, 1999). Software managers are therefore advised to give priority to critical assessments and subsequent improvement of software metrics programs (Mendonça & Basili, 2000). The presented approach for assessing a metrics program stresses the organizational completeness of the metrics program (Applegate, 1994) and the underlying control rationale (Kirsch, 1997). Software managers are encouraged to adopt this inexpensive and comprehensive scheme to critically assess improvements in metrics programs. Our experiences suggest that the research community should give priority to research that emphasizes metrics programs as organizational systems rather than isolated technological tools for measuring software processes. Future research could further explore these opportunities by adopting well-established approaches to assess metrics programs as management information systems (see for example DeLone & McLean, 2003). Additional research is also required to assess and further develop the presented approach to assess improvements in metrics practices for practical use.

REFERENCES

Andersen, P.B. (1990) A Theory of Computer Semiotics. Semiotic Approaches to Construction and Assessment of Computer Systems. Cambridge University Press, Cambridge.

Applegate, L.M. (1994) Managing in an Information Age: Transforming the Organization for the 1990s In R. Baskerville, S. Smithson, O. Ngwenyama, and J.I. DeGross, Eds., Transforming Organizations With Information Technology, (A-49), 15-94. Amsterdam: Elsevier.

Avison, D., F. Lau, M.D. Myers and P.A. Nielsen (1999) Action research, Communications of the ACM, Vol. 42, No. 1, January, pp. 94-97.

Berry, M. and R. Jeffery (2000) An instrument for assessing software measurement programs, presented at Empirical Software Engineering Conference, Staffordshire, UK.

Checkland, P. (1981) Systems Thinking, Systems Practice. John Wiley.

Checkland, P. and J. Scholes (1990) Soft Systems Methodology in Action. John Wiley.

Chidamber, S.R. and C.F. Kemerer (1994) A metrics suite for object oriented design, IEEE Transactions on Software Engineering, Volume 20, no. 6, page 476-493.

Coghlan, D. and T. Brannick (2001) Doing Action Research in Your Own Organization, Sage, London.

Dekkers, C.A. (1999) The secrets of highly successful measurement programs, Cutter IT Journal 12(4), 29-35.

deLone, W.H. and McLean, E.R (2003) The deLone and McLean model of information systems success: A ten-year update, Journal of Management Information Systems 19(4), pp. 9-30.

Fenton, N.E. and M. Neil (1999) Software metrics: Successes, failures and new directions, The Journal of Systems and Software (47:2-3), pp. 149-157.

Frederiksen, H.D. and L. Mathiassen (2002) Diagnosing Metrics Practices in a Software Organization. In New Perspectives on Information Systems Development: Theory, Methods and Practice. Kluwer Academic, New York, USA.

Frederiksen, H.D. and L. Mathiassen (2003) Diagnosing metrics practices in a software organization. Submitted for IEEE Transactions on Engineering Management.

Frederiksen, H.D. and J. Rose (2003) The social construction of the software operation, Scandinavian Journal of Information Systems, Vol. 15.

Galliers, R.D. (1992) Choosing Information Systems Research Approaches. In: R. Galliers, editor. Information Systems Research: Issues, Methods and Practical Guidelines. Blackwell Scientific Publications, Oxford.

Grady, R.B. (1992) Practical Software Metrics for Project Management and Process Improvement, Prentice Hall.

Hall, T. and N. Fenton (1997) Implementing effective software metrics programs, IEEE Software 14(2), 55-64.

ISBSG www.isbsg.org.au

Iversen, J. and K. Kautz (2001) Principles of metrics implementation, in Improving Software Organizations: From Principles to Practice, L. Mathiassen, J. Pries-Heje, and O. Ngwenyama, Eds., pp 287-305, Addison-Wesley, Upper Saddle River, New Jersey.

Iversen, J. and L. Mathiassen (2003) Cultivation and engineering of a software metrics program, Information Systems Journal, (13:1), January, pp. 3-20.

Jones, C. (1996) Applied Software Measurement, McGraw-Hill.

Kirsch, L.J. (1996) The Management of Complex Tasks in Organizations: Controlling the systems development process, Organization Science, Vol. 7, Issue 1, January-February, pp. 1-21.

Kirsch, L.J. (1997) Portfolios of control modes and IS project management, Information Systems Research, Vol. 8, No. 3, September, pp. 215-239.

Kitchenham, B., S.L. Pfleeger, and N. Fenton (1995) Towards a framework for software measurement validation, IEEE Transactions on Software Engineering 21 (12), 929-944.

Kock, N. & F. Lau (2001) Introduction. Information systems action research: serving two demanding masters, Information Technology & People, Vol. 14 no. 1 (2001), pp. 6-11.

Mathiassen, L. (2002) Collaborative practice research, Information Technology & People (14:4), pp. 321-345.

McFeeley, B. (1996) IDEAL A User's Guide for Software Process Improvement, CMU/SEI-96-HB-001.

McKay, J & P. Marshall (2001) The dual imperatives of action research, Information Technology & People, Vol. 14, no. 1, pp. 46-59.

Mendonça, M.G. and C.R. Basili (2000) Validation of an approach for improving existing measurement frameworks, IEEE Transactions on Software Engineering 26 (6), 484-499.

Niessink, F. & Vliet H. van (1999) Measurement should generate value, rather than data, in Proceedings of the Sixth International Software Metrics Symposium, pp 31-38, Florida, USA.

Pedersen, M.K. (1996) A Theory of Informations. Samfundslitteratur, Copenhagen.

Susman, G. and R. Evered (1978) An assessment of the scientific merits of action research, Administrative Science Quarterly, Vol. 23, pp. 582-603.

Wasserman, A.I. (1996) Towards a discipline of software engineering, IEEE Software, Vol. 13, no. 6. November, pp. 23-31.

STANDARDIZING SOFTWARE PROCESSES–AN OBSTACLE FOR INNOVATION?

Ivan Aaen[1] and Jan Pries-Heje[2]
[1]Aalborg University, Denmark; [2]The IT-University of Copenhagen, Denmark

Abstract: Over the last 10 years CMM has achieved widespread use as a model for improving software organisations. Often CMM is used to standardise software processes across projects. In this paper we discuss this standardisation of SPI in relation to innovation, organisational size and company growth. Our discussion is empirically based on years work and experience working with companies on SPI. In the concrete our discussion is enhanced by vignette stories taken from our experience. As a result we find that standardisation focussing on process, metrics, and controls may jeopardize innovative capabilities and company growth. We conclude by identifying and describing four mechanisms that can used to avoid disaster.

Key words: Innovation, Standardization, Software Process Improvement, organizational change, company growth

1.   INTRODUCTION

A standard is "something established and generally accepted as a model, example, or test of excellence ..." or "something established as a measure of reference of weight, extent, quantity, quality, or value" (Websters 1997). The effort that software producing organizations are making to become better, more efficient, faster, or improved in other ways have come to be known as Software Process Improvement (SPI). Some SPI efforts are carried out in an ad-hoc based way but quite many has used the Capability Maturity Model as a map to improve (Aaen *et al.* 2001).

The Capability Maturity Model - or just CMM - is a framework characterizing a 5-step path for software process improvement (Paulk et al. 1995). The path describes key processes at each of five levels. The description includes a number of goals at each level. An organization has to meet the goals at one level to reach the next. I.e. to go from the basic level 1 where behaviour is characterized by being ad-hoc and intuitive to level 2 you need to achieve the goals incorporated in six key process areas: requirements management, sub-contractor management, project planning, project tracking, quality assurance, and configuration management.

In fact CMM has become a generally accepted model for software process improvement. Several thousand organizations have undertaken SPI using CMM. In September 2003 the Software Engineering Institute reported (SEMA 2003) that 11823 projects in 2150 organizations had been appraised using CMM since 1987, and 1342 of them within the last four years. Thus it is fair to call CMM a standard for SPI.

But is it a good idea to standardize SPI using CMM? A focus on process, on metrics, and on controls may jeopardize innovative capabilities. 10 years ago CMM proponents such as Bill Curtis (cf. 1994) argued that the CMM enables innovation by removing some of the process-based impediments for exploring-designers, giving them time to think, rather than inhibiting them with a chaotic environment that limits their time and opportunities to exchange ideas. On the other hand critics such as James Bach (1994) claimed that "CMM is dangerous to any company founded upon innovation. Because the CMM is distrustful of personal contributions, ignorant of the conditions needed to nurture non-linear ideas, and content to bury them beneath a constraining superstructure, achieving level 2 on the CMM scale may very well stamp out the only flame that lit the company to begin with."

So Bach claimed that a standard such as CMM was a major obstacle to innovation. He based this argument on (among other things) personal observations in the two companies Apple and Borland; both producing IT products in Silicon Valley. We set out to research whether that is (still) a true statement in practice – in a Danish context? Or whether it is possible to standardize without sacrificing innovative capability in software process improving organizations?

2.   RESEARCH METHOD

The first step in answering the research question was to survey what others had written on innovation, standards and SPI. We found quite many references to each of these three areas but very few to the combination of 2 or 3 of them. In the sections below we give an account of what we found.

The second part of our research was a qualitative study. Over the last seven years since 1997 we have been involved as both neutral observers and action researchers in three companies. Furthermore we carried out an interview study in a smaller company. Together this gave us a large amount of data that we could use to analyze the impact of CMM as a standard in practice. Details about the companies are given in **Table 1**. We have chosen not to reveal the true identity of the companies.

Table 1. Facts about the case studies drawn on in this paper

Case	Size of organization at the time of the interviews	Industry and what offered	Where situated	Employees interviewed	Organizational roles represented in interviews
Mini	20-25 employees.	E-commerce, systems integration and web applications	Denmark, Copenhagen	3 in 2001, and 3 in 2003	CEO; software architect and developer
Midi	More than 500 employees – approximately 1/3 develops software	IT applications for the financial sector	World wide	More than 10 in each year 2003	CEO; SPI-manager, project manager; process user / developer
Maxi	More than 1000 IT-related employees	IT applications for use in the financial sector	Scandinavia	More than 10 each year in 1997-99 and again 2001-03	CEO; depart-ment managers, project managers, software developers
Mega	1980s; More than 10000 IT-related employees	Tele-communi-cations	World wide	More than 10 in each year 1997-1999	CEO; project manager; software architect/ developer

It would be a paper in itself to give an account of each of the companies. Instead we have chosen to include so-called "vignettes" where it is relevant in our analysis. In that way our empirical data is used to illustrate and substantiate important points.

3.   INNOVATION

Innovation is both a term of importance but also a quite ambiguous term that many researchers have tried to understand and give an overview of (cf. Nelson & Winter 1977, Dosi 1988). A core distinction seems to be between *innovative* product development and *routine* product development (Christensen 2002). Innovative product development is based on an idea that is perceived as new by experts or buyers/users (the latter is equal to Rogers' (1995) definition of innovation). The development process is associated with uncertainty, either technical (can it be done?) or use-related (will they accept it?), and the process itself is often carried out as a distinctive and targeted searching process. Whereas the result of routine product development to experts or users often is obvious, there is no substantial uncertainty, and the process can often be characterised as just ordinary "seek and find".

Vignette #1 illustrating routine product development

Mini is a software house that among other things develops database intensive web applications. Every application is new, the customer is new, and to the user it probably appears that the final application is new. But Mini has a standardized development process that they use every time. It included making 2-3 prototypes, showing them to the user, getting feedback, and incorporating the feedback in the next prototype or the final application. By having this process in place that among other things reduce ambiguity and uncertainty Mini has reduced an innovative development process to a routine development process.

Our question is whether both or only one type can be standardized? Certainly the process of coping with uncertainty can be standardised. One can also standardise the seek-and-find process involved in routine product development. But the good idea – the "new" thing – cannot be standardised.

Another important distinction seems to be between:

– Product innovation where the results of the innovation is sold on a market and thereby used outside the company
– Process innovation where the results (better processes) are used internally

Vignette #2 illustrating the schism between standardisation and innovation

In the mid-90s Mega Headquarter decided to require that all the subsidiaries in different countries should use CMM. After a few years all subsidiaries were certified at level 2 or higher. Then, however, Mega ran into financial troubles. At this point Mega was heavily criticized in the press for developing products that were not innovative enough but over-engineered.

CMM is clearly aimed at process innovation, but with a belief that good processes will lead to good products (we will return to that belief in the next section). In practice it can also be done – by some at least. For example the Danish Software House Systematic, that are certified at CMM level 3 was appointed the most entrepreneurial company in Denmark in 2003. And the CEO pointed to CMM and good processes as a major reason for that achievement.

4. INNOVATION VS. STANDARDIZATION – HOW VS. WHAT & WHY

In this section we shall return to our main question: How can we standardize without sacrificing innovative capability in software process improving organizations? How can we avoid throwing out the baby with the bath water when doing SPI?

In *Managing the Software Process* (Humphrey 1989) Watts Humphrey defines the fundamental ideas embedded in the CMM. Here he defines a standard as a "rule or basis for comparison that is used to assess size, content, or value", i.e. as a product standard. He combines this with a process-standard by defining a procedure as a "defined way to do something, generally embodied in a procedures manual" and by way of illustration he likens this definition with the procedure concept used in programming (Humphrey 1989). Based on IEEE-STD-610 the CMM combines Humphrey's two notions into one and defines standard as (Paulk, Curtis et al. 1993): "mandatory requirements employed and enforced to prescribe a disciplined uniform approach to software development".

CMM based SPI focus on improvement via standardization and formalization as exemplified by the following two process concepts key to CMM based SPI (Paulk, Weber et al. 1993):

– *Organization's standard software process* - The operational definition of the basic process that guides the establishment of a common software process across the software projects in an organization. It describes the fundamental software process elements that each software project is expected to incorporate into its defined software process. It also describes the relationships (e.g., ordering and interfaces) between these software process elements.

– *Project's defined software process* - The operational definition of the software process used by a project. The project's defined software process is a well-characterized and understood software process, described in terms of software standards, procedures, tools, and methods. It is developed by tailoring the organization's standard software process to fit the specific characteristics of the project. (See also *organization's standard software process, effective process,* and *well-defined process.*)

To Watts Humphrey SPI targets a process that can be controlled, measured, and improved. Improvements to the process will only stick if they are reinforced by careful introduction followed by periodical monitoring (Humphrey 1989): "The actions of even the best-intentioned professionals must be tracked, reviewed, and checked, or process deviations will occur. If there is no system to identify deviations and to reinforce the defined process, small digressions will accumulate and degrade it beyond recognition."

Here and elsewhere focus is on establishing and following a defined software process. Innovation and innovative capability is not a foremost concern in CMM based SPI. As defined and practiced in many organizations the CMM tends to emphasize operating norms and focus on building an ability to detect and correct errors in relation to such norms. Despite these efforts many organizations experience that such operating norms are often not adhered to, even after having invested heavily in developing standards and educating the work force in them.

CMM presents itself as TQM applied to Software developing organizations. Citing prominent TQM authors such as Deming, Juran, and others Watts Humphrey acknowledges the heritage from TQM, and Mark Paulk of the Software Engineering Institute states this relation directly: "CMM-based improvement is an application of TQM principles to software" (Paulk 1996).

In light of CMM being TQM applied we would expect a similarity between the two in the perceptions of standardization but this is not the case. There is a marked difference, and the difference seems not to be due simply to dissimilarities between the software industry and the industries traditionally using TQM, i.e. the car industry and similar manufacturing industries.

Kondo – a prominent writer on TQM – describes work standards broadly as consisting of three items: (1) aim of the work, (2) constraints on carrying out the work, and (3) means and methods to be employed in carrying out the work (Kondo 2000). In a major book Creech – another prominent writer on TQM – not even mentions let alone defines the term (Creech 1994).

Applying TQM to software teams Zultner also use a relatively lax definition of standards. To him a standard (Zultner 1993) "… may be a set of instructions, macros, drawings, and so forth – anything that allows the team to do something in a consistent way".

The traditional CMM-SPI approach is centralist and top-down: (1) understand the current status of the process relative to CMM, (2) create a vision of the desired process relative to CMM, (3) establish a list of required actions, (4) produce a plan for accomplishing these actions, (5) commit resources required to execute, and (6) start over at step 1 (based on (Humphrey 1989)). Each loop is expected to contribute to the set of standards enforced in the organization. Due to the time frame involved in one loop this tends to create a gulf between SPI efforts and everyday practical processes.

In line with traditional TQM Zultner focus on teams in his description as opposed to CMM's *centralist* perspective on function. Creech describes centralism as one major difference between TQM and mainly process-oriented type of improvement approaches such as CMM.

Another difference between CMM and TQM is scope. According to Paulk et al. the CMM essentially is a subset of TQM: "The CMM focuses on the process aspects of a Total Quality Management effort" (Paulk, Curtis et al. 1993).

Commenting on the "quality movement" Gareth Morgan describes how many TQM programs have got caught in old bureaucratic patterns and cultural norms, leading to disastrous failure rates (Morgan 1986). According to Morgan the philosophy of promoting continuous improvement has done much to institutionalize the practice of challenging taken-for-granted norms and practices at an *operational* level. Morgan describes the power of the Japanese concept of Kaizen like this (Morgan 1986):

- Employees are asked to dig beneath the surface of recurring problems and uncover the forces that are producing them.
- They are encouraged to examine existing modes of practice and find better ones.
- They are encouraged to create "languages," mind-sets, and values that make learning and change a major priority.

Challenging operating norms and assumptions in this way, Morgan argues that quality approaches create information, insights, and capacities through which a system can evolve to new levels of development.

CMM-based SPI builds on formalization by process specialists and thus on separating thinkers from doers. This is at odds with Kaizen insisting that ongoing improvement be the province of every member of the organization and that alignment be ensured throughout the organization as well.

Even if we accept to focus only on the process aspects of TQM there are serious problems in CMM-based SPI. Zultner describes ongoing improvement (Kaizen) in TQM as a product of variation – all real-world processes vary and this variation forms the very locus of improvement. Thus ongoing improvement requires a combination of

- Standardization – standardize-do-study-act – to reduce variability and lock in later gains,
- Improvement – plan-do-study-act – at the team level.
- Innovation – mainly breaking through bottlenecks for dramatic process improvement.

CMM being a subset of TQM – the process part - leads to a focus on universal goals rather than goals derived from the market situation, and this again tends to separate CMM-based SPI from practice and to a focus on process perfection for the sake of the process itself, and aiming to build the ideal software process for the organization.

Separated from everyday concerns such process perfection will likely be relatively static failing to seize opportunities for improvement in practical work. In fact we can expect the process standard to become an obstacle to change. As Kondo puts it: "since standardized working means and methods have been formulated after careful consideration of all the angles, they must be the most productive and efficient means and methods possible, regardless of who uses them – at least the people who drew up the standards think so" (Kondo 2000).

TQM is much more than Kaizen, and Kaizen is more than the CMM equivalent of SPC. Zultner on Quality function deployment (QFD) and Policy deployment (PD): These are completely absent in CMM – leading to only fickle focus on horizontal integration, on customers, on vertical alignment, and on advancing the organization for long-term survival.

4.1   Threats to innovation in the CMM approach

The above discussion has identified a number of important characteristic traits in the CMM approach to TQM:

1. *A strict process orientation.* According to Creech builds on five pillars: Product, Process, Organization, Leadership, and Commitment, but CMM basically only address the process-pillar (Creech 1994). Following Creech this means that SPI is not a holistic approach to quality.
2. *A centralist approach.* SPI is often practiced as a committee-oriented approach, alternatively as something being done by a specialist staff function. Either way this means a separation of thinkers from doers with a main focus on the organization-wide process. A focus on conformance rules, inputs, and procedures.
3. *A focus on function.* SPI projects are predominantly internal to the software organization. The aim is to improve internal work procedures based on the assumption that good products come from good processes.
4. *A focus on standard adherence.* The goal is to minimize variation and thereby improve process reusability and learning from previous projects.

What then are the threats to innovative capability coming from these traits?

A strict process orientation: The lack of customer orientation may lead to goal deflection in the improvement effort. Such goal deflection might result in employees seeing process adherence as a goal rather than a means. Process development might overshadow customer satisfaction leading to a waste of creative strengths on problems of less importance to the customer. A strict process orientation might also lead to process gold plating building ever more comprehensive and universal processes. Finally and perhaps worst: employees might "misuse" creativity on building process workarounds when organizational processes stand in their way.

Vignette #3 illustrating a very rigid and centralised process orientation

In Midi it was decided to use CMM to improve the organisation. An intensive effort led to the achievement of CMM level 2 in 2001. This effort was led by a centralised SPI-group. However, developers were complaining that following CMM added too much bureaucracy and that innovation was hurt. To cope with this Midi intensified the focus on a "bottom-up" SPI approach and focused improvements initiatives on the individuals in the organisation. A new de-centralised improvement effort was implemented to supplement the effort in the centralised SPI-group. The core idea was that responsibility for the core of both development and improvement needed to be decentralised to work.

A centralist approach: Centralism traditionally favours the deployment of solitary methods – i.e. that only one method is provided for a particular type of task. Kondo describes how solitary methods might lead people to feel that they are not responsible for non-conformance of product quality. Restrictive centralism thus might lead to employee alienation. Separating thinkers from doers means a split between insiders and outsiders in software projects. How can opportunities be seized and how can process innovations come to life? The split in itself might lead to stifled motivation at both ends. Turf guarding is also a likely outcome at both ends if careers accidentally collide in a clash between specific interests in the project and general interests in the staff function or committee. Lastly organization-wide process standardization might lead project members to turn on the autopilot or drop into apathy. None of these problems will benefit creativity or innovative capabilities neither at the level of software project nor at the organizational level.

Vignette #4 illustrating the danger of a centralised development method

In Maxi there was a development method in place describing phases and milestones, and making available templates and examples. An interview study in the organisation uncovered that the method wasn't being used due to three things: (1) Difficult to find things. (2) Build on one-size-fits all concept. (3) No help to tailor method for specific project. A project was established focusing on process tailoring (a CMM level 3 key process by the way). But before the project could report their recommendations top management had decided on a new centralized project model that every project should follow. In the new model there was a core and a number of options. So the tailoring project was stopped and the new project model was implemented. Everyone followed the new project model because it was enforced – but in interviews it was found that the three problems originally identified were still there.

A focus on function. Despite the undisputed assumption that good products come from good processes the focus on function implies not focusing on the end product of the software project. This lack of a product focal point might lead to conflicts in installing a sense of purpose and achievement to project members. The focus on function is one step away from the machine bureaucracy where routines may be routinized for the sake of routinization alone. Such focus on formalization and routinization might easily stand in the way for creativity.

Vignette #5 illustrating too much focus on function

At Mega the Headquarter had had enforced everyone to use CMM. So a lot of effort was put into implementing processes and following them strictly. In hindsight it is clear that the effort should have been directed towards the product instead. Because competitors took over several important markets while Mega was honing their processes.

A focus on standard adherence. Where adherence to norms becomes and end in itself fundamentalism lurks in the shadows. Fundamentalism might take the form of dogmatism, of mind control, or of goal deflection. This is when rules and regulations come to stand in the way for creativity and achievement. Another and less conspicuous threat coming from standard adherence is a blocking of variation. If successful SPI efforts lead to software projects becoming ever more similar in process the end result might be, that the very variation that fosters creativity and innovation is obliterated.

Vignette #6 illustrating too much standard adherence

This is a story that was part of the culture at Maxi. About 10 years ago there was a development model that was quite lengthy and very detailed. Everybody was supposed to follow it from page 1 to page 2000. The saying was that you didn't have to bring your brain to work; you could just start on page 1 and follow the process. Furthermore it was controlled whether you followed it, and warnings were given if you didn't. But as it always happens in organisations there came a catastrophic project. But the project leader claimed that he had followed the process – a good example of too much standard adherence. And so it was decided to give up the very detailed development model.

5. STANDARDISATION OF SPI IN RELATION TO COMPANY GROWTH

Innovation literature talks about two "engines" that lies behind technological innovation in organisations. One is the entrepreneur model the other is the big company model. Traditionally it has been argued that bigger is better. If you are big you can invest more, you are less vulnerable to failure, and you have the resources to attract the best. An entrepreneurial organisation on the other hand is a small independent company that can react fast. Acs & Audretsch (1988) and Rothwell & Zegweld (1982) argue that the entrepreneurial organisation has four advantages compared to the big company:

1. Management and decision structures are lighter and faster in small companies
2. Small companies are often headed by dynamic innovative people involved in the innovation process themselves
3. Innovation often involves communication across functions. In small companies that is simple, whereas in larger companies that type of communication can involve crossing departments, borders, countries, culture etceteras
4. Larger companies rewards their best and most innovative people by lifting them up to a management position – and thereby effectively taking them away from innovation

So can one standardize the advantages of a small entrepreneurial company? Let us take the advantages one by one. Small and light decision structures are at odds with CMM. In fact CMM specify a number of key processes that will add structure and will enforce more decisions to be made. Standardisation will not influence the involvement of dynamic managers in the innovation process. Whom you promote is neither influenced. But how you communicate is at the core of CMM. In fact CMM enforces a specific repeatable and measurable way of communicating (at least if you move up the levels). However, there is nothing in CMM that enforces slow or cumbersome communication. Probably the continuous improvement idea built into CMM will surmount some of communication disadvantages of enterprise growth.

Vignette #7 illustrating how rewarding may counterbalance innovation

At Maxi a local project manager was known all over the company for having saved the organisation from an embarrassing failure. An off-shore outsourcing project was again and again delivering code that could not be used. Finally the local project manager stepped in, broke all the rules (also a kind of process innovation), and managed to deliver the new innovative product on time as promised to the customers. She was soon promoted to second-in-command in that development unit, and thereby effectively out of the loop when talking about the day-to-day work on innovations – but probably with more influence on innovations from a management viewpoint.

Vignette #8 illustrating communication barriers

In Mega three set of stakeholders were trying to set the agenda. Headquarter were pushing CMM. Another department at Headquarter had helped the Danish subsidiary identify a number of so-called Vital Few Actions. And locally a number of improvement areas had been decided. Communication between the three functions was lacking. The result was paralysis. Nothing happened until the local management cut through and made a decision on which one of the three agendas should be the one to be followed.

Companies start as small and possibly entrepreneurial, and if they are successful they may expand. Larry Greiner (1972, 1988) has developed an evolutionary model of company growth.

In his model the organization starts its life in the Creative Phase shown in the lower left corner. In this phase, the founders of the company are typically entrepreneurs, communication among the people in the organization is informal, long work hours are normal, and the feedback from the market is immediate, as well as the reaction from management.

As the company grows in size and matures, it reaches its first crisis. Dubbed the Crisis of Leadership this is when informal communication no longer will suffice. The dedication, long hours and small salaries of the first hired "pioneers" are no longer adequate motivation. Furthermore new procedures are needed to exploit economies of scale and to provide better financial control. To solve the leadership crisis a strong manager is required. Often the owner/founders will not have the necessary skills and knowledge, and so will "hate to step aside even though they are ... unsuited to be managers" (Greiner 1972, 1998).

Having survived the leadership crisis the organization will "embark on a period of sustained growth under able and directive leadership" (Greiner 1972, 1998). In this Direction Phase hierarchies develop within the organization, the upper levels take responsibility for the direction of the organization, and communication becomes more formal. Also in this phase, formalized systems for accounting, incentives, work practice, and job specialization will arise.

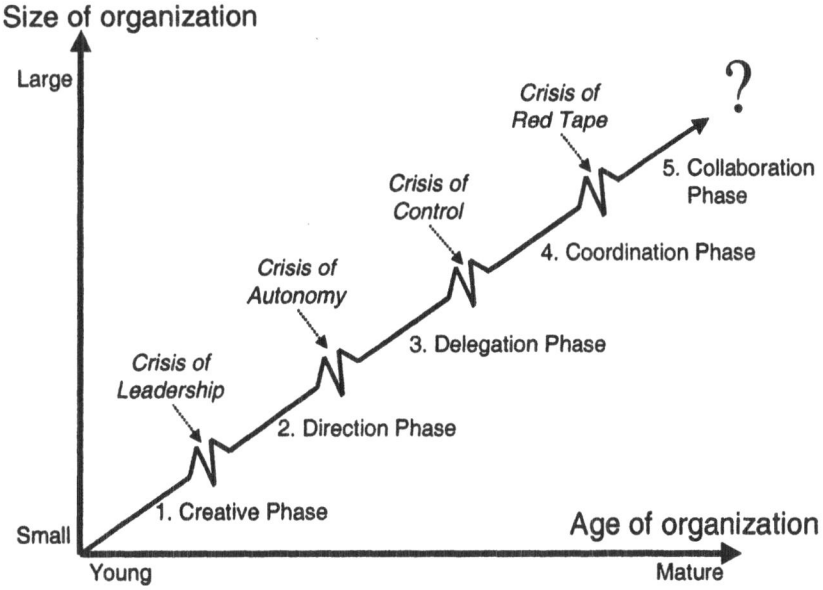

Figure 1.

In this second phase of organisational growth a standard can be useful to formalise systems and communication.

The second crisis then is the Crisis of Autonomy. Middle-level managers come to see the centralized decision structure of the second phase organization as a burden, and some autonomous field unit middle managers will start running their own shows. Often the reaction by top management is an attempt to restore centralized management. At this point some organisations stick to standardisation but face a number of the disadvantages discussed in section 4 such as a too centralist approach or an overly strong focus on standard adherence. To solve the autonomy crisis a more decentralized organization structure is needed offering greater responsibility and autonomy to middle management. This structure marks the start of the Delegation Phase.

Vignette #9 illustrating the usefulness of CMM

In Mini the owner and CEO wanted to appear and actually work more professional to be able to survive in the marketplace (with many big competitors) and maybe even grow from the 20 employees they had had for along time. They decided to use CMM as inspiration. CMM was presented to the employees. They discussed it and voted on what they found most useful. The outcome were three new processes that were then adapted to the organisations and implemented as the local "standard" – this is how we do things here. After the implementation the CEO described CMM as highly useful in the professionalisation of Mini.

The Delegation Phase ends in the Crisis of Control where top management realizes that they have lost "control over a highly diversified field operation." Again using standardisation to centralise and to focus on standard adherence is not the answer. This third crisis is overcome by using coordination techniques such as formal planning, by organising product groups as investment centres, and by initiating staff functions that controls and reviews for line managers. This way the company enters the Coordination Phase.

The next crisis then is the Crisis of Red Tape where the line is looking at staff functions with more suspiciousness, and distrustfulness evolves between headquarters and the field. Overcoming this fourth crisis then leads the organization into what Greiner calls "the last observable phase in previous studies." In this last phase strong interpersonal collaborations are established to overcome the crisis. A more flexible and behavioural management approach is implemented through the use of teams. Staff functions are reduced in number, and the motivational structure is geared more to team performance than to individual achievements.

6.   DISCUSSION

We have now analysed the standardisation of SPI in relation to innovation, organisational size and company growth. We have found that:

a) Standardisation can mainly be used for routine product development – and coping with uncertainty in innovative product development
b) Standardisation using CMM is clearly aimed at process innovation
c) Using CMM for standardisation easily leads to process orientation that is too strict
d) Using CMM for standardisation can easily be too centralized
e) Using CMM for standardisation can lead to much focus on function
f) Using CMM for standardisation can lead to too much focus on standard adherence
g) CMM can be useful in the first phases of a company's growth life-cycle, especially in the direction phase
h) CMM can be dangerous to use in the later phases of a company's growth life-cycle

So what can be done? We have four suggestions:
1. A holistic quality approach
2. A decentralist approach building on teams at the operational level
3. A project-oriented approach (as opposed to a focus on function)
4. A focus on situational discretion (as opposed to a focus on standard adherence)

But let us take the discussion one at a time.

A holistic quality approach (as opposed to a strict process orientation). This is probably the most controversial of our propositions. To our mind CMM-based SPI seems fundamentally flawed by its focus on process, metrics, and controls. TQM is a systemic or holistic approach involving process but also product, organization, leadership, and commitment. Perhaps software process improvement (SPI) really should be software total quality management (STQM). We believe that innovative capabilities may be strengthened by a focus on customers and on products. For project members this leads to an emphasis on goals and achievements relative to the project.

A decentralist approach building on teams at the operational level (as opposed to a centralist approach). A decentralist approach could mean methodological pluralism where projects experience degrees of freedom towards means combined with clarity with respect to ends. When organizational level means and methods are required they could be split in two groups: one for novices and one for experienced workers. Kondo argues in favour of such a split (Kondo 2000). Novices can use manual describing essentials of a basic process that should be known by every employee. Experienced workers can share special tips and tricks in the other group.

A decentralist approach would build on personal involvement and responsibility. The variations involved in any real-world software project would provide good opportunities to seize opportunities for process innovations.

A project-oriented approach (as opposed to a focus on function). Such an approach would entail a focus on the team rather than the function. Teams and individuals situated in practice allows for innovative solutions to processes and products. The project team is the ideal unit for handling improvisations possibly leading to process and not least product innovations.

A focus on situational discretion (as opposed to a focus on standard adherence). Situational discretion requires that authority and accountability be localized in the same team. Kondo argues in favour on less emphasis on process standardization and more on product standardization (Kondo 2000). This implies more focus on the aim of the work and on providing freedom in the use of means and methods, and less focus on process enforcement to allow for creativity. Following Kondo our proposition is to limit process standards to a minimum and thereby to focus less on "dos" and more on "don'ts" when procedural regulations are needed. The will result in fewer and leaner instructions replaced when needed with constraints. An ideal incubator for innovation.

7.   CONCLUSION

In this paper we have discussed the standardisation of SPI in relation to innovation, organisational size and company growth. Our discussion is empirically based on years work and experience working with companies trying to improve their software development. In the concrete our discussion was substantiated by vignette stories from our experience. As a result of the conclusion we find that standardisation with a main focus on process, on metrics, and on control may jeopardize innovative capabilities and company growth. However, we suggest to use the following four mechanisms thoughtfully to allow for a culture of innovation: (1) A holistic quality approach, (2) A decentralist approach building on teams at the operational level, (3) A project-oriented approach, and (4) A focus on situational discretion

ACKNOWLEDGEMENTS

The authors contributed equally to this article; the order of their names reflects alphabetical order.

REFERENCES

Aaen, I., Arent, J., Mathiassen, L. and Ngwenyama, O. (2001). A Conceptual MAP of Software Process Improvement. *Scandinavian Journal of Information Systems,* Special Issue on Trends in the Research on Software Process Improvement in Scandinavia, Vol. 13, pp. 123-146.

Acs, Zoltan J. & David B. Audretsch (1988). Innovation in large and small firms: An empirical analysis. American Economic Review, vol 78, no. 4, 678-690.

Bach, James (1994). The immaturity of the CMM. *American Programmer,* 7 (9), 13-18.

Christensen, J.F. (2002). Produktinnovation – process og strategi ("Product Innovation – process and strategy"). 2nd Edition. Handelshøjskolens Forlag.

Creech, B. (1994). The five pillars of TQM: how to make total quality management work for you. New York, Truman Talley Books/Dutton.

Curtis, Bill (1994). A mature view of the CMM. *American Programmer,* 7 (9), 19-28.

Dosi, Giovanni, Christopher Freeman, Richard Nelson, Gerald Silverberg & Luc Soete (Eds.) (1988). Technical Change and Economic Theory. Pinter Publishers.

Greiner, Larry (1972). Evolution and revolution as Organizations grow. *Harvard Business Review*, (50), 37-46.

Greiner, Larry (1998). Evolution and revolution as organizations grow. *Harvard Business Review*, May-June, pp. 55-64.

Humphrey, W. S. (1989). Managing the Software Process. Reading, Massachusetts, Addison-Wesley Publishing Company.

Huy, Quy Nguyen (2001). Time, Temporal Capability, and Planned Change. *Academy of Management Review*, 2001, Vol. 26, No. 4, pp. 601-623

Kondo, Y. (2000). "Innovation Versus Standardization." TQM Magazine 12(1): 6-10.

Morgan, G. (1986). Images of Organization. Newbury Park, Cal., Sage Publications.

Nelson, Richard & Sidney Winter (1977). In search of useful theory of innovation. *Research Policy* (6) 1, pp. 36-76.

Paulk, M. C., Weber, C., Curtis, B., and Chrissis, M. B. (1993a). Capability Maturity Model for Software, Version 1.1, Software Engineering Institute.

Paulk, M. C., Weber, C., Curtis, B., and Chrissis, M. B. (1993b). Key Practices of the Capability Maturity Model, Version 1.1, Software Engineering Institute.

Paulk, M. C., Weber, C., Curtis, B., and Chrissis, M. B. (1995) The Capability Maturity Model: Guidelines for Improving the Software Process, Addison-Wesley, Reading, Mass.

Paulk, M. C. (1996). Effective CMM-Based Process Improvement. 6th International Conference on Software Quality, Ottawa, Canada, SEI.

Rogers, Everett M. (1995). *Diffusion of Innovations.* 4th Edition. Free Press, New York.

Rothwell, Roy & Walter Zegweld (1982). *Innovation and the small and medium sized firm.* Frances Pinter, London.

SEMA (2003). Software Engineering Measurement and Analysis Initiative. Process Maturity Profile, Software CMM, CBA IPI and SPA Appraisal Results, 2003 Mid-Year Update, September 2003. Downloaded November 2003 from www.sei.cmu.edu.

Websters (1997). The new International Webster's Concise Dictionary of the English language. International Encyclopedic Edition. Trident Press International.

Zultner, R. E. (1993). "TQM for technical teams." Communications of the ACM 36(10): 79-91.

ORGANISATIONAL DYNAMICS IN THE SOFTWARE PROCESS IMPROVEMENT: THE AGILITY CHALLENGE

Anna Börjesson[1] and Lars Mathiassen[2]
[1]*Ericsson and IT University of Gothenborg, Sweden;* [2]*Georgia State University, USA*

Abstract: It is a challenge to keep up momentum in Software Process Improvement (SPI) during organizational changes. SPI initiatives get interrupted, are side-tracked and progress slowly when changes occur in the target organization. This paper explores this crucial relation between SPI initiatives and the dynamics of the organization to be changed. The study builds on longitudinal data from the introduction of a new approach to requirements management into a software unit within Ericsson. Our focus is on the challenges involved in managing the SPI initiative as a sequence of organizational changes occurs in software development. We discuss the findings from this study in relation to the SPI literature and the literature on organizational agility and software agility. Our research indicates that SPI can benefit from agility ideas if the innovations are integrated well with other agility initiatives within the software organization. We therefore suggest that there is a need to coordinate and align the software agility movement with SPI issues to arrive at a comprehensive and holistic understanding of how software organizations can respond effectively to dynamics in their environment.

Key words: Software Process Improvement, Agility, Organizational Dynamics, SPI Implementation Success

1. INTRODUCTION

Software Process Improvement (SPI) has been adopted by many organizations as a strategy to enhance their capability to deliver qualitative software (Humphrey, 1989; Grady, 1997; Mathiassen et. al, 2002). Although very successful cases of improvement have been reported (Humphrey et al., 1991; Haley, 1996; Diaz and Sligo, 1997; Larsen and Kautz, 1997) there is also a critical debate on this approach to improve software performance (Bach, 1995; Bollinger and McGowan, 1991; Fayad and Laitinen, 1997; Humphrey and Curtis, 1991). The most recent report from the Software Engineering Institute puts the rate of failure at around 70% (SEMA, 2002). A growing amount of research offers explanations and guidance to improve the success rate of SPI initiatives. McFeeley (1996) discusses the steps and activities an SPI initiative should include and presents the IDEAL model for SPI. Mashiko and Basili (1997) and Ravichandran and Rai (2000) examine the relationship between SPI and software quality management. Fichman and Kemerer (1997) discuss the organizational barriers towards adoption of software process innovations. Abrahamsson (2000, 2001) discusses the need to manage commitment from different stakeholders. Nielsen and Nørbjerg (2001) emphasize the need to focus on social and organizational issues in SPI. Aaen (2002) argues for helping software developers help themselves rather than having change agents drive SPI. Börjesson and Mathiassen (2003a) argue that SPI initiatives benefit from spending more of their effort in the later phases of the IDEAL model where focus is on deployment. While these contributions offer a broad range of perspectives on how to improve the SPI success rate and guide the SPI initiatives towards success, very little attention is paid to one of the key challenges faced by SPI initiatives: the dynamics of the software organization they target.

Software organizations constantly need to react to market dynamics, new customer requirements, technological innovations, and mergers between software companies. The degree and pace of the implied organizational dynamics have increased over the past years as indicated by the notions of fast-moving software organizations (Baskerville et al., 2001; Holmberg and Mathiassen, 2001) and radical IT-based innovations (Lyytinen and Rose, 2003). SPI initiatives are therefore increasingly faced with the challenge of developing and implementing improved processes into an organizational context that is constantly changing. To be successful they must be organized, managed, and executed in ways that take these organizational dynamics into account and address them effectively. The software engineering community has over the past years adopted agile philosophies and approaches to improve development practices in response to the increasing dynamics faced in software development (Abrahamsson et al., 2002). The theme of our research is to extend this notion of software agility by studying challenges related to improvement of software practices, i.e. SPI in dynamic organizational contexts.

The study is exploratory in nature. It combines an empirical study of SPI practices with insights from the literature on organizational agility and software agility. The purpose is to understand the challenges SPI initiatives face and the possible strategies and tactics they can adopt to deal with the dynamics the software organization is facing. We present data from a longitudinal case study during the period 2000 to 2003. The initiative was carried out in a product development unit within the Ericsson Telecom Company in Gothenburg, Sweden. The initiative focused on introducing a new requirements management approach and tool, Requisite Pro. We study the activities involved and the impact the new requirements management approach had on configuration control, baseline control, and traceability over time, while the software unit was subject to a number of organizational changes. Subsequently, we interpret the experiences from this initiative by relating them to the agility concept. The paper is structured around three main questions:

- What was the impact of organizational dynamics on the SPI initiative?
- In what ways could the initiative have benefited from adopting more agile SPI tactics?
- What are the implications for development of agile approaches within software engineering?

The argument is presented as follows. Section 2 presents the literature on SPI and organizational dynamics together with the literature on organizational agility and software agility. Section 3 describes the research approach. Section 4 presents data from the longitudinal case study with particular focus on the implications organizational dynamics had on the change initiative. Section 5 discusses the contributions of our research and implications for both practice and research. We conclude our findings in section 6.

2. THEORETICAL CONTEXT

Initially, we review the SPI literature with particular focus on how the issues related to organizational dynamics are addressed (Section 2.1). We then look closer at the literature on organizational agility that specifically aims to address challenges related to responding effectively to changes in the environment (Section 2.2). Finally, we look at how these ideas have been adopted to develop agile approaches to software development and open up for future research in the light of the organizational dynamics issues reported here (Section 2.3).

2.1 Organizational Dynamics in the SPI Literature

The relation between organizational dynamics and SPI success is not well investigated and understood in the core SPI literature. Parts of the SPI literature touch upon the challenges involved in dealing with organizational dynamics in SPI, but very little is said directly and the managerial tactics offered are few and vague. The Capability Maturity Model – CMM (Paulk et. al., 1995) is one of the most widely spread and used SPI models. CMM is structured in five levels of maturity and defined by a number of key process areas (KPAs). The idea is to work with the KPAs on the next level to address problems and improve practices in the organization. Very little is mentioned about organizational dynamics. The model assumes a relatively stable organizational environment while addressing KPAs to achieve higher CMM levels.

Humphrey (1989) discusses 'The Six Basic Principles' for SPI and one of them is 'Ultimately, everyone must be involved'. If improvement initiatives appear threatening to people, they will likely cause resistance to change. This implies that key stakeholders need to be involved to address the implications of organizational growth or the inclusion of new people into the organization. In the same way, McFeeley (1996) argues that management needs to be involved. He also specifically describes in the IDEAL model a step called 'Revise Organizational Approach'. The goal is to insure sponsorship for SPI from managers. This implies that SPI sponsorships need to be evaluated and negotiated when the software organization is re-organized. Grady (1997) emphasizes along the same lines the importance of 'Organizational Readiness', i.e. how much and how widespread the enthusiasm for change is. Weinberg (1997) argues in general that growth has a natural negative impact on quality. One of the reasons is that special initiatives are needed to ensure dedication to software development and commitment to change initiatives as new people become involved as a consequence of re-organizations.

Zmud (1982) discuss diffusion of new software practices. These practices, for instance configuration management and structured review, are typical examples of improvement areas that an SPI initiative drives. Zmud argues that the success depends on the organizational context and the resources in the organization whose behaviors need to change. Factors found to influence the diffusion of new software practices in one organizational context may be seen to have little or no impact in another organizational context. Fichman and Kemerer (1997) discuss these difficulties from an organizational learning perspective. Successful diffusion of software process innovations depends on the ability to facilitate organizational learning. This is especially relevant for complex organizational technologies such as object oriented programming languages.

More recent SPI literature (Holmberg and Mathiassen, 2001) discusses the challenges involved in combining a capacity for radical innovations while at the same time facilitating evolutionary improvements in software organizations. Holmberg and Mathiassen studied a fast-moving software organization and concluded a number of lessons learned to cope with a dynamic environment while simultaneously trying to improve professional practices. First, the authors argue that there is a reciprocal relationship between innovations and improvements that everyone in the organization must understand. Second, it is easy to ignore improvement work in favor of the more hype innovation work and the improvement culture must therefore be protected. Third, improvements must be conceived as relevant and useful in the organization. Finally, SPI practitioners must understand the core processes of the business they work in. Holmberg and Mathiassen argue that SPI is particularly important to integrate into fast-moving software organizations if they are to survive in the long run.

2.2   Organizational Agility

The agile movement is led by the Agility Forum at Lehigh University. The forum was formed in 1991 to implement the vision of a new system for production of goods and services through adoption and implementation of the agile organizational paradigm. Gunneson (1997) argues that even though there are different interpretations of organizational agility there are specific elements that are generally agreed upon. Agility is concerned with economies of scope, rather than economies of scale. Where lean operations are usually associated with efficient use of resources, agile operations are related to effectively responding to a changing environment while at the same time being productive. The idea is to serve ever-smaller niche markets and individual customers without the high cost traditionally associated with customization. The ability to respond is hence the essential and distinguishing feature of the agile organization (Dove, 2001). The reason that agility has emerged over the past decade as a key challenge for organizations is the increasing pace and unpredictability of changes in the economical, technological, and political environment (Gunneson, 1997; Dove, 2001).

We have chosen to focus on Dove's "Response Ability – The Language, Structure, and Culture of the Agile Enterprise". First, it emphasizes the complementary challenge of today's organization: "It must generate at least as much fuel as it consumes (profitability) and it must continuously adapt as necessary to changing environmental conditions (adaptability)" (Dove, 2001, p. 3). Second, it describes a set of elements that are required to successfully develop agile practices. Third, it offers explicit criteria by which one can evaluate the current organization as part of moving towards more agile practices.

Agility requires "the ability to manage and apply knowledge effectively, so that an organization has the potential to thrive in a continuously changing and unpredictable business environment" (Dove, 2001, p. 9). The two key elements that are required to practice agility are hence *response ability* and *knowledge management*. Response ability is achieved through change proficiency (process) and flexible relationships (structure) that are reusable, reconfigurable and scalable. Knowledge management in turn requires collaborative learning (process) and knowledge portfolio management (structure) including the identification, acquisition, diffusion, and renewal of the knowledge the organization requires strategically. Each of these elements are detailed and discussed in Dove's approach (2001) to the agile organization.

The core feature, response ability, is understood and assessed as illustrated in Figure 1. The opportunistic organization has high reactive proficiency, but it has weak proactive abilities. It follows best practices, listens to the customer, and is good at improving current capabilities. The innovative organization has high proactive proficiency, but it has insufficient reactive abilities. It can quickly introduce new technologies, services, strategies, and concepts to adapt to changing conditions in its environment. The organization is fragile when it has insufficient reactive ability to identify and explore opportunities related to its current capabilities as well as insufficient proactive ability to innovate as required by the environment. When market pressures are high and the environment is turbulent the ideal is the agile organization that combines high levels of reactive and proactive abilities.

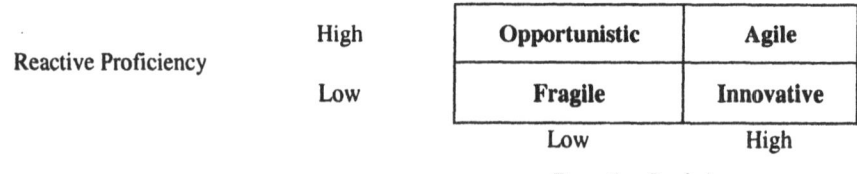

	High	**Opportunistic**	**Agile**
Reactive Proficiency			
	Low	**Fragile**	**Innovative**
		Low	High

Proactive Proficiency

Figure 1. Response Ability States (Dove, 2001)

2.3   Software Agility

The agility concept has been adopted by the software engineering community and a recent review of the literature is provided by Abrahamsson et al. (2002). The introduction of extreme programming (Beck, 1999) is acknowledge as the key step in this development, but there are also other published methods like Crystal Methods (Cockburn, 2000), Feature-Driven Development (Palmer and Felsing, 2002), and Adaptive Software Development (Highsmith, 2000). The key idea is "the recognition of people as the primary drivers of project success, coupled with an intense focus on effectiveness and maneuverability. This yields a new combination of values and principles that defines an *agile* world view" (Highsmith and Cockburn 2001, p. 122). Several of the key actors involved have expressed these values in the Agile Software Development Manifesto (Beck et al. 2001):

- Individuals and interactions over processes and tools.
- Working software over comprehensive documentation.
- Customer collaboration over contract negotiation.
- Responding to change over following a plan.

The strategy adopted to practice these values is, according to Cockburn (2002), the use of light-but-sufficient rules of project behavior combined with the use of human- and communication-oriented rules. Agile processes are both light and sufficient. Lightness helps a project to remain maneuverable, responding effectively to changes. Sufficiency is a matter of staying in the game, responding effectively to customer needs. Cockburn more specifically proposes a number of tactics to implement such a strategy: two to eight people in each room to facilitate communication and community building; onsite user experts to establish short and continuous feedback cycles; short increments of one to three months to allow quick testing and repairing; fully automated regression tests to stabilize code and support continuous improvement; and experienced developers to speed up development time.

3. RESEARCH APPROACH

This research is part of a collaborative practice study (Mathiassen, 2002) carried out at one of Ericsson's system development centers with more than 20 years of experience developing packet data solutions for the international market. This particular Ericsson organization has grown from 150 employees in 1995 to 900 in 2001 and been re-organized and downsized during 2001 to 2003. SPI has during this dynamic period become an increasingly important area in order to ensure quality deliveries.

The two authors represent industry and academia in close cooperation to secure relevant data and an appropriate theoretical framing of the study. The overall purpose of the research collaboration was two fold. We wanted to improve SPI practices at Ericsson while at the same time contributing to the body of knowledge on SPI.

The paper addresses the implementation of a new requirement management approach during the turbulent period from 2000 to 2003. Though the studied SPI initiative is successful in the end it suffers from initial weak impact and too slow progress. One of the authors was working in and been responsible for the initiatives. The potential bias and subjectivity is handled both through the collaborative research methodology and through discussion and interviews with engineers and researchers both within Ericsson and from the research community. The research is based on a case study (Galliers, 1992; Yin, 1994) with a focus on process implementation and use to assure SPI success. The SPI unit collected the basic data during the initiative with help of the Requisite Pro tool, SPI reports, and SPI project specifications.

The identified problems were collected from SPI reports and discussions with practitioners who used the new way of working. These discussions were made possible through project participation by one of the authors. The evaluations are based on questionnaires with follow-up interviews and discussions with software line and project managers and carried out in collaboration with those involved in the initiatives. Eight line and project managers were asked to answer questions about the implementation and use of the Requisite Pro tool within different development projects (see Table 1). Follow-up interviews were made in five cases to clarify previous answers.

4.   CASE

The studied case is about introduction of a new requirements management approach in one product development unit at Ericsson, Gothenburg, Sweden. The focus of the study is on the implementation and use of Requisite Pro and the related approach to requirements management. The use of this approach provides the organization with requirements configuration control, requirements baseline control, and traceability from requirements to test cases. The organization needed to improve requirements management to secure higher quality in the software parts of their products. There is a well-known potential in improving requirements management to secure better software quality (Humphrey, 1989).

4.1   The SPI Context

In late 1999 the SPI unit started to organize their work in dedicated initiatives supporting one software engineering project at a time (Börjesson and Mathiassen, 2003a). Projects that ran in partly overlapping sequence inherited the improved way of working from the previous project. In some cases the organization ran parallel software engineering projects that only partly differed in software and functionality. In these cases the SPI initiatives were coordinated through an SPI initiative that consisted of people from the two dedicated initiatives.

Each SPI initiative consisted of 7-8 part time process engineers that supported approximately 150 software engineers. Each initiative focused on several different software engineering practices such as requirements management, configuration management, project management and test management. 1-2 process engineers within each SPI initiative were dedicated to work with requirements management.

4.2   Adoption of Requirements Management

In late 1999, there was a management decision to adopt Rational Unified Process (RUP) (Krutchen, 2000) within the organization. The decision was a result of a major investigation comparing state-of-the-art software processes with respect to content, potential and coverage. One part of this decision aimed to change requirements management practices. Requisite Pro, a new way of describing requirements, use cases, and a new way of tracing requirements both to design and test was introduced. We have structured the introduction of the new requirements management approach into four phases (see Figure 2) corresponding to increased adoption of the approach. Each arrow represents a software engineering project that delivered one release of the product based on the approach. The phases are further explained in Table 1 (implementation success) and Table 2 (characteristics of the phases).

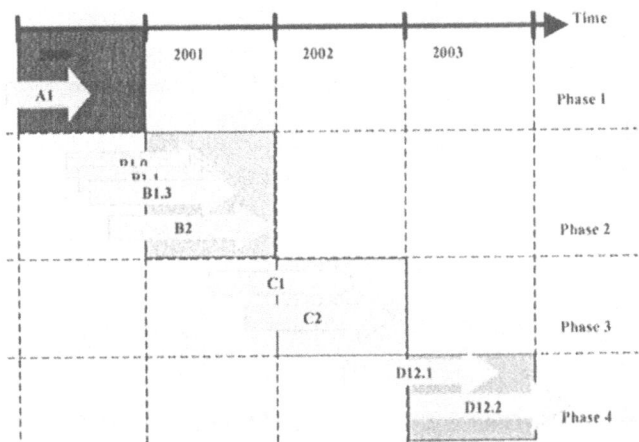

Figure 2. Improvement phases for new requirements management approach

Successful implementation of the requirements management approach is defined as the extent to which the tool was used to support control of requirements, i.e. baseline control, configuration control and test case traceability. Baseline control implies freezing a set of requirements artifacts as a baseline for all subsequent requirements activities. No changes can be made to those artifacts without a formal decision amongst the involved stakeholders. Configuration control means keeping track of versions of requirements artifacts, whether they are approved, and other status information. To have baseline and configuration control implies that all requirements artifacts are stored and visible at a place known to everyone needing them. Test case traceability means that all test cases can be traced to a specific requirement. Table 1 shows the implementation success of the requirements management approach in each phase with respect to baseline control, configuration control and test case traceability. We focus on the extent to which Requisite Pro supported requirements management. There were previously various manual controls of the requirements based on tools such as Excel.

Table 1. Implementation success of Requisite Pro (Project numbers refer to Figure 2)

Phase	Achieved SPI result	Configuration and Baseline control	Test case traceability
1	No focused and dedicated SPI initiative was organized. Several improvements in the requirements management area were made, e.g. adoption of spread sheets to control requirements manually.	A varying number of spread sheets stored within project areas or on local discs. No tool support to control requirements.	0%
2	All requirements in B2 were stored in the Requisite Pro tool, but only part of the test cases. B2 adopted tool support for configuration control of requirements.	Requisite Pro introduced and started to be used in B2. The time pressure in the B1.x releases made the introduction slower. Project management did not use the tool for follow-up.	20% (based on interviews with responsible managers)
3	The requirements management approach was improved based on experiences from B2. The C2 project managed to use the tool to support most test cases.	Requisite Pro further diffused and used. More features were adopted from Requisite Pro to provide more information. Many engineers did still not understand why they should provide information to Req Pro.	50% (based on interviews with responsible managers)
4	D12.1 managed to achieve baseline and configuration control as well as full support of test case traceability. D12.2 managed to get full control from the start of the project.	More or less 100% use of Requisite Pro by all engineers. The tool used by management for active follow-up.	99,7% (based on measures in Req Pro)

4.3   The Change Process

Each software engineering project described in Figure 2 and Table 1 benefited from the SPI initiative, but the speed and effectiveness of the change process is questionable. Why were the software projects in phase 2 or 3 not capable of achieving the level of tool-support that was reached in phase 4? To understand in more detail how the SPI initiatives worked during the different phases, Table 2 shows data for each phase: what the main problems were, how the SPI initiative was organized, which practitioners that were involved, and how results were achieved.

Table 2. Characteristics of the SPI initiative

Phase	Problem	SPI initiative	Involvement	Result
1	• Requirements documents stored on local discs • Adoption of ad-hoc spread sheets • Low tool support to cover test	• No formally organized initiatives. Several attempts were made by the software project to get all spread sheets under control.	• Requirements engineers • Test engineers	• Poor tool support for baseline control • Poor tool support for configuration control • Poor tool support for traceability
2	• Low Requisite Pro competence in software projects • Few competent users made administration of Requisite Pro hard • Major resistance to change • Low involvement from requirements engineers • Technical difficulties with Requisite Pro	• Process engineers and external consultants defined the requirements strategy, held courses, and supported the software projects.	• Requirements engineers • Test engineers • Process engineers • External consultants	• Requisite Pro was introduced and used by a few dedicated individuals. Project B2 achieved tool support for configuration control of requirements. Still no baseline control. • •Test case traceability was improved, but still low.
3	• New staff involved • New managers • Major resistance among the new people	• Senior engineers were involved to a higher degree • Management became involved and took responsibility for tool adoption	• Requirements engineers • Test engineers • Process engineers • Managers	• Tool supported configuration control • Tool partially supported base line control • Improved support of test case traceability
4	• New SPI staff involved • New technical difficulties • New category of engineers involved	• High involvement from engineers • The tool was introduced as a management tool to support control of requirements	• Requirements engineers • Test engineers • Process engineers • Managers • Design engineers	• Tool supported configuration control • Tool supported base line control • Tool supported test case traceability

No major improvements were implemented based on the new requirements management approach in the first phase. All focus was directed towards getting manual configuration control of requirements and test cases. The implementation success of the new requirements management approach improved in a stepwise fashion in phases 2, 3 and 4. Figure 3 illustrates this change process graphically related to the four phases. The general expectation was that the organization should benefit from the implementation more progressively when the requirements management initiative was launched. This is illustrated by the dotted line. The question is why this didn't happen and what barriers could be reduced and enablers improved to speed up the change process. What were the main reasons for not reaching higher success in phase 2 or 3?

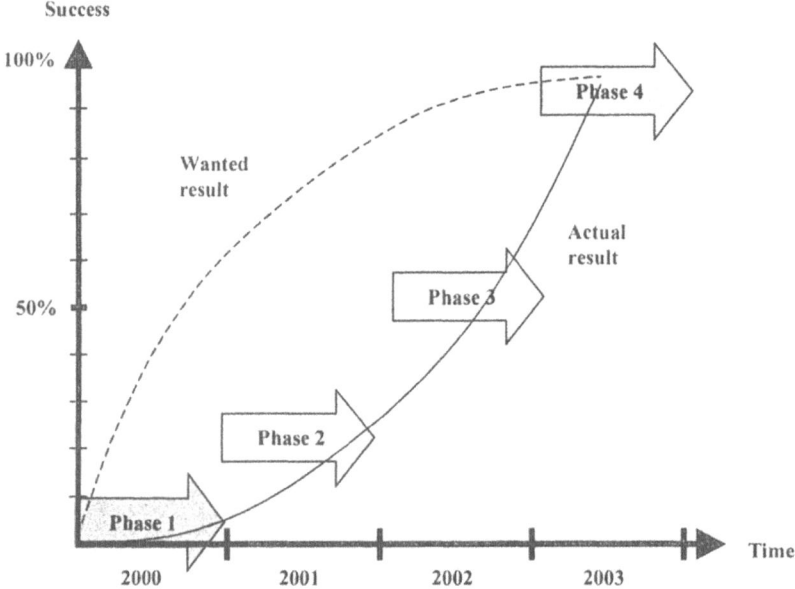

Figure 3. The adoption of the new requirements management approach

4.4 Organizational Dynamics

SPI initiatives are exposed to various forms of change reactions (Humphrey, 1989; McFeeley, 1996; Fichman and Kemerer, 1997; Grady, 1997). Weinberg suggests that SPI initiatives should be seen as sequences of reactions to attempts to change engineering practices from old status quo to new status quo (1997). The change process will go through a chaotic period and different attempts to integrate the new process into current practices.

During these events practices might return to old status quo or they might reach a sustainable, new status quo. This understanding of the change process is illustrated in Figure 4 and adopted to our case in Figure 5. Figure 4 also shows a simplified change curve to visualize the possible change reaction in Figure 4 and 5.

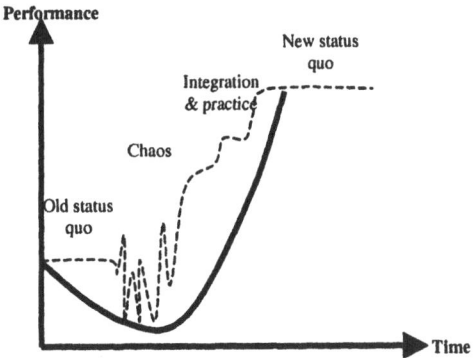

Figure 4. The simplified change curve

In each phase, Weinberg's reaction to change occurred. The underlying reasons are partially revealed in Table 1 and Table 2: degree of practitioner involvement, tool difficulties, and degree of management commitment. These are all reasons acknowledged in the SPI literature (Humphrey, 1989; Grady, 1997; Abrahamsson, 2000). These characteristics, however, only partially explain why the change process progressed slowly and with little effect. Figure 5 includes an additional explanation: the organizational dynamics of the target software organization. The Y-axis 'Success' in Table 5 and 'Performance' in Table 4 both indicate the same thing, i.e. getting positive effects from the change.

Two major re-organizations occurred during the change process and interfered with the improvement initiative as it was progressing well. In late 2001, two local development companies were merged together to save money by sharing administrative units and concentrating development on future strategic products for Ericsson. In late 2003, two divisions within the development company were merged to achieve higher integration between different parts of the product by working closer together and using the same processes. The idea was also to reduce overhead from line and project management. Of the 200 new employees that were added to the software organization between phase 2 and 3, approximately 50 were affected by the new requirements management approach, and of the 150 new employees that were added between phase 3 and 4, approximately 100 were affected. Many integration activities took place such as general training in the product and the RUP processes, the development tools used, and the market that was targeted by the developed software. These integration activities were in the first merger (late 2001) managed by a special initiative that planned all general training. In the second merger (late 2002), an initiative was started to align the software processes used by the two organizations, which resulted in a fairly well described process. This process was, however, not sufficiently deployed. Furthermore, no attention was directed towards the ongoing SPI initiatives that at those points in time still weren't fully accepted and seen as a natural part of the daily practices in the software projects.

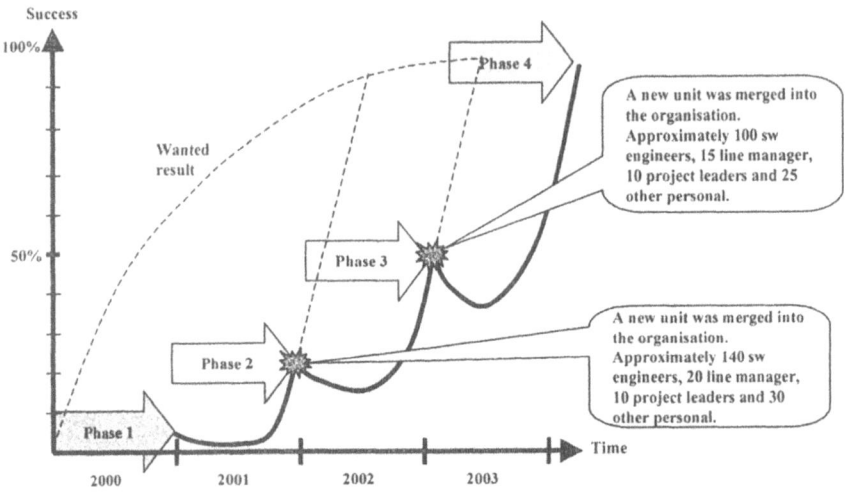

Figure 5. The interrupted change curve

5.   DISCUSSION

In the following, we discuss and analyze the case and relate it to relevant literature on SPI and agility. The section is structured as a response to our research questions followed by a discussion of the limitations of the findings.

What was the impact of organizational dynamics on the SPI initiative?

SPI research has identified a number of relevant issues that correlate with the findings from this case study. Humphrey (1989), McFeeley (1996), Grady (1997) and Abrahamsson (2000) all argue for the need of management commitment to assure successful SPI. Table 1 reveals that management was insufficiently involved to facilitate successful implementation of the requirements management approach at an earlier point in time than phase 4. Moreover, senior engineers were not involved before phase 3 and the SPI work was driven by process engineers and process experts in the first phase. Involvement of senior engineers and close collaboration between software and process engineers is important to create a good mix of practice pull and process push (Börjesson and Mathiassen, 2003b). Iteration has also been recognized as a useful approach to SPI where learning from previous iterations can leverage successful implementation in later iterations (McFeeley, 1996; Grady, 1997; Börjesson, 2003). One interpretation of the case is that successful implementation of the new requirements approach simply required a number of iterations involving different software engineering projects. As the SPI work iterated, lessons were learned, more features of the tools and templates were used, more roles became involved, and the initial requirements approach was as a consequence improved. Interpretations like these are well in line with the literature on SPI and organizational change and the case can in this way be said to confirm existing knowledge.

The most interesting aspect of this case is, however, that two major organizational changes occurred in the software unit at Ericsson during the SPI initiative, see Figure 5. Between phase 2 and 3 approximately 50 new employees and between phase 3 and 4 approximately 100 new employees were affected by the requirements management approach. From the data in Table 2, we see that no major attention was paid towards integrating these individuals and the different engineering traditions that they represented into the ongoing improvement work. The SPI initiative itself demonstrated no ability to sense and respond to these organizational changes, only organization-level responses were launched. During the first merger there was general training in the product and RUP, in the development tools, and in the market that was targeted by the developed software. During the second merger, a dedicated initiative was launched to align the different software processes used by the two organizations. The latter initiative resulted in a fairly well described process; but the process was never fully deployed.

The two mergers and the related organizational changes had serious implications for the SPI initiative because new engineers with quite different backgrounds had to accept, learn and adopt the new requirements approach. We know that organizational dynamics affect personal behaviors and the ability to execute improvement initiatives successfully (Zmud, 1982; Weinberg, 1997; Holmberg and Mathiassen, 2001). The new employees that were added as a result of the mergers were, however, not integrated into the SPI work so they could influence and contribute to the improvement work and the quality of the results. They therefore reacted passively or negatively to the SPI initiative (Weinberg, 1997; McFeeley, 1996). At the same time, no response was launched from the SPI initiative targeting this challenge. This task was left to general initiatives that were launched independent of the SPI project. The independent initiatives did, however, not effectively resolve the issues implied by the two mergers. The ongoing SPI initiative was as a result not capable of addressing the new employees and the organization as a whole was not capable of managing organizational mergers and improvement initiatives in parallel (Holmberg and Mathiassen, 2001). The organizational dynamics of the targeted software units had for these reasons quite a negative impact on the SPI initiative resulting in loss of momentum and a very slow and ineffective implementation process.

In terms of response ability, see Figure 1, the proactive proficiency demonstrated by the SPI organization was in this case low. There were no sense and respond mechanisms in place to help the SPI initiative identify, assess, and cope with the two mergers. The reactive proficiency of the SPI organization was at most medium. Activities did take place in response to the two mergers, but they were not integrated with the SPI initiative and they had little positive effect. These interpretations point in direction of a fragile or at best opportunistic SPI response ability in the presented case.

In what ways could the initiative have benefited from adopting more agile SPI tactics?

Dove (2001) defines the two key elements required to practice agility as response ability and knowledge management. Response ability requires change proficiency and flexible relationships and knowledge management is facilitated by knowledge portfolio management and collaborative learning (Dove, 2001). Table 3 offers an analysis of how the Ericsson SPI initiative could have benefited from more agile tactics along each of these dimensions. This analysis shows how the analyzed SPI initiative, viewed in isolation, could have benefited from higher proactive proficiency, greater flexibility towards inclusion of new employees, better and more explicit presentation of the RUP knowledge management facilities, and increased collaborative learning. While such tactics would likely have helped *the SPI initiative* to cope more effectively with the two mergers, the question remains of how to better align and coordinate across different activities within *the organization as a whole* to increase its response ability to changes in the environment.

Table 3. Analysis of organizational agility dimensions

Response Ability	Change Proficiency (the ability to proactively and reactively respond)	The SPI organization had weak to medium reactive proficiency (capability to react to the two mergers), but weak proactive proficiency (capability to sense, assess, and create responses to changes). Dove calls this state fragile or opportunistic. Such organizations do not innovate processes and tools as a proactive response to emerging changes and future needs. The lack of being able to identify such dynamics gives the SPI organization limited ability to respond effectively to change. *The SPI organization would have benefited from proactively including issues related to the mergers into each of its ongoing improvement initiatives. The SPI initiative would have benefited from explicitly analyzing and responding to the risks and issues that resulted from the two mergers.*
	Flexible Relationships (scalable and reusable relationships)	The process infrastructure and relationships between improvement initiatives did easily accommodate new people into the organization and thereby facilitate their learning and gaining from their experience and knowledge. There were, however, no sense mechanisms in place to make the SPI initiative aware of the emerging problems. Moreover, the SPI initiative had no readiness to involve additional competencies and assure commitment from new colleagues. *A proactive and more open attitude towards new colleagues and an ability to identify them as valuable resources rather than as problems would have benefited the ongoing SPI initiative.*
Knowledge Management	Knowledge Portfolio Management (the organization's management of core competencies)	Ericsson had an up-to-date infrastructure for managing knowledge. Their adaptation of RUP was available, supported and managed. However, the strategy for how to benefit from the RUP adaptation was not integrated into the SPI initiative to support the new employees, who came from other cultures with other methods and tools. *The organization would have benefited from integrating the RUP strategy and the related knowledge management practice better into the SPI initiative to help new employees understand and make use of the RUP adaptation.*
	Collaborative Learning (the support for collaborative learning net-works and events)	Fichman and Kemerer (1997) argue that successful diffusion of software process innovations depends critically on the ability to facilitate organizational learning. The general and independent knowledge management activities that took place in the software units had little or no effect on the SPI initiative. More importantly, no systematic attempt was made to link the knowledge already embedded into the ongoing SPI initiative with the knowledge of the new software engineers. *A focused and strong involvement of the new engineers into the ongoing implementation of the new requirements management approach would have resulted in additional learning and commitment to the new practices from the new software engineers.*

What are the implications for development of agile approaches within software engineering?

The agility concept has so far been adopted by the software community as summarized in (Abrahamsson et al., 2002) and expressed in the Agile Software Development Manifesto (Beck et al, 2001). One of the values, to respond to change over following a plan, is an expression of the response ability of agile organizations (Dove, 2001). This research suggests, however, that the current adoption of the agility concepts within software engineering is restricted in a couple of ways. First, the focus is mainly on responses to changes in customer demands. Other forms of dynamics realted to technological innovations, market changes, and organizational mergers are not explicitly addressed. The presented case shows, however, a need to adress software practices, SPI, and organizational mergers in an integrated fashion. Second, the ability to respond is tied directly to the operational level of developing software, i.e. to the software project level. There are few attempts to link software project agility to issues related to SPI or to other forms of organizational dynamics than changes in customer demands.

Similarly, there is within the SPI literature plenty of focus on issues related to change. Humphrey (1989) discusses the need of change agents and change management skills to successfully execute SPI and Weinberg (1997) has introduced the change model illustrated in Figure 4 and emphasizes the need for change artistry in SPI. There is, however, as we have argued no explicit attempts within SPI to address issues related to organizational dynamics in SPI. The presented case from Ericsson demonstrates the impact that organizational changes can have on SPI innitiatives and the possible advantages of adopting agile tactics within SPI to cope effectively with changes without severe loss of momentum and progress.

Our research suggests, however, also that organizational dynamics can seldom be dealt with effectively in isolation. Organizational disruptions typically penetrate many activities within the organization and they call for coordination and alignment of responses across functions and organizational levels. We suggest therefore, that agile software development and agile SPI are important contributions to dealing with the dynamics that software organizations face; but the overarching challenge is for the software community to rethink and develop its approach to software agility. The present focus on agile software development needs to be developed and integrated with the challenges related to agile SPI and both needs to be conceived as elements in agile software organizations more generally, see Table 4. We suggest - in line with the organizational agility idea (Gunneson, 1997; Dove, 2001) - that such a holistic and integrated software agility concept is needed if software organisations are to cope effectively with the increased speed of innovation, the need to continously improve current practices, and, last but not least, to respond effectively to emerging customer demands through delivery of quality software. More research is needed to develop this integration between streams of thinking and practice and to arrive at a holistic and comprehensive understanding of the agile software organization.

Table 4. The agility challenge

Organizational focus	Agility Challenge
Software Development	• To respond to changing customer demands • To participate in and adopt software process improvements • To sense and respond to technological innovations and market dynamics
Process Improvement	• To sense and respond to changes in software development • To sense and coordinate with other change initiatives • To provide a flexible process infrastructure
Software Organization	• To balance and coordinate development, improvements and innovations • To develop appropriate infrastructures and knowledge management practices • To develop response ability, organizational learning and metrics and in support of agile practices

The presented study has limitations. Most importantly, the case has focused on the relationship between SPI and organizational dynamics without including other dynamics between software organizations and their environment. There are also other factors, like culture and leadership, influencing any change effort than the ones reported here. Furthermore, case study research always implies biases from the specific environment in which it is conducted. It is therefore important to stress that the results from this research provide suggestions and indications rather than firm conclusions. Future research into organizational agility within the software industry will hopefully increase our understanding for how software projects, SPI initiatives, and software organizations at large can benefit from more agile strategies and tactics.

6.   CONCLUSIONS

This study has analyzed longitudinal data from the introduction of a new approach to requirements management into a software unit within Ericsson during 2000-2003. Our focus has been on the challenges involved in managing the SPI initiative as organizational changes occurred in the unit.

The research has made an effort to address the questions: What was the impact of organizational dynamics on the SPI initiative? In what ways could the initiative have benefited from adopting more agile SPI tactics? What are the implications for development of agile approaches within software engineering?

A number of already identified findings such as the necessity of commitment to SPI, the involvement of senior engineers, and the benefits of iterative approaches are further supported by this study. New insights into the relationship between SPI and organizational dynamics suggest, however, that limited agile capabilities have a severe and negative impact on the momentum and progress of SPI initiatives. The SPI initiative in question was not able to effectively respond to the need for integrating new employees. The SPI organization's response ability was fragile or at best opportunistic with low to medium reactive proficiency, and low proactive proficiency.

It is likely that the organization would have been capable of reaching SPI success faster if it had used more agile SPI tactics, such as addressing the organizational changes directly as part of the SPI initiative by involving and educating the new employees in the ongoing SPI work. Our study suggests, however, that it is important to approach organizational dynamics issues in a coordinated and coherent fashion. Further research is therefore needed to include SPI issues and other challenges related to managing organizational dynamics into the software agility movement. The software industry is advised to rethink and more actively adopt agile strategies and tactics to respond effectively to changes in their environments and to manage ongoing SPI work in parallel with other improvement and innovation initiatives.

REFERENCES

Aaen I. (2002) Challenging Software Process Improvement by Design. Proceedings of ECIS 2002 The European Conference on Information Systems, Gdansk, Poland, June 6-8.

Aaen, I., Arent, J., Mathiassen, L. and Ngwenyama, O. (2001) A Conceptual MAP of Software process Improvement. Scandinavian Journal of Information Systems, Vol. 13, 123-146.

Abrahamsson, P. (2000) Is Management Commitment a Necessity After All. In: Software Process Improvement? Euromicro '00, Maastricht, The Netherlands, IEEE Computer Society, 246-253.

Abrahamsson, P. (2001) Rethinking the Concept of Commitment in Software Process Improvement. Scandinavian Journal of Information Systems Vol. 13, 69-98.

Abrahamsson, P., Salo, O. and Ronkainen, J. (2002) Agile Software Development Methods – Review and Analysis, Oulu, Finland: VTT Electronics, Publication #478.

Baskerville, R., Levine, L., Pries-Heje, J., Ramesh, B. and Slaughter, S. (2001). How Internet Software Companies Negotiate Quality. IEEE Computer, Vol. 14, No. 5, 51-57.

Bach, J. (1995) Enough About Process: What We Need are Heroes. IEEE Software, Vol. 12, No. 2, 96-98.

Beck, K. (1999) Extreme Programming Explained: Embracing Change. Reading, Massachusetts: Addison-Wesley.

Beck, K., Beedle, M., van Bennekum, A., Cockburn, A., Cunningham, W., Fowler, M., Grenning, J. Highsmith, J., Hunt, A., Jeffries, R., Kern, J., Marick, B., Martin, R., Mellor, S., Schwaber, K., Sutherland, J. and Thomas, D. (2001) Manifesto for Agile Software Development. www.AgileManifesto.org.

Bollinger, T.B and McGowan, C. (1991) A Critical Look at Software Capability Evaluations. IEEE Software, Vol. 8, No. 4, 25-41.

Börjesson, A. (2003) Making SPI Happen: Iterate Towards Implementation Success. European Software Process Improvement Conference 2003.

Börjesson, A. and Mathiassen, L. (2003a) Making SPI Happen: The IDEAL Distribution of Effort. Hawaiian International Conference on System Science.

Börjesson, A. and Mathiassen, L. (2003b) Making SPI Happen: The Road to Process Implementation. Proceedings of the Twenty-sixth Information Systems Research Seminar in Scandinavia.

Cockburn, A. (2000) Writing Effective Use Cases. The Crystal Collection for Software Professionals. Reading, Massachusetts: Addison-Wesley.

Cockburn, A. (2002) Agile Software Development. Boston, Massachusetts: Addison-Wesley.

Diaz, M. and Sligo, J. (1997) How Software Process Improvement Helped Motorola. IEEE Software, Vol. 14, No. 5, 75-81.

Dove, R. (2001) Response Ability – The Language, Structure, and Culture of the Agile Enterprise. New York, New York: Wiley.

Fayad, M. E. and Laitinen, M. (1997) Process Assessment Considered Wasteful. Communications of the ACM, Vol. 40, No. 11, 125-128.

Fichman, R. G. and Kemerer, C. F. (1997) The Assimilation of Software Process Innovations: An Organizational Learning Perspective. Management Science, Vol. 43, No. 10, 1345-1363.

Galliers, R. D. (1992) Choosing Information Systems Research Approach, R.D. Galliers, 1992, in: R. Galliers, editor. Information Systems Research: Issues, Methods and Practical Guidelines, Blackwell Scientific Publications, Oxford.

Grady, R. B. (1997): Successful Software Process Improvement. Upper Saddle River, New Jersey: Prentice Hall.

Gunneson, A. O. (1997) Transitioning to Agility – Creating The 21st Century Enterprise. Reading, Massachusetts: Addison-Wesley.

Haley, T. J. (1996) Software Process Improvement at Raytheon. IEEE Software, Vol. 13, No. 6, 33-41.

Highsmith, J. (2000) Adaptive Software Development: A Collaborative Approach to Managing Complex Systems. New York, New York: Dorset House Publishing.

Highsmith, J. and Cockburn, A. (2001) Agile Software Development. The Business of Innovation. IEEE Computer, Vol. 34, No. 9, 120-22.

Holmberg, L. and Mathiassen, L. (2001) *Survival Patterns in Fast-Moving Software Organizations.* IEEE Software, Vol. 18, No. 6, 51-55.

Humphrey, W. S. (1989) Managing the Software Process. Reading, Massachusetts: Addison Wesley.

Humphrey, W. S., Snyder, T. R. and Willis, R. R. (1991) Software Process Improvement at Hughes Aircraft. IEEE Software, Vol. 8, No. 4, 11-23.

Humphrey, W. S. and Curtis, B. (1991) Comments on "A Critical Look at Software Capability Evaluations". IEEE Software, Vol. 8, No. 4, 42-46.

Krutchen, P. (2000) Rational Unified Process An Introduction Second Edition. Addison-Wesley Professional.

Larsen, E. and Kautz, K. (1997) Quality Assurance and Software process Improvement in Norway. Software Process – Improvement and Practice, Vol. 3, No. 2, 71-86.

Lyytinen, K. and Rose, G. M. (2003) The Disruptive Nature of Information Technology Innovations: The Case of Internet Computing in Systems Development Organizations. MISQ, Vol. 27, No. 4, 557-595.

Mashiko, Y. and Basili, V. R. (1997) Using the GQM Paradigm to Investigate Influential Factors for Software Process Improvement. Journal of Systems and Software, Vol. 36, 17-32.

Mathiassen, L. (2002) Collaborative Practice Research. Information, Technology & People, Vol. 15, No. 4, 321-345.

Mathiassen, L., Nielsen, P. A. and Pries-Heje, J. (2002) Learning SPI in Practice. In: Mathiassen et al. (Eds.) Improving Software Organizations. From Principles to Practice. Upper Saddle River, New Jersey: Addison-Wesley.

McFeeley, B. (1996) IDEAL: A User's Guide for Software Process Improvement. Pittsburgh: SEI. Handbook, CMU/SEI-96-HB-001.

Nielsen, P. A. and Nørbjerg, J. (2001) Software process maturity and organizational politics. In Russo, N. L., Fitzgerald, B., and J. I. DeGross (Eds.) Realigning Research and Practice in Information Systems Development: The Social and Organizational Perspective. Boston. Massachusetts: Kluwer.

Palmer, S. R. and Felsing, J. M. (2002) A Practical Guide to Feature-Driven Development. Upper Saddle River, New Jersey: Prentice-Hall.

Paulk, M. C., C. V. Weber, B. Curtis, and M. B. Crissis (1995) The Capability Maturity Model: Guidelines for Improving the Software Process. Reading, Massachusetts: Addison-Wesley.

Ravichandran, T. and Rai, A. (2000) Quality Management in Systems Development: An Organizational System Perspective, MISQ, Vol. 24, No. 3, 381-415.

Weinberg, Gerald M. (1997) Quality Software Management Volume IV – Anticipating Change. New York, New York: Dorset House Publishing.

Yin, R. (1994) Case Study Research. Newburry Park, California: Sage Publications.

Zmud, R.W (1982). Diffusion of Modern Software Practices: Influence of Centralization and Formalization. Management Science, Vol. 28, No. 12, 1421-1431.

MAPPING THE INFORMATION SYSTEM DEVELOPMENT PROCESS

Richard Vidgen[1], Sabine Madsen[2], and Karlheinz Kautz[2]

[1]School of Management, University of Bath, UK; [2]Department of Infomatics, Copenhagen Business School, Denmark

Abstract: Action research is used to gain understanding of how an IS development methodology emerges in practice and how it can contribute to value creation in organizations. The Multiview framework is used to guide the action research project. A graphical notation for mapping the unfolding of IS development projects is developed and applied to the project. Reflection on the project leads to a number of lessons being drawn about the organization of the IS development process, addressing themes such as vision, time pacing, and the role of architecture. The paper concludes with ideas about how the theoretical underpinnings of IS development might be bolstered by complex adaptive systems.

Key words: Action research, Multiview, IS development, Process mapping, Complexity

1. INTRODUCTION

A number of methodology authors recommend that the development process is tailored to fit the contingencies of the particular situation (Avison et al., 1998; Jacobsen et al., 1999) and empirical studies show that in practice the methodology is uniquely enacted for every development project (Stolterman, 1994; Fitzgerald, 1997). System developers adapt and apply methods and methodologies in a pragmatic way. A recent study indicates that this is also the case in web-based IS development (Baskerville & Pries-Heje, 2001), but so far little research has addressed the issue of how a local methodology is constructed, how it emerges in practice, and how effective it is in terms of creating value for the host organization. This paper presents an action research project in which the Multiview/WISDM framework (Vidgen et al., 2003) was used to guide the development of a market research data repository for a UK-based small to medium enterprise (SME). The aim of the paper is to show how the method emerged in practice and to reflect on that emergence in order to draw lessons about system development practice and theory.

2.   RESEARCH DESIGN

2.1   Action research

Action research was chosen as the primary research method in order to learn from first-hand experience how an IS development methodology (ISDM) emerges, i.e., the practice rather than the formalized descriptions given in text books and by method authors (Avison et al., 1999). The roots of action research can be traced back to Lewin's (1948) work on social change and social conflicts, through the Tavistock Institute's work on socio-technical theory (Emery and Trist, 1960), Checkland's (1981) view of human activity systems, to the Multiview/WISDM framework (Avison and Wood-Harper, 1990; Vidgen, 2002).

In their exposition of canonical action research, Davison et al. (2003) present arguments for and against the principle of theory. The counter arguments to using theory in action research are, firstly, it is difficult for the researcher to know what theory will be used or developed (particularly at the start of a project) and, secondly, there may be social issues arising for which no paradigmatic model exists. However, Davison et al. (ibid.) conclude that theory plays an essential role in guiding project activities and thus in distinguishing action research from action learning. Checkland (1991) also claims that the definition of a framework of ideas is important if action research is to be rigorous and have validity – it also helps differentiate action research from consultancy (Baskerville and Wood-Harper, 1996). Multiview/WISDM is used in this research to provide a structure for guiding action and as a basis for describing the unfolding of the project.

2.2   Multiview

Conventional systems analysis approaches, such as structured systems analysis, data modelling, and object-oriented methods, emphasize 'hard' aspects of the problem domain, that is, the certain and the precise (Avison and Fitzgerald, 2002). A hard approach is prescriptive and might be applied fairly consistently between organizations and within organizations. Checkland (1991) argues that systems analysts need to apply their craft to problems set in the ill-structured, fuzzy world of complex organizations. What makes organizations so complex and different is people - very different in nature from the data and processes that are emphasized in conventional IS development methods. People have different and conflicting objectives, perceptions and attitudes; people change over time. This may explain some of the dissatisfaction with conventional information systems development methodologies, i.e., they are not addressing the reality of organizational life.

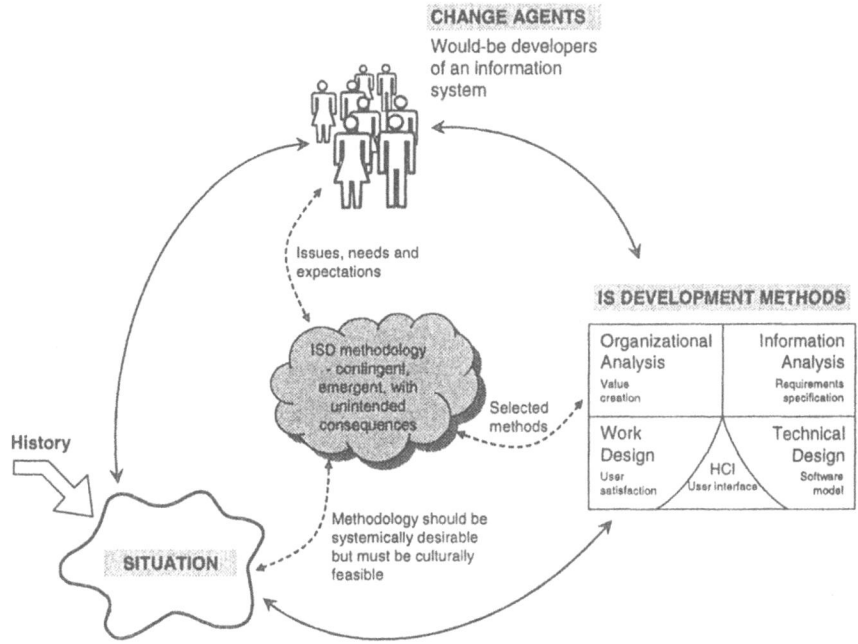

Figure 1. The Multiview framework (adapted from Vidgen et al., 2003).

Multiview is a systems development framework that has been defined and redefined over many years. It was espoused originally in Wood-Harper et al. (1985), developed further in Avison and Wood-Harper (1990) and Avison et al., (1998) with its latest reincarnation to be found in Vidgen et al. (2003). Multiview proposes that information systems development is a process of mediation between five elements: organizational analysis, information analysis, work design, technical design, and human-computer interface design (figure 1). The IS development activity is undertaken by human change agents with individual and shared interests in a particular organizational context with a distinct culture and history. Multiview draws on different tools and techniques into a blended approach. Each of these tools and techniques is used contingently, that is, as appropriate for each different problem situation. Even after some twenty years of refinement, Multiview is best described as a framework, not a step-by-step methodology, and its use as an 'exploration of information systems development' rather than the application of a prescriptive approach. Thus, according to the Multiview approach an IS development methodology is emergent only in practice through the complex interaction of change agents (including system developers) and development methods in an organizational setting. Multiview is, therefore, more usefully seen as a metaphor that is interpreted and developed in a particular situation, rather than as a prescriptive description of some real-world activity (Watson and Wood-Harper, 1995).

2.3   The client organization

Founded in 1991, Zenith International Ltd is a business consultancy specializing in the food, drinks and packaging industries worldwide. The main business activities are market analysis, strategic advice, technical consulting projects and conference organization. In spring 1998, the strategic aims of Zenith were to create a global presence, to broaden the product range, and to develop complementary skills (e.g., a synergy of market intelligence reports and consultancy). In September 1998 Zenith launched its first Web site, a typical cyber-brochure with limited interaction facilities.

The success of the company Web site launched in 1998 gave Zenith the confidence to explore e-commerce and the online delivery of market research content. In October 1999 Zenith and the University of Bath established a two-year joint project under the Teaching Company Scheme (TCS) with the objective of building an e-commerce system for selling research data on global drinks consumption (see Vidgen, 2002 for a full account).

The area of application for the action research reported here was a second TCS initiative, running for two years from October 2001, with the aim of creating a market research data repository (MRDR) that would contain details of companies and production volumes in the drinks industry. The primary action researcher worked in the role of Academic Supervisor for the project and was involved in hands-on development in the early stages of the MRDR. A second researcher was involved as an 'action case' researcher (Braa & Vidgen, 1999) for six months on the project, contributing primarily to the information analysis activity. A third researcher acted as an independent observer and conducted interviews with employees of the case organization, as well as with the action researcher and the action case researcher. This supplementation of action research by case study was chosen to bring more interpretive rigour to the project.

3.   THE INTERVENTION

The setting for the MRDR project was Zenith's market research department, which consists of six full time employees, including the Market Research Director. Each year the department produces a number of market reports, with the two most important ones being the 'Bottled Water' and 'Water Coolers' reports. The reports are based on data gathered from as many companies as possible in a line of business, such as bottled water. The reports are then sold to companies in the drinks industry, such as manufacturers (who provided the original detail data), packagers, and distributors. From initiation to publication, each report takes around three to four months to produce. Each report is led by a single market researcher who does the bulk of the work and gains a deep insight into the data and manages the structure of the report. A large volume of data has to be collected, stored and processed and information overload is the norm: "We've got loads of information on paper, on Excel files, all floating around" (Market researcher). A number of different software tools, such a Excel, Powerpoint and Word are employed in creating and formatting the reports:

> We use Microsoft Office essentially, Excel predominantly, PowerPoint to create charts and it is an extremely laborious process... For the 10 to 11 years the company has been going, we have produced company profiles in Excel format, we have linked each profile to overview tables and have created charts manually in PowerPoint, because that was the best presentation method at that time. Charts then pasted from PowerPoint, pictures into Word and put it all together using these packages. Obviously, this is extremely time-consuming. (Market Researcher)

The vision for the project, as documented in the project proposal, was clear from the outset: "to develop a unified market research data repository (MRDR) for the benefit of internal and external users encompassing data entry, data analysis, and report generation". The Market Research Director envisaged efficiency savings of up to 50% in market research report production time through the removal of labour intensive activities such as data entry and report formatting. The saving on clerical time would allow researchers to spend more time on data analysis and commentary, thus adding value to the basic report data (business effectiveness in addition to efficiency). In the larger context, a unified MRDR would allow data to be analyzed across reports, to create new revenue streams, to support consulting assignments, and to generally create a platform for greater organizational flexibility.

The development approach specified in the project proposal was incremental prototyping followed by pilot implementation of a single market report encompassing data collection, data analysis, and report production. The MRDR would then be rolled out for further market reports and external access to reports and ad hoc queries would be made available to clients via the Internet. Technically, the project was expected to make use of the new generation of content management software (CMS) (Goodwin and Vidgen, 2002). The incremental prototypes would be produced on a six week time-boxed basis, allowing frequent delivery of product to users and thus building credibility and keeping communications open. The plan was for the content of the deliverables to be constrained by the time available, as proposed in agile software development (Highsmith, 2002).

3.1 The project organization

The two year TCS programme (October 2001 – October 2003) was intended to develop the skills of the Teaching Company Associate, who was a recent MSc Computer Science graduate, to transfer knowledge from the academic institution via the Academic Supervisor to the company, and to achieve a tangible business benefit for the company. The Associate was employed on a full-time basis by the University for the two-year life of the project, but based in the company's offices taking day-to-day direction from the Industrial Supervisor, who was the Market Research Director at the company. These three – Associate, Academic Supervisor, and Industrial Supervisor – formed the core of the MRDR project team. The Market Research Director at Zenith also took the role of programme facilitator and was thus the business sponsor and internal champion for the project.

The formal organization of the MRDR project under TCS guidelines required regular steering committee meetings and technical meetings. Steering committee meetings were held every four months and attended by the Chairman of the company, a representative of the Teaching Company Directorate, and all the members of the project team. Technical meetings were held monthly by the project team to review progress and to document actions and decisions. The trust between Zenith's Chairman and the project team is illustrated clearly:

> Once a month is a working technical meeting, which I receive the minutes of. But I don't have an influence on that. And the reason why I don't, is that they are all very good, they know what they are doing and I can't contribute. I can contribute at the broader level and just reassure myself that the project is going well, but because they are a particularly good team, they are getting on with it and I'm happy with that. (Chairman, Zenith).

3.2 The unfolding of the project

The project phases are summarized in table 1. Narrative has been used to present the case, but it is difficult to discern a pattern to the unfolding of the development methodology from text. Langley and Truax (1994) note that longitudinal process research generates large quantities of data and that Miles and Huberman (1984) recognize the usefulness of visual displays. The presentation in table 1 is, of course, a vastly simplified presentation of the project; a severe condensing of interviews, research diaries, project documents, and so on. One way of presenting the unfolding of this project and to discern the shape from a morass of detail is to present the research using a process map.

The notation used in figure 2 takes elements from Langley and Truax's (1994) process flow chart and Thorp's (1998) results chain modelling, with the addition of further notation to make an explicit connection between the Multiview framework in figure 1 and the process map. The notation used in figure 2 is as follows.

Square-edged boxes represent initiatives. Initiatives are labelled according to the Multiview methods matrix: OA = organizational analysis, IA = information analysis, WD = work design, TD = technical design, HCI = human computer interface design. SD is used to indicate software development. *Rounded boxes* represent outcomes. *Solid horizontal arrows* represent precedence and influence, but not necessarily cause and effect. Three weights of line are used to show different strengths of connection – the thicker the line the more influential the connection. A *zig-zag line* is used to show an initiative that has been terminated or is temporarily in abeyance.

A *lozenge* symbol is used to indicate a reorienting impact. *Hexagons* represent assumptions that drive initiatives. *Ovals* represent perturbations – events that are outside the control of the system (part of the environment). *Dotted vertical arrows* connect the assumptions and events to relevant initiatives. A plus sign (+) indicates that the connection is facilitating and a minus sign (-) that it is inhibiting.

From the initiatives and outcomes in figure 2 it is possible to identify the major phases that emerged in the project: exploration; development of database; investigation of business process/job satisfaction; design of software architecture and development of core operational software; live use of software to produce market reports; extension of software (e.g., Internet access by clients to MRDR). The project used the Multiview framework to guide the project activities but the detailed content and the timing were emergent from the interplay of the actors, the situation, and the methods. The phases were only identifiable with the benefit of hindsight, i.e., as a result of the process mapping exercise.

Table 1. The MRDR project activities

Time period	Activity
Oct 2001 – Jan 2002	*Initiation:* developer trained in technology used by Zenith; review of content management software (CMS) conducted resulting in decision to custom build solution.
Feb 2002 – Jul 2002	*Database modelling:* database is recognized as core to MRDR. The plan allowed for six weeks elapsed time to design database on the assumption that a CMS would be implemented – due to the complexity of the MRDR data structures analysis and design took five months to reach a stabilized database.
Apr 2002 – May 2002	*HCI development 1:* an early prototype was developed to provide the users with a tangible output, allowing feedback on look and feel and a first test of the database structure.
May 2002 – Jun 2002	*Formal requirements analysis:* the informal notes and analysis of business processes were written up using flow charts and UML use cases.
Jul 2002	*Job satisfaction investigation:* application of the Multiview framework suggested that attention be given to job satisfaction of market researchers. The ETHICS questionnaire was rejected by the human resources manager (see Vidgen and Madsen (2003) for a full account). A revised questionnaire combining job satisfaction and use cases was developed and this highlighted that users felt they spent too much time collecting and formatting market data as opposed to analyzing, summarizing, and commenting.
Aug 2002 – Dec 2002	*Technical architecture:* the original three-tier architecture was superceded by a four-tier architecture based on XML. This was a response to the complexity of the MRDR application and the desire to build a flexible platform for data sharing.
Sep 2002 – Jan 2003	*Development of company detail reports:* the emphasis of the project was on Web delivery, but Zenith's Chairman wanted the MRDR to produce an exact facsimile of the current paper reports. This required the introduction of a more sophisticated formatting technology, XML-FO (formatting objects), to deal with page headers, page breaks, etc. for output in PDF format.
Jan 2003 – Feb 2003	*HCI development 2:* the Market Research Director of Zenith needed a deliverable from the MRDR project to sustain interest and credibility within Zenith. If data company detail data were entered into the database then a directory of companies in the water cooler industry could be generated. To support data entry of company detail data the user interface was redeveloped.
Mar 2003 – May 2003	*Water cooler company directory:* Company data entered into the database, the water cooler directory report produced automatically in PDF format, and marketed and sold to clients.
Jun 2003 – Jul 2003	*Market summary analysis:* detailed company volume data summarized into market overviews (e.g., top 50 bottled water companies in Europe).
Jul 2003 – Sep 2003	HCI development 3: testing of interface with users identifies extensive modification needed to support needs of market researchers in the production of live reports.
Oct 2003 – Jan 2004	*Market report production:* the first full market report, West Europe Bottled Water, is produced using the MRDR.
Feb 2004 onward	*Extension:* further reports produced from the MRDR, external access via the Internet, and new business initiatives.

4. LEARNING FROM THE INTERVENTION

The explicit specification of learning through reflection is where action research can make a contribution to practice and to theory. The key lessons that have arisen throughout an on-going process of reflection and a subsequent analysis of the process map (figure 2) are now summarized.

4.1 Projects need a vision

With regards to organizational analysis and value creation (figure 1), the research underlines the need for information system development projects to have a clearly articulated vision that is couched in terms of the value to be created for the organization (Vidgen, 2002). Highsmith (2000) calls this the project mission – "anything that helps the project team define the desired results of the effort at a summary level" (p. 44). Success is then judged on the basis of how well the vision is achieved, not by how well the plan was implemented. With regard to the MRDR, the project description on the cover page of the project proposal gave a clear statement of the vision:

> To create an enterprise repository for Zenith's research data, using a web content management software solution, to support internal knowledge development and external sales of research data. (TCS Project Proposal)

This vision remained constant throughout the project, because it focussed mainly on 'what' was to be achieved in broad terms, and provided a sense of direction whenever there was a danger of the project drifting or becoming mired in technical issues. However, by including a statement of 'how' this would be achieved, i.e., through the use of content management software, the clarity of the vision in the project proposal was weakened. At any given level of recursion the vision should address the 'what' rather than the 'how'. A similar approach is suggested in more mainstream strategy planning by Kaplan and Norton (1993) who proposed the balanced scorecard. The balanced scorecard contains a vision and mission statement together with a set of critical success factors and a small number (around sixteen) of relevant measures. The action research project was evaluated in financial terms (a requirement of TCS funding) but clearly a balanced scorecard approach could have been taken to broaden the evaluation criteria to include aspects such as job satisfaction.

4.2 Frameworks provide guidance, not prescription

By organizing around a vision, an IS project needs to be guided toward a desirable outcome rather than pursue a planned result. Although methodology as contingent and locally situated practice might be described as amethodical, this does not mean that it is a series of random actions (Truex et al., 2000). The project reinforced the value of the Multiview framework as an aid to *guiding* the system development process and influencing mental models rather than a prescription. For example, the value of using Multiview as a guide was illustrated in the project by the investigation of work life quality and the subsequent development of the use case job satisfaction instrument. Without a guiding framework it is quite likely that the sociotechnical aspect of systems development would not have been addressed at all.

Related to the idea of Multiview as guiding framework, we found that the outline project plan was perceived to be more valuable than the detailed Gantt chart. A Gantt chart was produced because it was required for the steering committee meetings, but the (over-)detailed plan was symbolic rather than of practical use in managing and controlling the project. Gantt charts may indeed be useful but they need to be specified at a level of detail suitable to the task in hand and then black-boxed for higher levels of the project organization structure; the steering committee was not interested in the fine-grain project detail and neither did the Gantt chart help the project team organize their day-to-day work.

4.3   Architectures help manage complexity

Technical architectures are useful in technical design as a way of managing complexity. Although agile software developers prefer to 'keep it simple' and prefer to avoid building infrastructure, this approach can lead to the developers being overwhelmed by the complexity of the software. In this project, the evolution of a four-tier architecture based in XML allowed separation of database, stored procedures to transform the data, components for encapsulation of business rules, and a presentation layer to deal with delivery to different platforms (e.g., web browser, paper, spreadsheets, and OLAP tools). However, the architecture should be a response to the complexity of the situation and not a fixed structure. We planned for a three-tier architecture but found that four layers were needed to cope with the inherent complexity of the application.

4.4   Time-pacing rather than time-boxing

The agile method Scrum comprises 30 day iterations called 'Sprints' (Highsmith, 2002). We found that such a pace was impossible to maintain, which can be attributed to an inexperienced developer (the TCS Associate), the introduction of leading edge technology (XML-FO in particular), the need to develop project infrastructure (the four-tier architecture), and a relatively complex database structure. However, we do not conclude from this that the agile principle of fast and regular delivery is inappropriate. Brown and Eisenhardt (1998) talk about 'time-pacing' as an internal metronome that drives organizations according to the calendar, e.g., "creating a new product every nine months, generating 20% of annual sales from new services" (p. 167). Time pacing requires organizations to change frequently but interestingly can also stop them from changing too often or too quickly. Time pacing is therefore not arbitrary; rhythm is used by organizations to synchronize their clock with the marketplace and with the internals of their business (although Brown and Eisenhardt give no indication as to how an organization might set the pace of the internal metronome). Although it is quite possible that the 30 day Sprints in Scrum are an institutionalization of best practice time-pacing of system development it is also possible that different situations with different developers, skills and organizational contexts will need to find their own rhythm. The idea of forced IS delivery through time-pacing is an interesting one with many implications for the IS development process and future research.

4.5   No 'one' is in control

With regard to the unfolding of the project as mapped in figure 2 it is difficult to say precisely how and why the IS development structure ended up as it did. It seemed that the project shape emerged from the interplay of the team members and their responses to external perturbations. Streatfield (2001) presents the traditional view of managers as 'in control' as selecting, designing, planning a course of action, correcting deviation, working in a stable environment with regular patterns, conformity, and consensus forming. Streatfield (ibid.) continues with the 'not in control' view, which sees action as evoked, provoked, emerging, amplifying deviation, and an unstable and unpredictable environment with diversity and conflict (figure 8.2, p. 134). Rather than accept one or other of these poles managers must work with the paradox of control. They have to accept the anxiety that is generated, using gesture-response as part of the subjective interaction in groups and help all organizational members look for meaning in the living present while perpetually constructing the future.

The implication of this view is that IS project managers should accept and even embrace the paradox of control, i.e., they are simultaneously 'in control' and 'not in control' and need the courage to live with the resulting anxiety. This view of control as emergent seems to describe well the experience of the MRDR project. On a day to day basis much of the control was in the hands of the Associate, but there were also hierarchical influences from the Academic Supervisor concerning the technology and development methods, and from the Industrial Supervisor with regard to business issues. The mix of control also changed over time as the Associate gained technical skills and confidence in her abilities. Thus, it is difficult to say who was in control of this project in a traditional managerial sense; control was an emergent property and outcome rather than a causal input.

5.   SUMMARY

The emergent IS development methodology used to build a market research data repository (MRDR) through action research was mapped using a graphical notation. This style of presentation allowed a two year project to be presented succinctly in a one page diagram (figure 2). Despite the danger of over-simplification, we believe this approach to presenting the results of action research provides a basis for comparing multiple projects and identifying common themes such that the dynamics of method emergence in practice can be better understood.

Using the process map reflection on the lessons from the action research project led to a number of themes: the need to articulate and organize around a vision, the use of Multiview as a guide to action, the role of technical architecture in managing technical complexity, time-pacing and the internal metronome of system development, and the paradox of control. An underlying theme to all these lessons is a response to complexity. Agile software development as proposed by Highsmith (2000) draws much on complex adaptive systems theory, key features of which are self-organization and emergent order, the dispersion of control, and the presence of internal mental models that contain implicit and explicit assumptions about the way the agent's world operates. An important area for further research is to investigate whether complex adaptive systems theory can provide a sound theoretical grounding for the lessons drawn in this research, for Multiview, and for IS development in general.

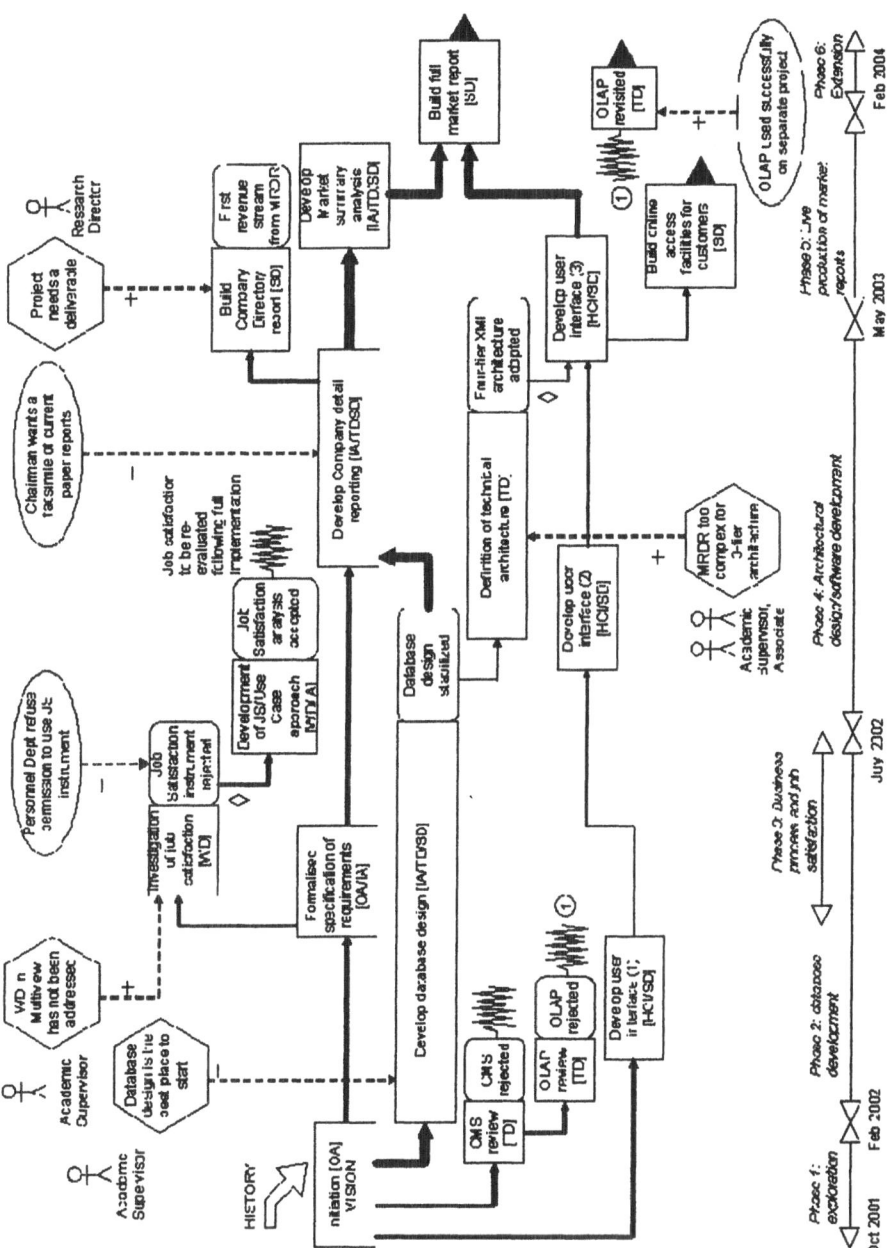

Figure 2. Process map of the unfolding MRDR development methodology

REFERENCES

Avison, D. E., Lau, F., Myers, M. and Nielsen, P. A. (1999). Action Research. *Communications of the ACM*, 42 (1), 94-97.

Avison, D. E. and Fitzgerald, G. (2002), Information System Development: Methodologies, Techniques, and Tools, McGraw-Hill, Maidenhead.

Avison, D. E. and Wood-Harper, A. T. (1990). *Multiview: An Exploration in Information Systems Development*, McGraw Hill, Maidenhead.

Avison, D. E., Wood-Harper, A. T., Vidgen, R. T. and Wood, J. R. G. (1998). A Further Exploration into Information Systems Development: the evolution of Multiview 2. *Information, Technology & People*, 11 (2), 124-139.

Baskerville, R. and Pries-Heje, J. (2001). Racing the e-bomb: How the Internet is Redefining Information Systems Development. In: *Realigning Research and Practice in Information System Development*, (Eds, Russo, L., Fitzgerald, B. and DeGross, J.), IFIP TC8/WG8.2 Working Conference, Boise, Idaho, USA, July 27-29

Baskerville, R. and Wood-Harper, A. T. (1996). A Critical Perspective on Action Research as a Method for Information Systems Research. *Journal of Information Technology*, 11 235-246.

Braa, K. and Vidgen, R. (1999). Interpretation, intervention and reduction in the organizational laboratory: a framework for in-context information systems research. *Accounting, Management & Information Technology*, 9 (1), 25-47.

Brown, S., and K. Eisenhardt (1998). *Competing on the Edge of Chaos*. Harvard Business School Press, Boston.

Checkland, P. (1981). *Systems Thinking, Systems Practice*, Wiley, Chichester.

Checkland, P. (1991). From Framework through Experience to Learning: the essential nature of Action Research. In: *Information Systems Research: Contemporary Approaches and Emergent Traditions*, (Eds, Nissen, H.-E., Klein, H. K. and Hirschheim, R.), North Hoalland, Amsterdam.

Emery, F. E. and Trist, E. L. (1960). Socio-Technical Systems. In: *Management Sciences, Models and Techniques*, Vol. 2 (Eds, Churchman, C. W. and Verhulst, M.), Pergamon, pp. 83-97, Oxford.

Davison, R., Martinsons, M., and Kock, N. (2004). Principles of Canonical Action Research. *Information Systems Journal*, 14, pp. 65-86.

Fitzgerald B. (1997), The use of Systems Development Methodologies in Practice: A Field Study, *Information Systems Journal*, 7(3), pp. 201-212.

Goodwin, S. and Vidgen, R., (2002). Content, Content, Everywhere ... Time to Stop and Think? The Process of Web Content Management. *IEEE Computing and Control Engineering Journal*, 13(2): 66-70.

Highsmith, J., (2000), *Adaptive Software Development: a collaborative approach to managing complex systems*, Dorset House, NY.

Highsmith, J. (2002). *Agile Software Development Ecosystems*, Addison-Wesley, Boston.

Jacobsen I., Booch G. and Rumbaugh J. (1999), *The Unified Software Development Process*, Addison-Wesley.

Kaplan, R., & Norton, D., (1993). Putting the Balanced Scorecard to Work. *Harvard Business Review*, Sept-Oct, 134-147.

Langley, A., and Truax, J., (1994). A process study of new technology adoption in smaller manufacturing firms. *Journal of Management Studies*, 31 (5): 619-652.

Lewin, K. (1948). *Resolving Social Conflicts*, Harper, New York.

Miles, M., and Huberman, A., (1984). *Qualitative Data Analysis*. Sage, CA.

Stolterman E. (1994), The 'transfer of rationality', acceptability, adaptability and transparency of methods, *Proceedings of the 2nd European Conference on Information Systems (ECIS)*, Nijehrode University Press, Breukeln, pp. 533-540.

Streatfield, P., (2001). *The Paradox of Control in Organizations*. Routledge, London.

Thorp, J., (1998). *The Information Paradox*. McGraw-Hill, Montreal.

Truex, D. P., Baskerville, R. and Travis, J. (2000). Amethodical Systems Development: the deferred meaning of systems development methods. *Accounting, Management and Information Technology*, 10 (1), 53-79.

Vidgen, R. (2002). WISDM: Constructing a Web Information System Development Methodology. *Information Systems Journal*, 12 247-261.

Vidgen, R., Avison, D.E., Wood, R., and Wood-Harper, A.T. (2003), *Developing Web Information Systems*, Butterworth-Heinemann.

Vidgen, R., & Madsen, S., (2003). Exploring the Socio-Technical Dimension of Information System Development: use cases and job satisfaction. In: *Proceedings of the 11th European Conference on Information Systems*, Naples, Italy, June 19-21.

Watson, H. and Wood-Harper, A. T. (1995). Methodology as Metaphor: The Practical Basis for Multiview Methodology (a reply to M. C. Jackson). *Information Systems Journal*, 5 (1), 225-231.

Wood-Harper, A. T., Antill, L. and Avison, D. E. (1985), *Information Systems Definition: the Multiview Approach*, Blackwell Scientific Publications, Oxford.

TAKING STEPS TO IMPROVE WORKING PRACTICE: A COMPANY EXPERIENCE OF METHOD TRANSFER

Björn Lundell[1], Brian Lings[2], Anders Mattsson[3], and Ulf Ärlig[3]

[1]*Department of Computer Science, University of Skövde, Sweden;* [2]*School of Engineering, Computer Science and Mathematics, Department of Computer Science, University of Exeter, UK;* [3]*Combitech Systems AB, Jönköping, Sweden*

Abstract: Methods are vital for systems development in companies, but their effective transfer into practice necessitates an understanding of the many factors affecting success. It can be extremely expensive to adopt a method in order to evaluate its suitability for a given context; what is required is a progressive commitment according to feedback. This paper documents the experience of progressive method transfer within one company.

Key words: method transfer, experience of transfer, working practice

1.   INTRODUCTION

Effective use of methods is vital for competitive advantage in systems development contexts. The changing nature of technologies and requirements necessitates flexibility with respect to method adoption, and organisations need a systematic approach to the continuous improvement of work practices. In this paper we consider the transfer of a method into Combitech Systems AB[1] (hereafter referred to as Combitech), an IT company within the SAAB group[2]. Combitech had become interested in the possible use of a method to support work practice improvement activities within the company.

---

[1] http://www.combitechsystems.com
[2] http://www.saab.se

The method under transfer is the *2G* method, a qualitative method evolved for use in socio-technical domains. It was originally developed for evaluation of CASE-tools in specific usage contexts, but its scope of application has subsequently been broadened – a potential indicated in the method documentation (Lundell and Lings, 2003).

A number of studies of transfer of the *2G* method have previously been undertaken in a number of different company contexts (e.g. Lings and Lundell, 2004; Rehbinder *et al.*, 2001). These have allowed us to theorise about effective processes of method transfer and influenced the approach taken to transfer at Combitech. In the rest of this paper we describe the progressive approach adopted, and reflect on the experience of transfer so far.

2.   ON THE METHOD BEING TRANSFERRED

There is an underlying assumption behind the *2G* method that evaluation is a socio-technical activity, and that a key early activity in any evaluation is the development of a reference framework. Two inter-related frameworks are developed during an application of the *2G* method. To produce them, interview and other data is analysed both from an organisational and a technological perspective, using a two-phase process. In the first phase, the focus is on organisational need; in the second phase the focus shifts to how needs might be met through current technology (where technology is interpreted broadly, for example to include IS development methods). The *2G* method is not prescriptive, but gives clear guidelines on the use of different kinds of data source, and how frameworks may be evolved. It is important to the method that both phases in its application take place in the organisational setting in which the technology under investigation would be used. The method is intended as a general method, scalable according to context.

3.   ON METHOD DEVELOPMENT AND TRANSFER

We can usefully consider seven steps in the development and transfer of a method, namely:
1. Method development and documentation
2. Method application by the method developer in a controlled context
3. Method application by the method developer in a company context
4. Method transfer to a method user within a controlled context

5. Method transfer to a non-company method user within a company context
6. Method transfer to a company method user within a company context
7. Diffusion within a company already exposed to the method

From the point of view of transfer, the importance of the first three steps is to establish a clear conceptual framework for a method. This means that its underlying value systems must be transparent, and any assumptions about method use must be clear. It is important that they match those of any intended company context. The fourth step allows initial checks that the method is well enough established to be transferred to a user not previously exposed to it. This is a check that it can be learned and is usable at least in a controlled context. Both of these steps are necessary, but the real test of transfer comes only when a method is transferred into its intended context for usage. For example, the use of novices within a classroom setting may say little about an experienced practitioner's perception of a method within a company project.

The main concern of this paper is with steps 5 and 6, where a method is transferred in to a real company context. Moody (2002) notes, from experiences of use of a data modelling method in a large commercial organisation, that "it is very difficult to get new approaches, especially those developed in academic environments, accepted and used in practice." (p. 393) Shanks (1996) ascribes this to the inherent risks and expenses for the organisation involved. For successful method transfer, therefore, such risks and expenses must be minimised.

The final, diffusion, step is beyond the scope of this paper but is the motivation behind Combitech's interest in the current transfer studies: the *2G* method is seen to have wider applicability throughout the company.

4. THE COMBITECH EXPERIENCE

The method receivers are consultants working in Combitech, and the specific context for initial *2G* method transfer is a development project undertaken within the company.

Combitech is a medium sized, geographically distributed enterprise working with advanced systems design and software development, electronic engineering, process optimisation and staff training. It has approximately 230 employees and covers a broad spectrum of business areas such as defence, aviation, vehicles, medical industry and telecom.

The company has a long experience of systematic method work and model driven systems development. In several development projects UML is used (e.g. Mattsson, 2002), but other methods and techniques are used as well. Development tools, such as Rational Rose (from Rational) and Rhapsody (from Ilogix) are also used in some projects.

From the company's perspective, 2G was of interest because it offered a rich and systematic way of dealing with socio-technical phenomena. Managers had been made aware of earlier application of the 2G method with respect to contextualised CASE tool evaluation, and the company had a general interest in tool-assisted model-driven development.

The process started with round table talks about previous applications of the method, and potential contexts within the company where it might be applicable. This led to the specification of a pilot project, in which a university researcher would be seconded to Combitech for a four month period to act as a method user (step 5 above). This was seen as an efficient way in which to gauge stakeholder reactions to the method. It also gave the company access to a continuous dialogue about the method and its potential usefulness within the company.

A consultant software engineer elected, with encouragement and full support from management, to follow the application in detail in order to be able to apply it internally. The dialogue established was seen as a positive and lightweight, hands-on introduction to the method. There was a strong sense of involvement by the software engineer, and the 4 month duration was seen to offer time for reflection and critical thinking throughout the process. However, during weeks with high short-term demands and other commitments, it was more difficult to preserve the momentum on this longer-term activity. It was felt that the extra load on stakeholders would be lightened if method application could be pursued on a more contingent basis.

It was important that the method could be viewed with appropriate scepticism, as it was being undertaken by an 'outsider' with no management commitment to eventual deployment. This placed it as a limited study, its role being as much for professional development as for method evaluation.

On completion of the pilot, Combitech were then in a good position not only to assess stakeholder reaction to the method, but also to judge its potential within the company. In particular, discussions ensued on how to apply the 2G method in the company's ongoing work on method tailoring. From the company's perspective, there was a perceived competitive advantage offered by the 2G method in supporting activities in method tailoring in a structured way. The pragmatic and strategic frameworks central to the 2G method, interpreted as the *how* and *why* with respect to development methods, were seen as a means of contributing in two ways: firstly, for continuous improvement actions; secondly to assist in the initiation of new staff into company culture and working practices.

Having committed to deploying the method within the company, the next step (step 6 above) is to transfer the method fully to a company member – in this case the 'shadow' from the pilot project. The focus for this study is a tailoring of a development method, primarily influenced by RUP but also with other influences. At this stage of the process, mentoring is seen as the most effective mechanism to assist in method transfer. The mentor could be any method expert, but in this case is one of the method developers. This will be supplemented by stakeholder seminars, involving method experts and company members, for more reflective feedback and analysis.

A number of issues have been raised in the process of executing step 6. The major one concerned finding an appropriate project, which needed to be: representative of intended usage; internal; not too limited; and with reasonable planning time. It was perceived that there was no risk in applying the method, but it did demand stakeholder time – and this needed to be planned in the light of commitments. This is not straightforward for a consultancy company, with great variability in project sizes and lead times. The real need was to find a window for the intended method user just prior to the target project for analysis (in this case, method tailoring for a software development project). This window needed to be close to the project, large enough to evolve a useful framework, and at a time when interviews with relevant stakeholders could be scheduled. On the other hand, post-project analysis was felt to be less time-critical.

It is beyond the scope of the current project to consider step 7 in the process, which at the very least must await analysis of the experience of step 6. This is the stage at which costs could increase sharply, as other consultants are trained with the method. However, the cost would be ameliorated by the dissemination activities implied within step 6, and management would already have had clear indications of the method's suitability and acceptability to stakeholders – i.e. risks would have been considerably reduced.

5. SUMMARY

We have identified seven steps in the process of method development and transfer, from the development of a method through to its full deployment within a company. From a company perspective, the most important of these are steps 5 through 7: from a pilot application of the method using an external 'consultant', through transfer to member of the company and finally diffusion within the company. We have taken as an example the transfer of the *2G* method into Combitech Systems AB, noting how the multiple steps can reduce risk and increase the likelihood of stakeholder acceptance. The overall experience for the company has been positive. In particular, it is felt that exchanges between consultants based on tacit knowledge have been increased and given greater structure. This has also facilitated reflection and created greater consensus within the group, leading to easier acceptance of change and therefore increased adaptability. This experience is consistent with methods as a means of improving current work practices rather than as prescriptive ways of working.

ACKNOWLEDGEMENTS

The authors are grateful to Hanna Zaxmy, the external method user for step 5 in the transfer process.

REFERENCES

Lings, B. and Lundell, B. "On transferring a method into a usage situation," in *IFIP WG 8.2 IS Research Methods Conference: Relevant Theory and Informed Practice*, 15-17 July, Manchester, 2004, Kluwer *(accepted for publication)*.

Lundell, B., and Lings, B. "The *2G* method for doubly grounding evaluation frameworks," *Information Systems Journal* (13:4), 2003, pp. 375-398.

Mattsson, A. "Modellbaserad utveckling ger stora fördelar, men kräver mycket mer än bara verktyg," *On Time*, April, 2002 [www.ontime.nu] *(in Swedish)*.

Moody, D.L. "Validation of a method for representing large entity relationship models: an action research study," in *European Conference on Information Systems*, 6-8 June, Gdansk, Poland, 2002, pp. 391-405.

Shanks, G. *Building and Using Corporate Data Models*, Ph.D. thesis, Monash University, Melbourne, 1996.

Rehbinder, A., Lings, B., Lundell, B. Burman, R. and Nilsson, A. "Observations from a Field Study on Developing a Framework for Pre-Usage Evaluation of CASE Tools," in *New Directions in Information Systems Development: IFIP WG 8.2 2001 Conference*, N.L. Russo, B. Fitzgerald, and J.I. DeGross (eds.), Kluwer, Boston, pp. 211-220.

PART III

ASSESSING INNOVATION DRIVERS

EVALUATING INNOVATIVE PROTOTYPES
Assessing the role of lead users, adaptive structuration theory and repertory grids

Carl Magnus Olsson[1] and Nancy L. Russo[2]

[1]*Viktoria Institute, Gothenburg, Sweden;* [2]*Operations Management & Information Systems, Northern Illinois University, DeKalb, USA*

Abstract: Innovation evaluation approaches have primarily focused on studying the adoption and use of existing technologies. However, as development timelines and product life cycles continue to shrink, it is useful to be able to evaluate emerging innovative technologies. Recognizing this, the research outlined in this paper shows how lead users (Von Hippel 1988) can be combined with adaptive structuration theory (DeSanctis & Poole 1994) and the repertory grid technique (Kelly 1955) in evaluation of innovations at the prototype stage. A context-aware application, co-developed by the researchers and industrial representatives, is presented as an illustration of such a prototype. The paper demonstrates how IT researchers can take an active stance in the evaluation of on-going technological innovation designs.

Key words: Adaptive structuration theory, repertory grids, lead users, context-aware applications

1.   INTRODUCTION

In the diffusion of innovation (DOI) realm, we typically think of how the user will evaluate the innovation, such as a new technology, to determine likelihood of adoption and diffusion. Traditional measures of the "adoptability" of an innovation include ease of use, fit, trialability, and visibility (Rogers, 1995). The importance of the context of the innovation and other social issues has been recognized (Swanson 1994; Hedström, 2003; Lyytinen & Rose, 2003), as has the need to examine the innovation process from the appropriate viewpoint, be it individual, group, or network (Prescott & Conger, 1995; Lyytinen & Damsgaard, 2001).

However, as the pace of change accelerates, and organizations feel the pressure to introduce new product innovations ever more rapidly, there is a need to evaluate innovations that are still in the development stage to determine if the design is on the right track. As Lyytinen and Yoo (2002) discuss, this rapid pace of change is challenging the distinction between technical (developmental) research and behavioral (use) research, and thus IT researchers should be actively involved in studies where technologies are being built and tried out – not after the fact when they have entered the market

The type of IT innovation that is of particular interest in this paper is in the context of mobile, pervasive, and ubiquitous computing; in other words, systems that are being integrated into everyday activities through mobile phones, PDAs and other handheld devices, systems that are integrated non-obtrusively into everyday objects such as appliances and wearable computing devices. Lyytinen and Yoo (2002, p. 337) call this a "nomadic information environment" which is described as "a heterogeneous assemblage of interconnected technological, social, and organizational elements that enable the physical and social mobility of computing and communication services between organizational actors both within and across organizational borders." This nomadic information environment is identified by high levels of mobility, large-scale services and infrastructures, and diverse ways in which data are processed and transmitted.

This paper addresses a particular aspect of nomadic information environments: Context-Aware Applications (CAAs). Context-aware applications typically use location-based data to trigger predefined behavior (Schmidt, Beigle & Gellersen 1998). Examples of earlier context-aware applications included the notification agent, the meeting reminder agent, call-forwarding applications, and active badge systems (Siewiorek, 2002; Dey et al., 2001). Other contextual information includes human factors such as the user (habits, emotion, physical condition), social environment (co-location of others, social interaction, group dynamics), and tasks (spontaneous activity, engaged tasks, general goals). Aspects of the physical environment, such as conditions (noise, light, pressure) and infrastructure (surrounding resources, communication, task performance) are also relevant (Schmidt et al. 1998)

The purpose of this research is to examine a method for evaluating innovations in the prototype stage. A prototype of a particular context-aware application called CABdriver (Context-Aware Backseat Driver) is used to outline how this approach has contributed to the innovation process. CABdriver is a handheld, interactive in-car application developed in conjunction with a major European automaker. It is described in detail in the following section.

As Lyytinen and Yoo (2002) suggest, the technical and social issues involved in using this type of system cannot be separated easily. Advanced technologies, such as CABdriver, are more likely to impact social aspects than traditional business computing systems are (DeSanctis & Poole 1994). Therefore, our approach to evaluating CABdriver must take into account both social and technical aspects. The approach taken to evaluate the innovation includes the use of lead users (Von Hippel 1988), the repertory grid technique (Kelly 1955), and adaptive structuration theory (DeSanctis & Poole 1994). Adaptive structuration theory provides a useful framework to examine the use of CABdriver. The repertory grid technique is used as the method of capturing the structuration process through interviews with lead users. Because the innovation is in the prototype stage, evaluation by a large group of users is not feasible. Lead users, selected for their early experience with the innovation, can provide "early warning" evaluation perspectives.

After the description of CABdriver, the research approach is detailed, followed by a discussion of the analysis and contribution of the research. The concluding section summarizes what we have learned from this study and identifies future avenues of research in this area.

2.   CONTEXT-AWARENESS FROM AN IN-CAR PERSPECTIVE

Development of the concept CABdriver (Context-Aware Backseat Driver) was initiated in January 2003 as a joint project with Saab, Mecel and Vodafone. This first prototype of the CABdriver concept is a handheld game that is influenced by contextual information from three areas – the car, the immediate surrounding and the distant surrounding. In addition to giving the passenger something to do at the same time as contextual information affects the behavior of the CABdriver game, the intention is to supply this information in a way that allows the player to convey what may be interesting or even critical to the driver.

2.1   Background

From a research perspective, the main focus related to in-car use of IT has so far been on adapting traditional human-computer interfaces to the car in order to reduce risks of distracting the driver. For instance, Goose & Djennane (2002) look at WIRE3, an adapted HTML browser, Salvucci (2001) compares manual with voice controlled phone dialing, while Brookhuis, De Vries & Waard (1991), McKnight & McKnight (1993), Alm

& Nilsson (1994, 1995) and Redelmeier & Tibshirani (1997) focus on the most commonly talked about issue – how a driver's mobile phone conversation affects driving and the risk of having an accident. With respect to the safety of a car in motion, it is perhaps no surprise that the driver has been the primary focus of most research so far.

While the driver is preoccupied with the continuous negotiations of driving the car, passengers often have nothing specific to do. Handheld games and games available on mobile phones often serve the purpose of killing time during travel, especially among youngsters and teenagers. However, passengers still show considerable subconscious attention and interest for the changing surroundings (Appleyard, Lynch & Myer 1964).

2.2   CABdriver

Responding to this, the first prototype of the CABdriver concept is a handheld game for passengers. The game is a cross between the space shoot-em-ups of the eighties and more strategic games. The game runs solely on the PDA at the moment, although server like behavior in the Infotainment Center of the Saab 9-3 has been discussed. Contextual information is either supplied by a Saab 9-3 via an IP network over Bluetooth to a PDA, or via an IP network over Bluetooth or W-LAN from a computer to the PDA.

The game has two primary parts that the player spends her time in – one is the radar (Figure 1 and Figure 2) and the other is playing missions (Figure 3). On the radar, events within 5000 meters (approximately 3 miles) of the car – which represents the center of the radar – are shown. Missions are created on the radar based on real-world points of interest (gas stations, tourist attractions, hospitals, parking lots, shopping centers, airports, train stations, etc) and traffic messages (TMC data). These traffic messages are sometimes broadcasted live on radio (RDS-TMC), but those are only a small fraction of the thousands of traffic messages that the TMC standard supports. TMC information consists of data combined into a description of a specific situation that affects driving. The data includes accident types, closures and lane restrictions, carpool information, roadwork, dangerous situations, road conditions, temperature and precipitation, air quality and visibility, security alerts and so on. The gaming experience changes based on the content of traffic messages and car status. For example, special missions are automatically selected if the car is within close proximity of important events such as accidents, road blocks and animal warnings and if the car is low on gas, the game will only allow for missions close to the car (the center of the radar) to be started.

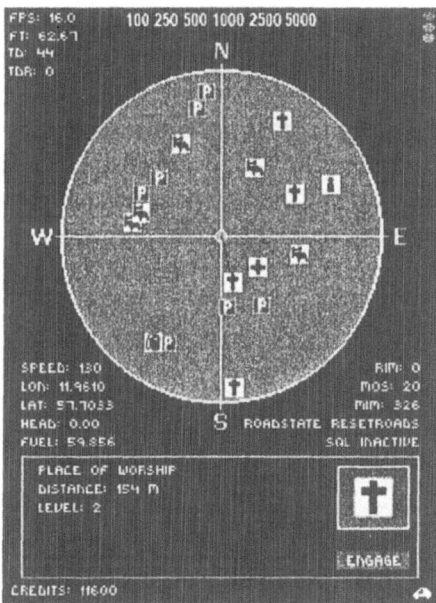

Figure 1. From the radar, a number of points of interest are shown in the surrounding area. Currently, the player is reading about a place of worship which is in danger of being invaded by enemy spaceships. The scale is shown above the radar (500 meters radius is chosen here).

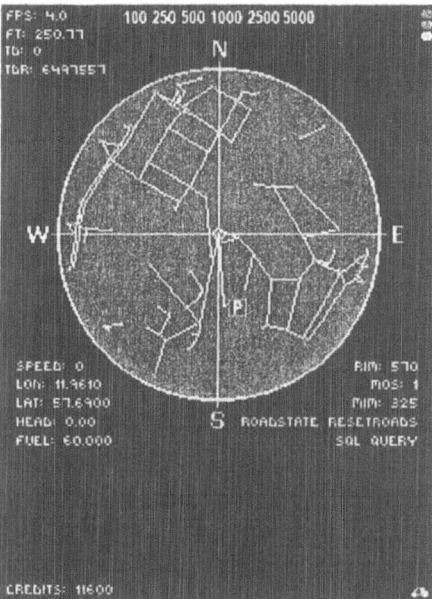

Figure 2. The road network is drawn on the radar with one parking space showing. A flashing car in the bottom right corner shows that the game is currently receiving data from the car.

Once a mission is generated (Figure 3), it is in turn influenced by information such as the actual car speed, speed limit, fuel consumption, outside temperature and driver work-load (a combination of anti-spin control and electronic stability program activity, breaking, hard acceleration, navigation system indicating a turn is close-by and recent use of the turn signal). Furthermore, certain missions may require the player to go to the trading section to upgrade her spaceship in order to stand a better chance of completing them.

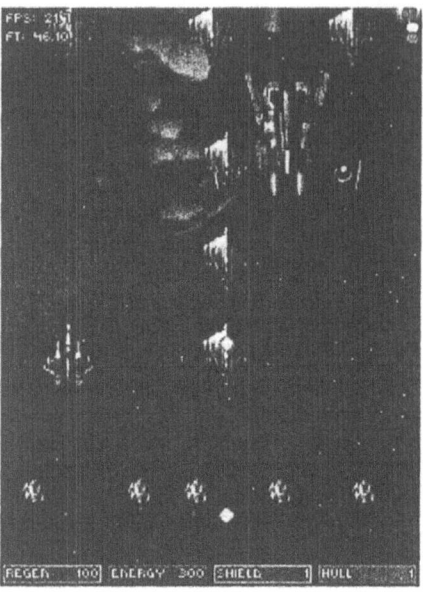

Figure 3. The dreaded driver work-load monster (indicating an occupied driver; big spaceship in upper right corner) enters a mission at the same time as space mines (string of small round balls at the bottom) from an excessive speeding violation can be seen. The game compares actual speed with speed limit for the current road; not shown here is the flashing blue and white line at the bottom of the screen that warns when the driver is speeding excessively. Driver work-load is at all times shown to the gamer (small ovals in the upper right corner). Fortunately, the driver still seems to drive environmentally sound (indicated by the full regeneration bar in the bottom left corner), something which affects the ability to recover from being hit by enemy shots, move your own space ship (mid-left, pointing up) and shoot back. The T-shaped formation of space ships is also enemy ships.

Future plans on how CABdriver could evolve include various support for multiplayer games such as in-vehicle (passengers interact with the game from different handheld devices), inter-vehicle (play with others who are traveling in cars close to you through an ad-hoc peer-to-peer network or 3G mobile network) or multiplayer internet (play with anyone connected to the internet, where the in-car player interacts with the game in a different way than those playing on stationary computers). Other game plans that have been discussed include never-ending simulation and evolution games (i.e. Sim City; SIMS) and quiz-games (i.e. Trivial Pursuit; Backpacker). Gender equality and other forms of pastime entertainment that attracts a wider range of ages are also under deliberation.

3. RESEARCH APPROACH

This research focuses on how adaptive structuration theory (DeSanctis & Poole 1994), in-depth qualitative interviews using the repertory grid technique (Kelly 1955) and carefully selected lead users (Von Hippel 1988) can be used in evaluation of innovations at a prototype stage. A context-aware application, CABdriver, is used as an illustration of such a prototype. Each of the components of the research approach is discussed below.

3.1 Adaptive structuration theory

This research uses adaptive structuration theory (AST), an extension of structuration theory (ST), of which the major constructs and propositions can be seen in Figure 4. At this point in the research process, AST is primarily used as a background for the two-month evaluation of CABdriver and is thus only touched on very briefly in this paper. [Readers are referred to Jones and Karsten (2003) for a description of the history and critique of the application of ST.]

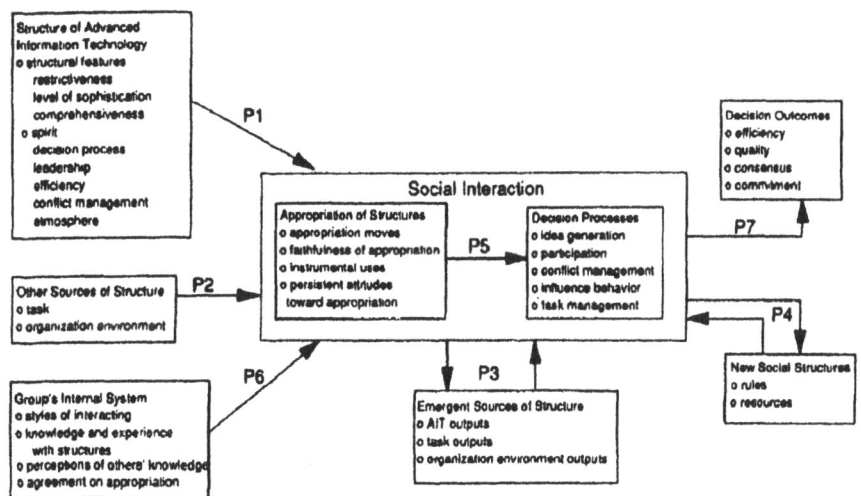

Figure 4. Major constructs and propositions of AST (DeSanctis & Poole 1994, p. 132).

AST was selected for this study because it offers a means of examining a variety of social and contextual factors that influence the use of a technology. As illustrated in Figure 4, AST takes into account the importance of interactions within the use situation, as well as different methods of use with varying degrees of fit with expected use patterns. This model provides a broad view of the innovation adoption and use-context, which was desired in this situation where it is not initially evident what factors are relevant to an evaluation of the prototype.

It should be noted that four aspects of this research differ from traditional use of AST. First, AST has previously been used primarily in organizational settings, often on groups using groupware, rather than on the individual level in an every-day situation as is the case with CABdriver. Second, CABdriver directly influences one user, rather than a group of users. This individual's use of CABdriver is intended to spawn attempts at influencing the behavior of the small group of non-users in the car. The interaction within the group is thus the result of one group member using the application rather than the common case within AST of multiple users of the same technology. Third, although the empirical results in this paper are used to discuss how lead user interviews can provide value, the upcoming every-day use evaluation will not be as long as traditional AST studies. However, AST is still an applicable theory in innovation processes; "...a second direction for this research is to directly test the explanatory and predictive power of AST" (DeSanctis & Poole 1994, p. 144). Fourth, we are bringing the technology to the group, rather than studying an already existing technology within that group. This means that we, as researchers and as part developers, have strong structural features we hope to see, which thus forces us to take extra care in not overlooking results that do not conform to our expectations.

3.2   Interviews: the repertory grid technique

Based on Tan & Hunter (2002), this research uses repertory grids – a cognitive mapping tool in personal construct theory (Kelly 1955) – to capture the structuration process of AST as well as to assist in not overlooking results that do not match our expectations. The repertory grid technique is used to empirically elicit and evaluate peoples' subjective perceptions. The technique is relatively straight-forward since a person, according to Kelly (1955), judges what she encounters in terms of dualities, such as *Fun—Boring*, *Nice—Rude*, and so on. The objects under investigation are called elements, and may be supplied by the researcher (Reger 1990) or elicited from the participants (Easterby-Smith 1980). Several rules apply for what makes valid elements (i.e. discrete – Stewart & Stewart 1981; homogenous – Easterby-Smith 1980; not evaluative – Stewart

& Stewart 1981; representative – Beail 1985; Easterby-Smith 1980). Constructs, i.e. the dualities for use as a grading scale, may also be either elicited or supplied. Elicitation is typically used (and is here as well), primarily by presenting the elements to the participants in combinations of three, a technique referred to as *triads*. The participant is asked to identify one of the elements in the triad that stands out in some way. This quality is then labeled by the participant, using her own terminology, after which she is asked to label in what way the other two elements differ from the first one in that aspect (thereby creating a duality). This process is iterated until all combinations of elements have been presented to the participant. Previous research has shown that seven to ten triads are sufficient to elicit all the participant's constructs in a particular domain (Reger 1990). Variants of the triadic process include using *dyads* (Keen & Bell 1980), combining elicited with supplied constructs (Easterby-Smith 1980), using a sample of the entire repertory of elements (Dutton et al. 1989) and asking for the opposite label rather than how the identified element differs from the other two (Epting et al. 1971).

After a set of constructs relating to a particular set of elements have been created by the participant, she is asked to grade each element on a scale – in this case set to be a five point scale, as recommended by Stewart & Stewart (1981) – with the dualities at each end of the scale (i.e. the construct *Fun—Boring* would mean Fun is 1 and Boring is 5). Techniques for analyzing the grids created include *content analysis* (Hunter 1997; Moynihan 1996; Stewart & Stewart 1981), *rearranging* (Bell 1990; Easterby-Smith 1980), transforming (Shaw & Thomas 1978) and *decomposing* (Bell 1990; Easterby-Smith 1980) repertory grids, and finally *analysis of content and structure* (Kelly 1955).

For this paper, the data has then been fed into Web Grid III, a frequently used and feature-rich tool for collecting, storing, analyzing and visually representing repertory grids. In particular, two statistical methods of data analysis have been used to transform and represent the data collected – the FOCUS cluster algorithm (Gaines & Shaw 1995; 1997; Shaw 1983), and PRINCOM mapping (Gaines & Shaw 1980; Slater 1976). An in-depth description of this process is given in section 4. One drawback with using only three participants is that the results cannot be considered to approximate a "universal result". For the results to be applicable in general, a sample of 15 to 25 participants is sufficient (Dunn, et al. 1986; Ginsberg 1989). However, in true qualitative fashion, we have not strived for "generally applicable" results, but rather chosen to focus our efforts on in-depth analysis of a few carefully selected lead users (section 3.3) and their perceptions of a particular in-car application and the surrounding environment (outlined as important by AST).

3.3   Identifying lead users

Von Hippel (1988, p. 107) characterizes how lead users are particularly useful for providing insight into evaluation of novel products, processes or services. "Although the insights of lead users are as constrained to the familiar as those of other users, lead users are familiar with conditions that lie in the future for most – and, so, are in a position to provide accurate data on needs related to such future conditions." Lead users are characterized as facing general marketplace needs months or years before the bulk of the marketplace encounters them. Typically, they are also positioned to benefit significantly by finding solutions to those needs.

Rather than adopting a quantitative approach to finding lead users, much like Von Hippel (1988) does in his illustration of this, we have taken a more subjective stance, as we hand-picked three individuals we had extensive experience from dealing with, in one way or another. Having lead user interviews as a part of this research serves three purposes: (1) it allows identification of any major design problems with CABdriver (and if need be, redesign of it) ahead of the more demanding every-day evaluation; (2) it permits us to discuss the research approach chosen (i.e. using a few carefully selected lead users for in-depth qualitative interview sessions to provide design feedback and preliminary results ahead of a more in-depth study of every-day use, in combination with repertory grids and adaptive structuration theory); and, (3), it will enable us to compare the results gained from lead users with the data we will receive in the every-day use evaluation (captured in forthcoming work).

Table 1. Lead users

Person H: Female, 28 years old.
H is a PhD student with less than 6 months to her defense. She is studying the development process of commercial off-the-shelf software, using a computer game development company as her case study.

Person M: Male, 30 years old.
M is a systems administrator with excellent technical skills. M actively plays many computer games and is well versed in development of applications on handheld devices, including for in-vehicle use.

Person J: Male, 32 years old.
J has a history with a major international computer consultancy company, where he worked with data warehousing for customers with large amounts of database information. From the various roles he held in projects, he gained much experience from how innovative solutions are born, developed and commercialized.

4. ANALYSIS AND DISCUSSION

Recorded, in-depth interviews (Wolcott 2001; Patton 2002) were performed individually with each of the three participants (section 3.3), using the repertory grid technique. In addition to the grids containing supplied elements, constructs were elicited from each participant, later used for grading all elements. *Laddering* (Reynolds & Gutman 1988) was further employed to allow participants to extensively explain and discuss their constructs as well as each grading of elements. In laddering, a series of general probing questions are asked (i.e. how and why), in order to stimulate participants to elaborate on their choices. Compared to open-ended questions prepared by the researcher, this technique enable participants to freely choose the aspects they themselves feel are most relevant for discussion. Each session lasted between two and three hours and participants were allowed to create as many constructs as they could think of, after being exposed to a selection of possible combinations of elements in each grid. This resulted in slight variations in the number of constructs for each grid, but it insured that all constructs that participants considered relevant were identified.

The intent of the analysis has been to identify differences and similarities between the elements used by the participants. The process of identifying these requires a number of stages during the analysis. During this analysis, the recorded sessions with clarifications, hesitation, confusion and elaboration expressed by the participants has helped in making qualitative judgments as to how to treat their constructs as well as grading of elements. In this sense, repertory grid analysis has served as a starting point for further analysis of the interview sessions rather than as any kind of "final results".

4.1 Stage one: data collection

At the first stage, each person was queried about three main areas – (1) different purposes of traveling by car, (2) different types of traffic situations one encounters in a car, and (3) the use of different applications related to CABdriver (with which all were familiar) in cars.

Elements in the first grid all focused on different aspects of car travel. The grid was labeled Travel, in order to clearly separate the grid from the other two grids. The elements (E) all relate to travel by car with the purpose of going to, for or on...:

Table 2. Travel elements

E1)	Vacation
E2)	Leisure pursuit
E3)	Work
E4)	Errand
E5)	Business

Elements in the second grid all focused on different aspects of car traffic. The grid was labeled Traffic, in order to clearly separate the grid from the other two grids. The elements all relate to travel by car in each of the following types of traffic:

Table 3. Travel type elements

E6)	Downtown
E7)	Highway
E8)	Rush-hour
E9)	Traffic jams

Elements in the third grid were all applications used during car travel. The grid was labeled Application, in order to clearly separate the grid from the other two grids. The elements all relate to in-car use during travel:

Table 4. Application type elements

E10)	Navigation system
E11)	RDS-TMC (traffic messages that interrupt the radio transmission)
E12)	CABdriver
E13)	Mobile phone game
E14)	Nintendo Gameboy

This means that each participant produced three grids. As the focus in this paper is how all three lead users perceive the three areas of investigation, all constructs and grades were put into one large grid for each domain of investigation (Travel, Traffic and Application). Below, we have chosen to show the stages of refinement only for the Application grid, as the process was the same for each grid. (Our intent is to show the feasibility of our evaluation approach rather than to conduct a full evaluation of CABdriver.) For the sake of completeness, the full Travel and Traffic grids are included in the Appendix.

4.2   Stage two: FOCUS cluster analysis

At the second stage, these elements were entered into Web Grid III for FOCUS cluster analysis. Here, we looked for relationships between the constructs, pondered contradicting results, consulted the recorded sessions and if need be, asked follow-up questions to the participants. Concluding this stage was a subjective re-labeling or re-use of constructs that seemed to represent the same aspect. This meant breaking up some of the links that the

FOCUS algorithm suggested. One example that can be seen of this (Figure 5) is the association of *No control—User control* with *Sound based—Screen based*. This had to do with the nature of our elements used in this grid, rather than what might be true in general. Among our elements, this was true, however. At this stage of analysis, we can also see how CABdriver shares more aspects with navigation systems and traffic messages through the radio (RDS-TMC) than the two more traditional games. It may also be noted that all participants, even the hard-core gamer, considered mobile phone games as the same type of element as a Nintendo Gameboy. When queried about this, M explained that the difference in hardware was not at all as important to him as the actual game plan of games on these two devices. Furthermore, contradictive results, such as the *Mobile—Stationary* construct which suggests that a Nintendo Gameboy and a mobile phone game are stationary rather than mobile was put down as accidental reversal of the scale by that participant (which he agreed likely when asked later).

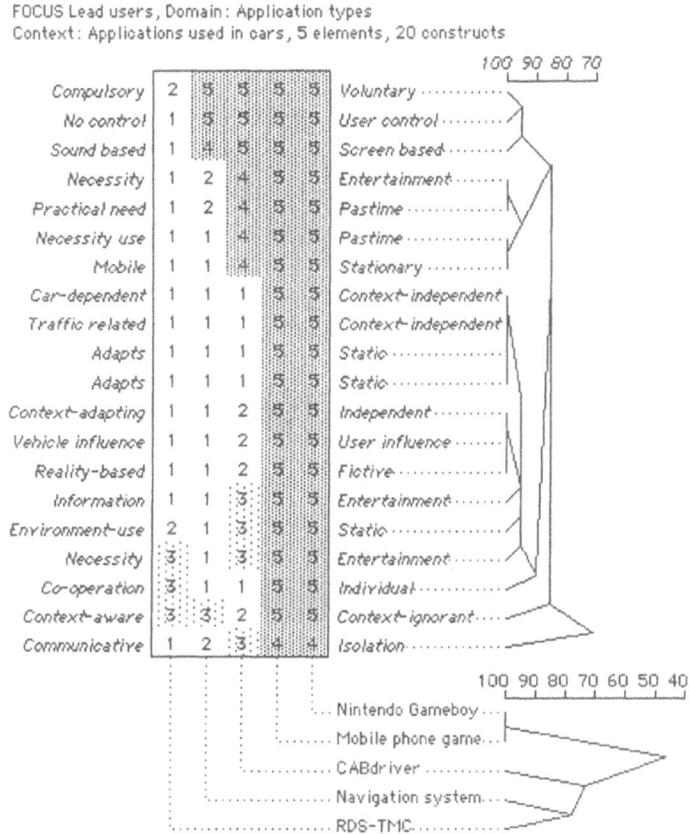

Figure 5. A FOCUS cluster analysis shows at which threshold the constructs relate to each other as well as the threshold at which the elements relate to each other.

Thus, after careful semantic review and interpretation, constructs were grouped and re-labeled when judged appropriate. Based on these new groups another FOCUS cluster analysis was performed (Figure 6), now displaying the vital differences. In this cluster, we can see the formation of a somewhat unexpected result from a designer perspective. CABdriver was, as expected, perceived as primarily a pastime entertainment. However, although recognized as reality-based and adapting to the context, the game remained worryingly close to a Nintendo Gameboy and a mobile phone game regarding in-car isolation. Actually, rather than making the player support the driver, the game was perceived as primarily pulling the driver into the game. This can be seen by the strong co-operation required by the game. Analysis of the interviews supported that this co-operation was strongly perceived by the participants, i.e. CABdriver influences driving. At design time, the game was intended to be entertaining, providing the player with a better understanding of the current environment (in-car, the immediate surrounding as well as the distant surrounding) and finally be able to provide some useful help in finding a parking spot nearby, the next gas station, traffic message warnings, and so on. Fortunately, this result is not all bad. As CABdriver is nearly impossible to play if the driver does not adapt a smooth, environmentally sound, driving style as well as stay away from excessive speeding, CABdriver seems to be very well positioned. At one time during the technical testing preceding these evaluations, a developer/player prevented the driver from passing a car in an area where he would have had to accelerate as hard as possible since the developer/player was out of energy already and he did not want a driver work-load monster entering his mission (let alone face the space mines that would come from excessive speeding).

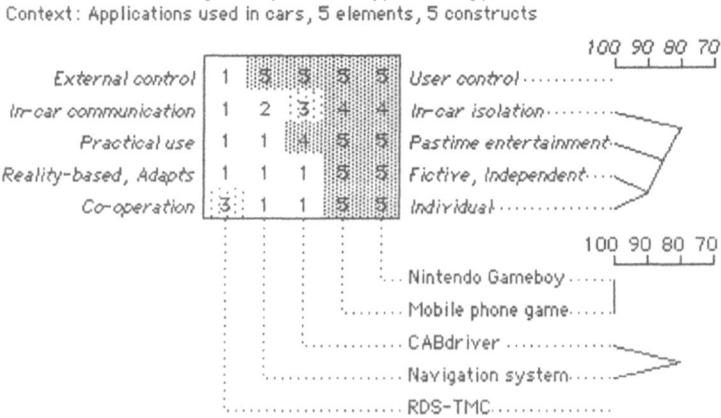

Figure 6. A FOCUS cluster analysis of the re-labeled and grouped constructs.

Furthermore, after grouping all constructs dealing with *Practical use—Pastime entertainment* together, CABdriver came out to share many aspects with a navigation system, something which can be explained by Figure 2 earlier. It may be a very crude navigation system, but turn-by-turn guidance is easily given by the CABdriver user when searching for that parking space or gas station.

4.3   Stage three: PRINCOM map illustration

At the third stage, a PRINCOM map was created (Figure 7). Care must be taken when analyzing a PRINCOM map since constructs have been rotated freely for maximum distribution in unlimited dimensions, while the elements have been placed on the map as if it was two dimensional. For instance, this explains why the construct In-car isolation—In-car communication is shorter than the others. As it is impossible for us to model an unlimited dimension space (although mathematically relatively simple to show using vector analysis), the very nature of a PRINCOM map requires qualitative judgment as to what to explain and what to avoid. Again, going back to the recorded sessions has helped this task. The model shown here is visualized in vector space to facilitate maximum separation of elements in two dimensions (Gaines & Shaw 1980; Slater 1976)

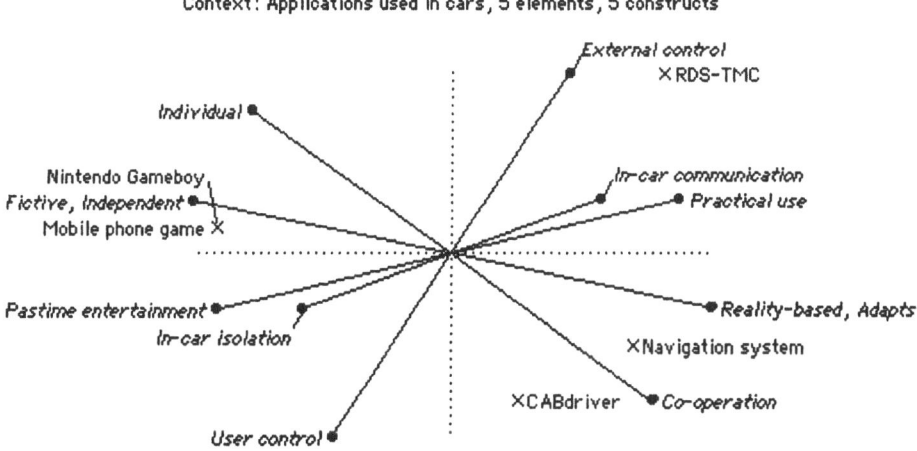

Figure 7. A PrinCom map provides an overview of how the lead users relate to different types of applications used during car travel.

Again, but possibly more clearly illustrated, we see how CABdriver and a navigation system are relatively similar in terms of the constructs. We can also see how *Pastime entertainment—Practical use* seems quite close to the construct *In-car isolation—In-car communication*. Careful review of the recorded sessions as well as mathematical analysis of these two vectors shows that the two constructs are only visually represented in this two-dimensional image as close, rather than having actual strong similarity.

4.4   Contributions

The contribution of this paper lies in illustrating how three established ways of conducting research can be brought together, forming a useful research approach in a new setting. In our case, this setting is the evaluation of innovative prototypes. Specifically, by using the approach on CABdriver, four valuable experiences emerge. First, the repertory grids in this paper contribute by showing how the structural features of technology, travel type and traffic situation are perceived by the lead users. These grids correspond to the existing sources of structures affecting the social interaction according to AST and demonstrate how repertory grids can be integrated with AST. Forthcoming every-day use evaluation will expand this integration by focusing on appropriation moves, emerging structure and decision processes. It is also noteworthy that four aspects of this research differ from traditional use of AST (which has been critiqued for being applied too mechanistically – Jones & Karsten 2003). This research differs by (1) having a non-organizational setting, (2) having an individual focus, (3) exploring the explanatory and predictive power of AST, and (4) bringing the technology to the group.

Second, the repertory grid technique provides a highly established tool for capturing individual cognition. This is of particular importance if strong hopes and expectations regarding the impact of the technology already exist. Using repertory grids and laddering grants a structured way for allowing the participants themselves to form the aspects of particular meaning to them, rather than having the preconceptions of the technology and intended impact guiding the research questions.

Third, in using the repertory grids as a starting point for further analysis of the recorded sessions rather than as "final results", we show an effective way to approach the recorded sessions. Using the grids as a first lens for analyzing cognitive data also helps meeting challenges from having strong preconceptions of desired outcomes.

Fourth, including lead user evaluations prior to the main evaluation is fruitful. These lead users were carefully selected based on extensive experience from dealing with them in one way or another, and thus represent our subjective point of view. Since the results were not intended to imply any general point of view representing an entire user group, the in-depth qualitative interviews with subjectively chosen users still apply. Using the participants to test the prototype design ahead of the every-day use evaluation, in search of any major design flaws previously missed, provided some unexpected results compared with what was intended at design time. In this case, the prototype did not require re-design based on these unexpected results since they still answered to CABdriver's purpose of influencing driving.

5. CONCLUSIONS

In this paper we have explored the use of a particular approach for evaluating innovative prototypes. We found the combination of adaptive structuration theory, the repertory grid technique, and interviews with lead users to be a useful method of evaluating a prototype. The use of this method on the context-aware application CABdriver has provided insights on four areas: (1) repertory grids can be a useful addition to AST, (2) repertory grids supported by neutral probing questions offers a valuable means to investigate cognition from an interviewee perspective rather than a researcher-driven perspective, (3) using the grids as a lens for approaching the collected data is an efficient way to start data analysis, and (4) by introducing carefully selected lead users, innovations can effectively be tested for design flaws and possible unexpected outcomes prior to any major evaluation or product launch.

Whereas this approach has been used with only one technology on a small level, nothing we found would prohibit its use with other types of innovative technologies. Forthcoming work will focus on a two month evaluation of every-day use of CABdriver in five families, and will also address the effectiveness of qualitative interviews with lead users in relation to the more demanding every-day use evaluation. This more extensive evolution process will not only provide more information on the usefulness of the evaluation method, but will also provide more detailed results regarding perceptions of the use of the technology itself.

REFERENCES

Alm, H. and Nilsson, L. (1994). Changes in Driver Behaviour as a Function of Handsfree Mobile Phones – a Simulator Study. *Accident Analysis & Prevention*, 26, 441-451.

Alm, H. and Nilsson, L. (1995). The Effects of a Mobile Telephone Task on Driver Behaviour in a Car Following Situation. *Accident Analysis & Prevention*, 27, 707-715.

Appleyard, D., Lynch, K. and Myer, J. R. (1964), *The View from the Road*, M.I.T. Press, Cambridge, Massachusetts, USA.

Beail, N. (1985). *An Introduction to Repertory Grid Technique. Repertory Grid Technique and Personal Constructs*, N. Beail (ed.), Brookline Books, Cambridge, MA, pp. 1-26.

Bell, R. C. (1990). Analytic Issues in the Use of Repertory Grid Technique. *Advances in Personal Construct Psychology* (1), pp. 25-48.

Brookhuis, K. A., De Vries, G. and Waard, D. (1991). The Effects of Mobile Telephoning on Driving Performance. *Accident Analysis & Prevention*, 23, pp. 30-316.

DeSanctis, G. and Poole, M.S. (1994) Capturing the Complexity in Advanced Technology Use: Adaptive Structuration Theory. *Organization Science*, Vol. 5, No. 2, pp. 121-147.

Dey, A. K., Abowd, G. D. and Salber, D. (2001). A Conceptual Framework and a Toolkit for Supporting the Rapid Prototyping of Context-Aware Applications. *Human-Computer Interaction*, Volume 16, pp. 97-166.

Dunn, W. N., Cahill, A. G., Dukes, M. J., and Ginsberg, A. (1986). The Policy Grid: A Cognitive Methodology for Assessing Policy Dynamics, in *Policy Analysis: Perspectives, Concepts, and Methods*, W. N. Dunn (ed.), JAI Press, Greenwich, CT, pp. 355-375.

Dutton, J. E., Walton, E. J., and Abrahamson, E. (1989). Important Dimensions of Strategic Issues: Separating the Wheat from the Chaff. *Journal of Management Studies* (26:4), pp. 379-396.

Easterby-Smith, M. (1980). The Design, Analysis and Interpretation of Repertory Grids. *International Journal of Man-Machine Studies* (13), pp. 3-24.

Epting, F. R., Suchman, D. J., and Nickeson, K. J. (1971). An Evaluation of Elicitation Procedures for Personal Constructs, *British Journal of Psychology* (62), pp. 512-517.

Gaines, B. R. and Shaw, M. L. G. (1980). New directions in the analysis and interactive elicitation of personal construct systems. *International Journal Man-Machine Studies*, 13, pp. 81-116.

Gaines, B. R. and Shaw, M. L. G. (1995). Concept mapping on the web. *Proc. Fourth International World Wide Web Conference*, O'Reilly.

Gaines, B. R. and Shaw, M. L. G. (1997). Knowledge acquisition, modeling and inference through world wide web. *Human Computer Studies*, 46, pp. 729—759.

Ginsberg, A. (1989). Construing the Business Portfolio: A Cognitive Model of Diversification. *Journal of Management Studies* (26:4), pp. 417-438.

Goose, S. and Djennane S. (2002). WIRE3: Driving Around the Information Super Highway, *Journal of Personal and Ubiquitous Computing*, Vol. 6-3, Springer-Verlag, London.

Hedström, K. (2003). The socio-political construction of Care-Sys: How interests and values influence computerization, *Proceedings of IFIP WG 8.6 conference*, Copenhagen.

Hunter, M. G. (1997). The Use of RepGrids to Gather Interview Data About Information Systems Analysts. *Information Systems Journal* (7), pp. 67-81.

Jones, M. and Karsten, H. (2003). Review: Structuration Theory and Information Systems Research. Research Papers in Management Studies, University of Cambridge, Judge Institute of Management.

Keen, T. R. and Bell, R. C. (1980). One Thing Leads to Another: A New Approach to Elicitation in the Repertory Grid Technique, *International Journal of Man-Machine Studies* (13), pp. 25-38.

Kelly, G. (1955). *The psychology of personal constructs*. Vol 1 & 2. Routledge, London, UK.

Lyytinen, K. and Damsgaard, J. (2001). What's wrong with the diffusion of innovation theory? The case of a complex and networked technology, in *IFIP 8.6 Conference Proceedings*, Banff, Canada.

Lyytinen, K. and Rose, G. M. (2003). Disruptive information system innovation: The case of internet computing. *Information Systems Journal*, Vol. 13, No. 4, pp. 301-330.

Lyytinen, K. and Yoo, Y. (2002). Research Commentary: The Next Wave of Nomadic Computing. *Information Systems Research*, Vol. 13, No. 4, pp. 377-388.

McKnight, A. J. and McKnight, A. S. (1993). The Effect of Cellular Phone Use upon Driver Attention. *Accident Analysis & Prevention*, 25, 259-265.

Moynihan, T. (1996). An Inventory of Personal Constructs for Information Systems Project Risk Researchers. *Journal of Information Technology* (11), pp. 359-371.

Prescott, M. B. and Conger, S. A. (1995). Information technology innovations: A classification by IT locus of impact and research approach. *Data Base Advances*, Vol. 26, No. 2 and 3, pp. 20-41.

Redelmeier, D. A. and Tibshirani, R. J. (1997). Association Between Cellular-Telephone Calls and Motor Vehicle Collisions. *The New England Journal of Medicine*, 336, 453-458.

Reger, R. K. (1990). The Repertory Grid Technique for Eliciting the Content and Structure of Cognitive Constructive Systems. *Mapping Strategic Thought*, A. S. Huff (ed.), John Wiley & Sons Ltd., Chicester, pp. 301-309.

Reynolds, T. J. and Gutman, J. (1988). Laddering Theory, Method, Analysis, and Interpretation. *Journal of Advertising Research*, February-March, pp. 11-31.

Rogers, E. M. (1995). *Diffusion of Innovations*. The Free Press, fourth edition.

Salvucci D. (2001). Predicting the Effects of In-Car Interfaces on Driver Behavior using a Cognitive Architecture. *Proceedings of Human Factors in Computing Systems*, ACM Press, New York.

Schmidt, A., Beigl, M. and Gellersen, H.-W. (1998). There Is More to Context Than Location. *Computers & Graphics* 23 (1999) 893-901.

Shaw, M. L. G. and Thomas, L. F. (1978). FOCUS on Education: An Interactive Computer System for the Development and Analysis of Repertory Grids. *International Journal of Man-Machine Studies* (10), pp. 139-173.

Shaw, M. L .G. and Gaines, B. R. (1983). A computer aid to knowledge engineering. *Proc. British Computer Society Conference on Expert Systems*, pp. 263—271.

Siewiorek, D. P. (2002). New Frontiers of Application Design. *Communications of the ACM*. Vol. 45, No. 12, pp. 79-82.

Slater, P. (1976). *Dimensions of intrapersonal space*. Vol 1. John Wiley, London, UK.

Stewart, V. and Stewart, A. (1981). *Business Applications of Repertory Grid*. McGraw-Hill, London, 1981.

Swanson, E. B. (1994). Information systems innovation among organizations. *Management Science*, Vol. 40, No. 9, pp. 279-309.

Tan, F. B. and Hunter, M. G. (2002). The Repertory Grid Technique: A Method For The Study of Cognition In Information Systems. *MIS Quarterly* Vol. 26 No. 1, pp. 39-57.

Von Hippel, E. (1988). *The Sources of Innovation*. Oxford University Press, New York, Oxford, USA.

APPENDIX

Princom and Focus of Travel and Traffic Grids

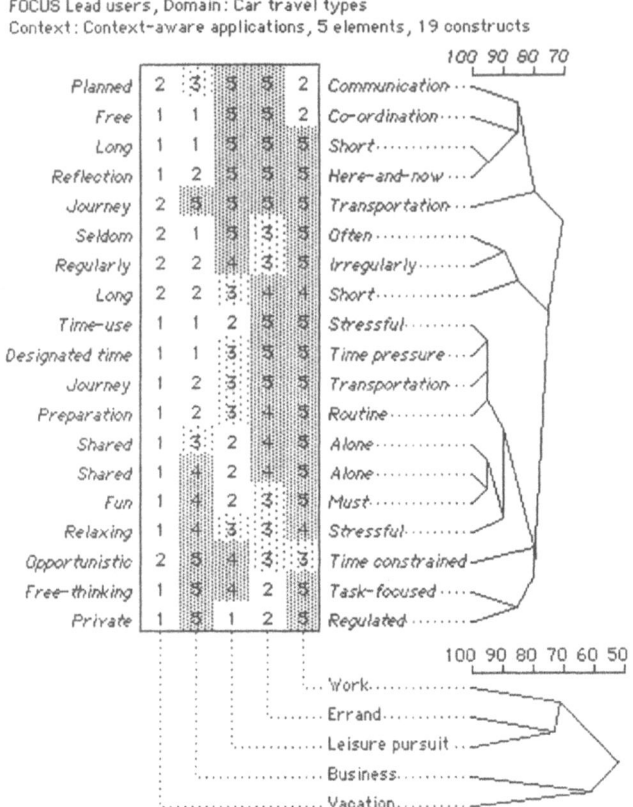

Figure 8. A FOCUS cluster analysis shows at which threshold the constructs relate to each other as well as the threshold at which the elements relate to each other.

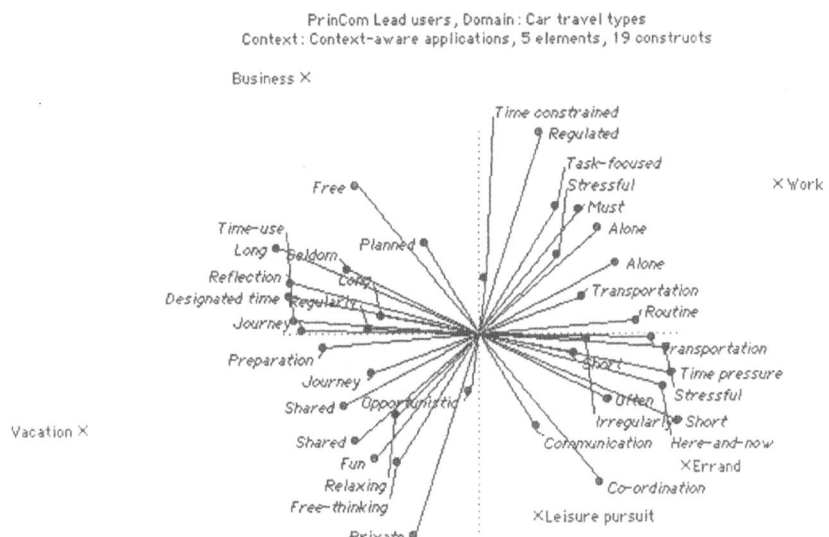

Figure 9. A PrinCom map provides an overview of how the lead users relate to different types of car travel.

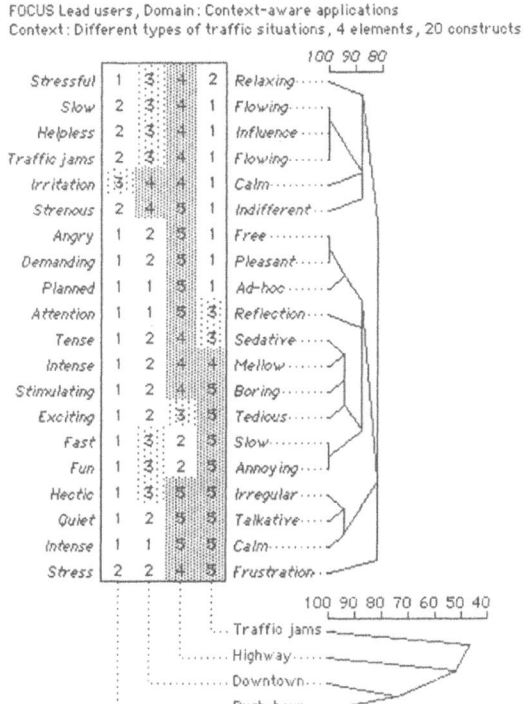

Figure 10. A FOCUS cluster analysis shows at which threshold the constructs relate to each other as well as the threshold at which the elements relate to each other.

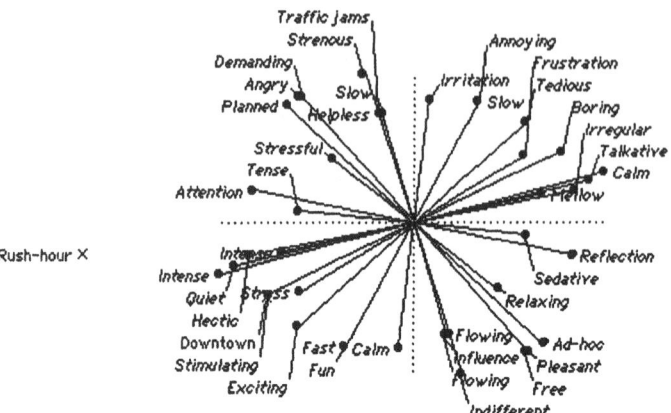

Figure 11. A PrinCom map provides an overview of how the lead users relate to different types of traffic situations.

APPLYING IT INNOVATION
An empirical model of key trends and business drivers

Keith Beggs
Intel Corporation

Abstract: To take root and proliferate in today's environment, innovation must serve a purpose. Although the various technologies behind IT innovation are incredibly diverse, the rationale for their adoption converge around six empirically derived themes. Each theme, or motive, is significant, valid and described with real-world examples. Two metrics characterize the extent to which the various motives are used to justify investments in innovative IT approaches and infer other trends. Instructions on ways others can further apply the business driver investment model are included in order to enable IT stakeholders to brainstorm new or underutilized approaches and better leverage each of the six business drivers of IT innovation adoption for their competitive advantage.

Key words: Innovation business drivers, empirical study, rationale for IT innovation investment

1.   INTRODUCTION

To take root and proliferate in today's environment, innovation must serve a purpose. In today's tough economic times, deployment of innovative IT solutions and capabilities does not occur in a vacuum or for it's own sake. Innovation is applied to provide distinctive new capabilities that in turn provide some form of competitive advantage or benefit. While innovation occurs through a myriad of technologies, methods and approaches, the *application* of IT innovation can be characterized along six key themes or business drivers. This paper will define these themes in a new analytical model, examine real-world application of them, and provide a conceptual framework for others to examine their discretionary IT investments and brainstorm new possibilities.

2.   THE BUSINESS DRIVER INVESTMENT MODEL

The business driver model (Figure 1) simplifies and categorizes the key business drivers behind adoption of IT innovation. Businesses use multiple paths to increase their adaptability and competitive advantage. The model can be applied broadly because it was derived empirically from real world IT consulting experience and analysis of over 600 pages of content from Fortune 500 IT consultants, product developers and practitioners.

Recent application of IT innovation occurs along six major themes or types of rationale. Three of these are more offensive or externally focused in nature. They focus on maximizing revenue, net income and competitive advantage through:
- New sales channels, methods and globalization efforts
- Deployment of new products and capabilities
- Positively differentiating existing products or services.

The benefits associated with these types of investment are typically visible above the gross income line on the balance sheet. Consequently, these investment motives are shown above the black line in the model.

Conversely, CIOs and stakeholders also strategically invest in IT to optimize internal operations and minimize expenses. These benefits are visible below the gross income line in the balance sheet. Hence, these three business drivers are displayed below the black line in the diagram:
- Cost and expense reduction
- Automation, productivity, and process redesign
- Strategic decision making and risk reduction assets.

To varying degrees in different types of industries, such as technology services and semiconductors, the themes on the left side of the diagram (increased sales, cost reduction) tend to have more quantifiable benefits, while the motives on the right (product differentiation, branding, or information for strategic decision making) tend to be more qualitative and harder to quantify.

The arrows in the diagram conceptually connect what is being innovated, IT innovations, with the six major types of rationale used to justify their adoption or use, the sides of the hexagon. They also demonstrate the relative importance and frequency of these investment motives using two metrics. White arrows describe how frequently this rationale was used to recommend adoption of IT innovations in a sample of web-based content across Fortune 500 IT consultants and practitioners (such as Accenture, IBM Global Services, Intel, and EDS). Black arrows correspond to the amount of actual discretionary investment observed within another sample of IT deployment initiatives. The length of the arrows represents relative importance of the driver and the percentage that rationale was cited or applied. Longer arrows are the predominate or most widely observed rationale. Together, these metrics provide some insight into the relative frequency and extent various rationale are used in IT decision making.

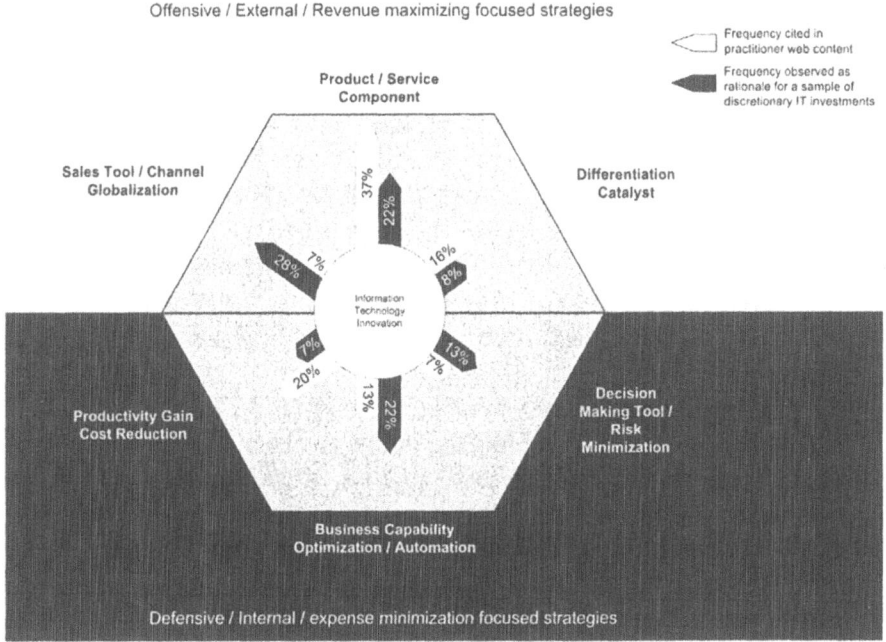

Figure 1. Key Drivers of IT Innovation Adoption

3. EXTERNALLY FOCUSED, INCOME GROWTH FOCUSED STRATEGIES

Tom Peters highlights the importance of investments to maximize growth and income by stating "Revenue enhancement (new products, innovation in general) is the ticket. While relative costs must remain under control, and fat kept to a minimum, it's the builders, in the long haul who will reap the rewards from Wall Street."[1] Andy Grove's comment: "We need to create waves …. of lust for our product"[2] further confirms the significance of growth-focused IT investments. Forward thinking teams recognize that IT innovation is a strategic asset to be leveraged, not just an expense to be minimized. Three major types of leverage points exist to do just that.

[1] Tom Peters, *The Circle of Innovation,* 1997, Knopf: New York, Toronto
[2] Alan Goldstein, *Dallas Morning News*, August 11, 1996, page 54.

- **New sales tools, channels and globalization efforts** – As globalization intensifies, we're clearly seeing IT capabilities (Internet, rich media, localization and internationalization, and other technologies) being used to produce additional revenue across domestic and international markets. IT capabilities provide global impact and efficiently reach new markets, particularly emerging markets or areas where a business has limited physical presence.
- **New service and product development components** – IT innovation is also being developed into new products or service capabilities, which provide additional revenue potential. As Moore's law plays out through time, various new IT capabilities continue to be the cornerstone of new products. Exciting new embedded, networking, server and microprocessor capabilities are being applied to enable new capabilities and usage models, and to help drive demand for replacement of legacy technologies.
- **Service and product differentiator catalysts** – In many cases, IT innovation is bundled with existing products or incorporated into service delivery to positively differentiate core products, even though the innovation itself is not the end product consumed. Insightful application of IT can positively differentiate other products. This is clearly evident with Progressive Insurance's approach to providing automated, real-time self-service insurance quotes.

4.   INTERNALLY FOCUSED, EXPENSE MINIMIZATION STRATEGIES

IT innovation is also a strategic asset with various internal benefits. However, the benefits span well beyond the obvious cost saving motive.
- **Cost reduction** – Replacing legacy technologies and approaches with new solutions and technology can reduce the cost of providing equivalent capabilities. This is clearly evident in trends like the replacing of expensive data centers based on proprietary standards with newer infrastructure, deploying wireless-enabled platforms, and replacing antiquated PC technology to reduce the total cost of ownership. New approaches to infrastructure services and usage, such as On Demand computing, also appear to offer promise.
- **Business process optimization, productivity and automation improvements** –These all provide new benefits and help increase competitive advantage. New database and web application innovations are helping to automate manual, labor-intensive processes and enable various process improvements.

- **Strategic decision making and risk mitigation capabilities** – If information is power, then IT innovation is a power grid that can illuminate the entire enterprise. IT capabilities such as OLAP, data mining, and decision support equip operations with strategic insights and help minimize risk. Although difficult to quantify, the value of these investments should not be underestimated. In certain cases, like HIPAA and the health care industry, compliance programs and legal mandates drive additional IT investment.

5. ANALYZING THE RESULTS

Looking at Table 1, we discover that while each investment motive is valid, they don't appear to be used to a uniform extent. Project data highlights the importance of web, localization and other related investments. However, their value is often self-evident, implied, and not explicitly mentioned in web content. It is not surprising that IT practitioners would consistently channel IT innovations into product development, perhaps more so than other industries. However, it initially appears that while frequently described, IT capabilities may not be fully utilized to differentiate other non-IT related products or services. Cost reduction is a frequent justification for internal investments. However, investments may often be speculative at first and focus on savings through new productivity or capability enhancing investments, rather that simply "like for like" technology replacements. It is also extremely encouraging to see IT decision makers invest in strategic decision making and information investments as well.

Table 1. Frequency and extent to which these motives are used

Motive category / theme	Metric 1: relative % of sampled IT practitioner web content that used that motive	Metric 2: relative % of sampled discretionary IT projects observed using that motive
New sales tools, channels and globalization efforts	7%	28%
New service and product development components	37%	22%
Service and product differentiating catalysts	16%	8%
Cost reduction (for *identical* capabilities)	20%	7%
Automation, productivity, process redesign	13%	22%
Strategic decision making and risk reduction assets	7%	13%
	100%	100%

The model we've discussed provides a fresh perspective on the major drivers or rationale for adoption of IT innovation. Although the various technologies behind IT innovation are incredibly diverse, the rationale for their adoption converge around six empirically derived themes:

Income, growth and externally focused
1. Sales capabilities and globalization efforts
2. Service and product development components
3. Differentiation catalysts

Internal capability and expense focused
4. Cost reduction
5. Automation, productivity, and process redesign
6. Strategic information and risk reduction capabilities.

Each theme or motive is significant, valid and described with real world examples. Two metrics which characterize the extent to which each rationale is used further validate the above "business driver investment" model. It is interesting to note that various rationale are not used in uniform frequency. CIOs and stakeholders must tailor and prioritize IT adoption and investments to various economic and situational constraints. Stakeholders should also evaluate their application of innovation by charting their major investments along each of the six edges of the business driver investment model hexagon. Working with other key stakeholders, one can then brainstorm new or underutilized approaches and identify ways to leverage each of the six business drivers of IT innovation adoption for their competitive advantage.

IT BUSINESS VALUE INDEX
Managing IT Investments for Business Value

Malvina Nisman
*Intel Corporation**

Abstract: Traditional financial and accounting measures often fail to properly quantify the business benefits of IT investments. Intel's Business Value Index (BVI) methodology helps us prioritize our IT investments through a common language and framework for assessing potential business value, impact on IT efficiency, and financial attractiveness.

Key words: Intel Information Technology White Paper, IT investments, business value

1. INTRODUCTION

IT investments are growing more complex, more strategic, and larger in scope. At the same time, traditional financial and accounting measures often fail to properly quantify the business benefits of those investments. Intel IT strives to invest in projects that bring both business value and IT efficiency, and are also financially attractive. To prove such bottom-line results, we developed the IT Business Value (ITBV) program at the end of 2001 to forecast the real value of an IT project and document the actual business value it delivers to Intel.

The Business Value Index (BVI) methodology helps to prioritize investment options, make data-driven decisions, and monitor progress. It encompasses business value, IT efficiency value, and financial criteria. Business value measures the impact of a project on our business strategy and priorities. IT efficiency value gauges how well the investment will use or enhance our existing infrastructure. The financial criterion measures the financial attractiveness of the investment, including time to return investment, cost/benefit ratio, and net present value (NPV) of a project.

*An earlier version of this paper was published by Intel on IT@Intel.

BVI complements analysis tools that measure return on investment (ROI). Using the BVI methodology, Intel's decision makers evaluate proposed IT investments based on potential business value, impact on IT efficiency, and financial attractiveness. Our decision makers also value early stage investments—managed as IT strategic options—using BVI.

2.   BACKGROUND

According to the META Group, more than 70 percent of IT organizations are still perceived as cost centers rather than value centers. Managing the real and perceived business value of IT is thus very difficult.

With tight IT budgets, an effective and strategic way of managing IT investments becomes more critical for the success of IT organizations. Faced with this challenge, Intel IT developed the BVI methodology, a new approach to identify and evaluate potential benefits from making investments. BVI evaluates factors beyond the traditional cost framework. This approach helps identify the potential strategic value of an initiative and effectively communicate it to our executive management.

3.   WHAT IS THE BUSINESS VALUE INDEX?

Some IT projects provide indirect benefits to a company, such as improving IT efficiency through optimized wide area network capacity while also delivering a new IT business capability to employees at a reduced cost. Other projects deliver direct benefits to a company by improving throughput, enhancing productivity, and increasing revenues. These projects deliver business value to the corporation but have no direct impact on the IT infrastructure. The Intel IT Business Value Index, shown in Figure 1, provides a way to conduct a relative comparison of these different IT initiatives to facilitate investment decisions that bring the most value to Intel and IT.

BVI is a composite index of factors that impact the value of an IT investment. It evaluates IT investments along three vectors: IT business value (that is, impact to Intel's business), impact to IT efficiency, and the financial attractiveness of an investment. All three factors use a predetermined set of defining criteria that includes customer need, business and technical risks, strategic fit, revenue potential, level of required investment, the innovation and learning generated, and other factors. Each factor's criteria are weighted according to the ongoing business strategy and business environment—changes in business strategy could change how criteria are weighted for different factors.

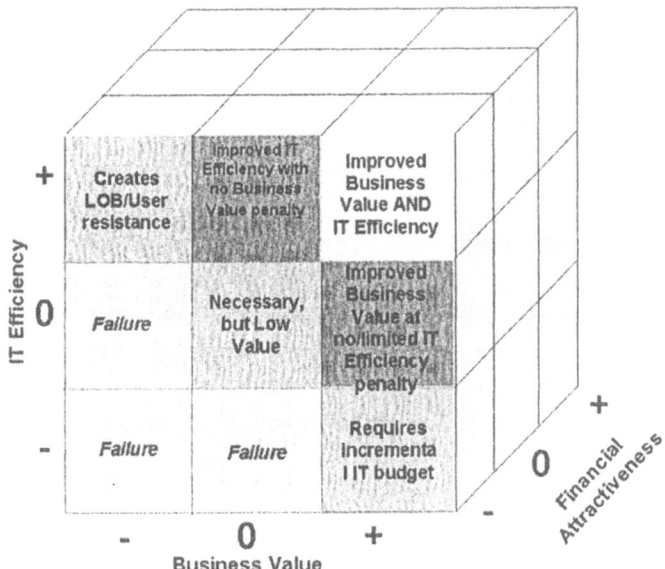

Figure 1. The Business Value Index

The predefined criteria include, but are not limited to:
– Customer need
– Business and technical risks
– Strategic fit
– Revenue potential
– Level of required investment
– Amount of innovation and learning generated

A crucial aspect of BVI is its ability to reveal the intangible benefits and strategic value of potential investments. It has a number of key characteristics that differentiate it from other prioritization methods. It provides a level comparison for multiple, strategic IT investment options. Weighted criteria support "what-if" analyses and rapid adjustments to changing business priorities. It tracks a potential investment's intangible benefits and strategic value, returning a series of decision points that depict the change in a project's relative value over a period of time.

BVI also highlights the most influential factors for assessing investments, indicates how each investment might add value to the company, and displays the results in a matrix that enhances comparative analysis of multiple investment opportunities. This approach supports options-based management of IT investments, helping us decide whether a project merits further funding.

Using the BVI methodology, our management compares and contrasts investments, and then decides what investments align best with our business priorities.

4.   HOW IS THE BUSINESS VALUE INDEX USED?

4.1   Criteria and weightings

Each vector in the BVI tool is comprised of a set of assessment criteria. Each criterion is weighted according to its importance in light of the business environment and current strategies. As project managers or program owners evaluate their proposed investments using the BVI tool, they score their project against these criteria on a scale of 0 to 3, depending how the IT investment will likely perform against a range of values set for a particular assessment criteria. Table 1 shows a range of assessment criteria for a particular type of investment.

Table 1. Sample Assessment Criteria and Scoring

Criteria	Weight	0	1	2	3
Customer pull/need	4	Low	Medium	High	Very high
Customer product cost reduction	3	Increase	No impact	Marginal reduction	Substantial reduction
Business strategic fit and impact	3	Low/NA	Medium	High	Very high
Customer performance improvement	3	Decrease	<5%	>5%	>10%

The products of the weighting and scoring of a particular investment's set of assessment criteria are summed to produce total scores for each of the three vectors—IT efficiency, financial attractiveness, and business value. After several projects have been scored, the Business Value Chart graphically depicts the three indices for each project. The position of a bubble relative to the horizontal and vertical axes shows each project's business value and IT efficiency values. The width of a bubble shows each project's financial attractiveness value. Figure 2, on the next page, depicts sample, evaluated projects.

The chart's graphical representation enables a quick assessment of investments relative to one another. The "win-win" circle shows that projects in the upper right corner of the chart have value both to IT and to Intel's business. Financial attractiveness is also important when evaluating the relative strength of various investments.

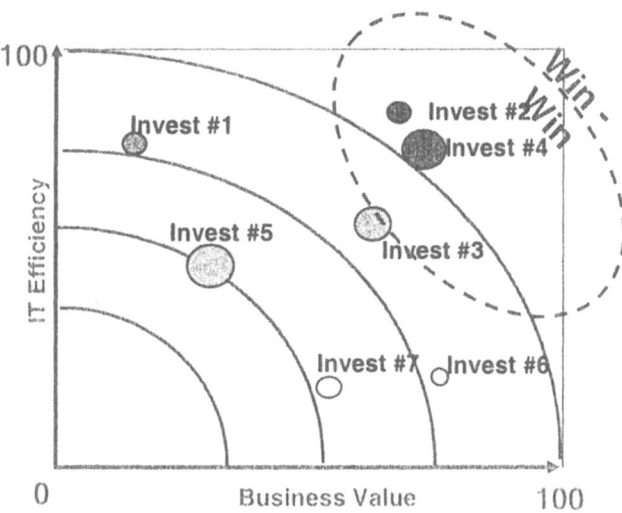

Figure 2. Sample Business Value Chart

Concentric circles identify projects with similar total BVI and IT efficiency index (ITI) scores. In Figure 2, for example, although investments 1 and 6 are on the same concentric circle, investment 1 is stronger in IEI, and investment 6 is stronger in BVI. The two investments have similar scores, but important differences exist. For that reason, business strategy and current priorities should guide the final prioritization of these projects, not solely the information on the chart.

Figure 3 shows a Project Comparison Matrix, which compares investments on a factor-by-factor basis. It shows the points accrued by each project in each factor minus the points lost. The matrix differentiates how the investments add value in different ways. Higher values in each section are shaded green or orange to visually depict the strengths and weaknesses of each project. The control values are shown at the bottom of the matrix.

This level of analysis—combined with knowledge of the group's current business strategy and environment—guides prioritization of investments. Differences in parts of the business or changes in business strategy can lead a team to adjust the weighting of different factors for analytical purposes.

Comparison Matrix																							
Business Value (corporate-wide Impact to Intel's business)																							
					Points Accrued											Points Lost							
Criteria	Weight	Max	Proi A	Proi B	Proi C	0	0	0	0	0	0	0	Prol A	Prol B	Prol C	0	0	0	0	0	0	0
Impact to Intel Revenue	5	14.3	14.3	0.0	4.8	0.0	0.0	0.0	0.0	0.0	0.0	0.0	0.0	14.3	9.5	0.0	0.0	0.0	0.0	0.0	0.0	0.0
IT Customer Pull/Need (i.e. Intel Biz group, or external company)	4	11.4	7.6	7.6	7.6	0.0	0.0	0.0	0.0	0.0	0.0	3.8	3.8	3.8	0.0	0.0	0.0	0.0	0.0	0.0	0.0	0.0
Impact to Intel's Business Risk	4	11.4	7.6	7.6	3.8	0.0	0.0	0.0	0.0	0.0	0.0	3.8	3.8	7.6	0.0	0.0	0.0	0.0	0.0	0.0	0.0	0.0
Level of alignment with Intel Strategic Objectives	3	8.6	2.9	5.7	5.7	0.0	0.0	0.0	0.0	0.0	0.0	5.7	2.9	2.9	0.0	0.0	0.0	0.0	0.0	0.0	0.0	0.0
IT Customers' Performance Improvement	3	8.6	5.7	2.9	2.9	0.0	0.0	0.0	0.0	0.0	0.0	2.9	5.7	5.7	0.0	0.0	0.0	0.0	0.0	0.0	0.0	0.0
Innovation or Enhanced Capability that solves a biz problem or creates competitive advantage	3	8.6	8.6	8.6	5.7	0.0	0.0	0.0	0.0	0.0	0.0	0.0	0.0	2.9	0.0	0.0	0.0	0.0	0.0	0.0	0.0	0.0
Customer Product Unit Cost	3	8.6	8.6	5.7	5.7	0.0	0.0	0.0	0.0	0.0	0.0	0.0	2.9	2.9	0.0	0.0	0.0	0.0	0.0	0.0	0.0	0.0
Impact on customer satisfaction via VOC results (click here to access reference on IT VOC data)	2	5.7	3.8	3.8	1.9	0.0	0.0	0.0	0.0	0.0	0.0	1.9	1.9	3.8	0.0	0.0	0.0	0.0	0.0	0.0	0.0	0.0
Confidence of Success Execution	2	5.7	3.8	5.7	5.7	0.0	0.0	0.0	0.0	0.0	0.0	1.9	0.0	0.0	0.0	0.0	0.0	0.0	0.0	0.0	0.0	0.0
Confidence in timeliness of solution delivery	2	5.7	3.8	5.7	5.7	0.0	0.0	0.0	0.0	0.0	0.0	1.9	0.0	0.0	0.0	0.0	0.0	0.0	0.0	0.0	0.0	0.0
Confidence that the solution will address the business need	2	5.7	3.8	5.7	5.7	0.0	0.0	0.0	0.0	0.0	0.0	1.9	0.0	0.0	0.0	0.0	0.0	0.0	0.0	0.0	0.0	0.0
Use of Intel Products	1	2.9	0.0	1.9	2.9	0.0	0.0	0.0	0.0	0.0	0.0	2.9	1.0	0.0	0.0	0.0	0.0	0.0	0.0	0.0	0.0	0.0
Tangible Benefits (benefits not captured above)	1	2.9	0.0	1.0	1.0	0.0	0.0	0.0	0.0	0.0	0.0	2.9	1.9	1.9	0.0	0.0	0.0	0.0	0.0	0.0	0.0	0.0

Control Values (changeable by user) Level 1: 5.0 Level 2: 10.0   Control Values (changeable by user) Level 1: 5.0 Level 2: 10.0

Figure 3. Sample Project Comparison Matrix

4.2 Ongoing assessment process

An ongoing Strategic Portfolio Management process continuously manages a team's investments, ensuring that investments are closely aligned with Intel IT's business strategies and priorities, and that IT resources are effectively utilized.

The Strategic Portfolio Management process is a closed-loop process using the BVI tool. The process starts with understanding IT's business strategies and then proposing projects to support those strategies. We access, prioritize, and rank proposals using BVI. Once the investment decisions are made, we commit resources based on the results of BVI and other qualitative data. The team builds monitors and indicators that track success at various stages of the project lifecycle.

We re-compute BVI on a regular basis to monitor the change in business value over time for the investments. Investment priorities can be adjusted based on the indicators and the BVI results. We track the actual results and benefits of the investment and measure them against the desired business results to provide feedback to the management process. This model's strength is that it enables continuous alignment between the IT investment portfolio and the dynamic business strategies and priorities. Actual results are measured against expectations set at the beginning of the investment. The model also enables data-driven decisions that may result in canceling projects or initiatives that fall short of expectations or no longer map to business priorities.

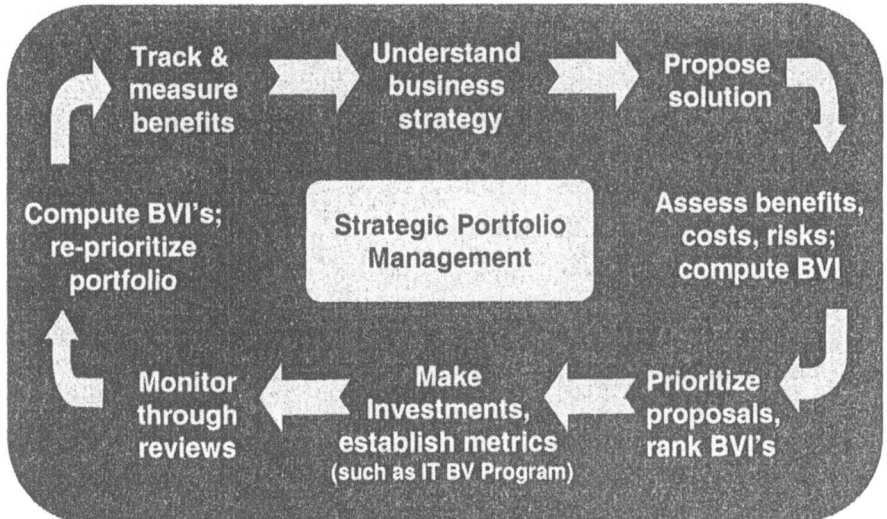

Figure 4. Ongoing portfolio management

5. RESULTS

BVI introduced a methodical, systematic approach to investment prioritization that supports data-driven decisions. Building on the BVI philosophy, we are implementing a strategic portfolio management framework that constitutes an integrated approach to investment decision-making and aims to identify and proactively manage a balanced investment portfolio mix. The framework encompasses common evaluation criteria (that is, BVI), investment categories and asset class definition, and visualization metrics.

BVI has assisted in decision-making activities over a wide variety of IT projects. For example, our Distributed Systems Management (DSM) team, People Systems group, IT Innovation organization, Customer Services group, and Global Infrastructure teams are using this tool to prioritize their project portfolios. We are using this approach to prioritize our IT Research and Development portfolio.

An example of a successful initiative prioritized through BVI is our Content Distribution System project, a peer-to-peer middleware services and knowledge management system. Using BVI, we prioritized the proof of concept (POC) for this project and gave it the go-ahead for development. Following the successful POC, this project has now emerged as a potential next-generation technology for IT.

Another example is our Electronic Content Distribution System (eCDS) project. We intercepted the development of eCDS before it was integrated into SKOOOL, a national educational portal in Ireland. The preliminary BVI evaluation proved accurate, resulting in a successful implementation. In parallel with SKOOOL, eCDS has also been integrated internally into an e-learning site, and externally at the University of Reading, as part of a pan-European, e-learning educational initiative.

5.1   BVI adoption: Example

The Telecom and Call Center Services (TCCS) group in our IT Global Infrastructure organization recently adopted BVI into their business process. TCCS is responsible for infrastructure projects funded by IT and for pay-per-view projects funded directly by IT customers. The group was looking for a method to help prioritize these diverse projects and make the right investment decisions. In light of the recent economic environment, prioritization became an even more important process in business management. TCCS turned to BVI as a potential tool to help solve their prioritization dilemmas.

A group composed of TCCS managers and project managers studied the BVI tool and started a pilot effort that evaluated whether BVI would fit their business requirements. The group selected a set of current and potential future projects for the pilot and used the BVI tool to conduct assessments.

As the pilot concluded, the group conducted a survey to gather feedback. The pilot team had found the tool valuable in helping them understand and assess the business impact of their projects and communicate that information more effectively to customers. The BVI tool also helped the team better prioritize their projects. The team commented that the time spent going through the process was well invested.

The group presented their findings to senior management, resulting in adoption of BVI as part of their ongoing business processes. Every project request is now associated with a BVI assessment before consideration for funding approval.

6.   CONCLUSION

BVI provides a common language and framework for discussing IT investments, assessing business value and IT efficiency contribution based on common criteria, and prioritizing diverse investments based on environment and IT strategy. The BVI process enables continued and proactive alignment of the IT project portfolio with corporate and IT business strategies.

Intel IT applies BVI to a variety of IT projects, which are then used by the ITBV metrics program. BVI creates an initial composite index to help gauge likely business benefits. The metrics program then looks more closely at the actual value delivered by a given project and assigns a real dollar value to the deliverables.

BVI is a decision support tool for IT investment decisions. It should not, however, be used as the only tool or factor in the decision making process. A thorough IT investment decision should be made based on the BVI results, the business drivers for different initiatives, the dependencies between projects (both within the group and between other groups), the impact to internal and external customers, the amount of investment, and other relevant business or environmental factors.

USING IT CONCEPT CARS TO ACCELERATE INNOVATION
Applied research and iterative concept development provide a vehicle for innovation and communicating a shared vision.

Cindy Pickering
Intel Corporation

Abstract: Innovation enables Information Technology to keep pace with the accelerating demand for computing power. To encourage innovation, Intel Corporation has implemented an initiative that cultivates innovation for innovation's sake. Emphasizing creative solutions rather than tangible results, Intel's IT Concept Car program encourages the "outside the box" thinking innovation requires. IT Concept Cars are based on the same premises and methods as automotive industry concept cars -- futuristic, novel, unique, and appealing vehicles used to try out new looks and capabilities. As in the automotive industry, an IT Concept Car is not necessarily fully functioning, but has just enough form and features to powerfully communicate the concept to the desired market segment and get a pulse on their interest and potential acceptance and/or sponsorship.

Key words: Information Technology; Concept Car; Innovation; Applied Research; Human Factors; idea; futuristic; vehicle; automotive industry; Usability

1. USING CONCEPT CARS TO DRIVE IT INNOVATION

In the auto industry, a "concept car" provides a gleaming view of innovation's potential. It does not represent linear progression instead it embodies a new vision beyond the next generation. The concept car's success is measured as much by the creativity it inspires as by the ultimate implementation of its details, which may significantly differ from the original concept.

Intel IT is taking steps to foster innovation as an adjunct to its mission to provide for the computing needs of Intel's more than 80,000 employees, by establishing its own Concept Car program based on methods similar to the automotive industry. Designed to promote and reward creative "outside the box" thinking, this Concept Car initiative encourages the same creative thinking within IT that drives Intel products to their position of industry leadership.

2.   IT CONCEPT CAR CHARACTERISTICS

An Intel IT concept car:
- Is novel and pleasantly surprising, and has immediately apparent future benefits.
- Provides a showcase for innovative market research and product development.
- Enables an experimental prototype or demo used to research consumer reaction, acceptance, and feedback.
- Signals a significant paradigm shift rather than an incremental change.
- Does not have to be fully functional.
- Integrates and may cleverly camouflage a collection of underlying technologies via intuitive, apparent human interfaces and blended environment.
An IT concept car differs from a research project in duration and focus:
- A concept car has a quick turnaround, often one quarter or less; funding and purchasing processes are streamlined. The emphasis is on the idea and the process.
- A research project is long-term, sometimes a year or longer, and follows a much more formal plan. It typically has a much broader scope, and emphasizes the result.

3.   TAKING THE IDEA TO REALITY

Several artifacts or outputs apply to IT concept Cars, and depending on the nature of the concept, not all of these will be created:
- Annotated sketch of idea to see if the concept makes sense
- Storyboards or more elaborate drawings and scripted descriptions to further define the concept
- Model or Mockup that enacts the story board
- POC Demo that is not fully functional but can be used to gather hands-on feedback / inputs

– Tradeshow Exhibit Quality Demo
  – posters, handouts, well-scripted demo
  – internal focus, small intimate settings
  – some well coordinated external venues
– Whitepaper collateral documenting lessons learned, usefulness, etc.
– Videos (web or VHS) that show use of the concept car, especially important for large physical concept cars that are not highly portable – but also serve as a permanent visual record

Limited production usage is not considered to be a concept car phase, but a concept car could transition to this, if successfully handed off to a technology development or implementation group.

The lifecycle for an IT Concept Car and how it maps to the artifacts above is described next.

3.1 Proposal creation and sponsorship

Anyone with a good idea can propose an IT concept car by submitting a completed IT concept car proposal form. In the proposal, the submitter briefly describes the concept, problem statement, benefits, usability testing plan, and can include the annotated sketch or illustration. The proposal also includes a bill of materials and the estimated cost and timeline. If a person with an idea does not want to build out his or her concept car to completion, a process is in place that seeks to find another person to work with the idea's originator—the prime objective is to encourage creative thinking, not to inhibit it with an overwhelming process.

3.2 Approval and funding

The proposal is reviewed and responded to within two weeks, either with approval, or with a request for changes and/or more information. A proposal might not get approval if the idea is not novel, or if the idea is out of scope of the Concept Car program, e.g., a limited production pilot.

3.3 Building out the Concept

Upon approval, the Concept Car owner begins the process of building out the concept via streamlined purchasing processes, and assisted by an internal contracting group if desired. This group has skilled resources for project management, storyboarding, design, engineering, development, and usability testing.

During build, the storyboarding, design, and development activities take place. Some concepts may involve hardware components that require some engineering to assemble them into a POC demo. In software oriented concepts, iterative design that progresses through storyboarding and interactive software mockups help to evolve the concept. Some concepts also include a physical setting or space.

3.4   Methods sharing, communication, and exposure

A concept car "community of practice", CoP, meets weekly to exchange concept car experiences, best known methods, prepare for upcoming events, and brainstorm new ideas. The program office has visibility to both internal and external opportunities for showcasing the Concept Cars and obtaining user feedback.

Example events where we have taken Concept Cars include:
- CTG Country Fair, an Intel internal event including posters and demos from Intel product and research groups, including IT
- Intel Developer Forum, an international external forum with an exhibit hall that includes Intel products and other commercial high tech vendors.
- IT Innovation Center, a world-wide collection of Intel IT spaces that include lab space, showcases for posters and demos, and digital conference room areas.

3.5   Usability Testing and User Feedback

By partnering with the IT Usability Design group, we have gained some expertise and common methods for usability testing across concept cars. The types of methods that we use include surveys, focus groups, interviews, observation, and heuristic analysis of a POC. Additionally, whenever we show a concept car at event, we get feedback from the audience on usefulness, usage scenarios, benefits, ideas for extensions and enhancements, and offers to test drive the concept at a later date. Following the event, the CoP discusses and documents this feedback in a post mortem.

3.6   Capturing and reporting key learning's

Concept car owners document what they learn so that others may benefit from their experience. The IT Research Core Team reviews concept car results and may approve further publication and additional follow-up research. Concept Car information is made available to Intel employees on an internal website; some will be presented the outside world in the form of white papers, demos, videos, and case studies. At least 2 Concept Cars have been captured as intellectual property candidates with 1 in the patent application process.

3.7   Completion and transitioning to the next phase

Concept cars are temporary. Some will be discarded after the study; others will be enhanced or even put into production. When the concept car is complete it is made available in virtual and physical "showrooms, such as the ITIC" so that others may experience the concept and benefit from its learning's. When we bring concept cars together in showrooms, we can also demonstrate how they can be used individually and together in different business scenarios.

4.   IT CONCEPT CAR EXAMPLES

4.1   Collaboration Vision

The Collaboration Vision was developed by the cross-functional Virtual Collaboration Research Team. For over 2 years they had been trying to paint a compelling powerful vision for virtual team collaboration, and when they decided to build a Concept Car, they found the vehicle to do it! The vision was featured at the 2003 IT Strategic Long Range Planning and Product Line Business Plan events where it inspired a 3 year program to drive towards the vision.

Unique characteristics of this concept car include the ability to see all "my" multi-team activities in one place, work without time and location boundaries, interact expressively with remote collaborators, and move effortlessly among applications and team spaces.

Usage scenarios include: Meet in real time across space and know who is participating, where they are and their time of day. Work asynchronously on a shared document or project tasks, while tracking progress. Coordinate responsibilities across projects and meetings, and manage diverse tasks.

Data from an Intel virtuality survey of 1264 employees supports the vision showing that: 2/3 employees multi-team (3-10 teams). Mobility and lack of shared practices and structure negatively affect performance while performance correlates positively to sociability, expressivity, and structure.

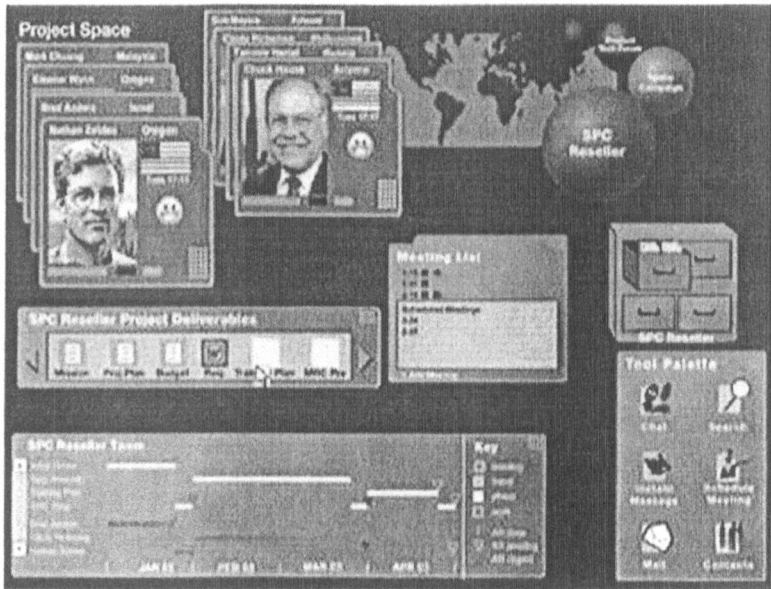

Figure 1. Virtual Collaboration Vision Concept Car

4.2   Senseboard

Senseboard is an Intel IT Research "Concept Car," identified through IT's partnership with MIT Media Lab's Tangible Media Group (Professors Hiroshi Ishii and Rob Jacobs). It is a special purpose electronic whiteboard that uses a sticky note metaphor, enabling users to physically manipulate and organize ideas while freely brainstorming and creating project task maps.

Figure 2. Senseboard: Electronic Sticky Note Whiteboarding

The primary features of Senseboard include: tactile interaction / involvement, saving session output electronically for future reference and remote participation in the group activity using a web interface that also supports audio/d video "social" interaction. Our research had the goal to explore usability and usefulness of this collaboration environment in both virtual and co-located team settings. When we took Senseboard to the Spring 2002 Intel Developers Forum, it inevitably elicited several new, beneficial usage models and scenarios in business, academic, and even home domains.

4.3   BACCRoom

The Intel BACC Room is a collaborative environment based on research and technology developed at the Center for Integrated Facility Engineering (CIFE) located at Stanford University, using multiple connected (most commonly 3) SmartBoard™ systems. An example for this concept helps illustrate its value: During review of major construction projects, status is tracked for many things - cost to date, cash flow, schedule status, facility status, etc. The Intel BACC Room facilitates this by showing many of these elements and their relationship over time! Suppose you have three BACC Room systems, with the project schedule on system 1, the current 3d model view of the facility on system 2, while system three shows the cost of the project to date. BACC Room adds the "glue" to synchronize the 3 views over time. As you move to a new date on the schedule, systems 2 and 3 auto-magically adjust to display the associated completion level for the 3d model view and cost for that new date in time.

Figure 3. Intel BACC Room Collaboration Environment

Additional uses for the Intel BACC Room are:
−   Outline work review, design review
−   Training
−   Maintenance
−   Remote scenarios for all the above!

4.4   Intelligent Dialog Interfaces

Intelligent Dialog Interfaces is a project to develop applications of speech interfaces and conversational systems in order to (1) assess the maturity of these technologies as natural user interfaces to computing; (2) assess ROI of conversational interfaces to prototype applications and report to the IT community; and (3) identify the most pressing research questions based on prototype deployment and inform the research community. The project is funded by the IT R&D Council and is in collaboration with Intel funded researchers at the MIT Media Lab and the Spoken Language Systems group at the Laboratory of Computer Science. Two application domains will be explored in the first year: services for Intel manufacturing technicians and Intel employee services.

Figure 4. Mobile Dialog Interface for Technician in the Factory

4.5   Instant Meeting Server

Attendees at business meetings held at off-site facilities frequently feel out of touch with the office during the meetings. The hotel may provide a wired broadband connection, but unless IT support staff is provided, few presenters could manage the tangle of network hubs and cable required to provide wired connections to all attendees. While some hotels may provide even wireless LAN connectivity, typically each attendee is required to go through a web based authorization process and is isolated from other attendees while on the network. Instant Meeting Server provides easy set-up of a self-contained, portable wireless LAN for meeting attendees to share a broadband internet connection and local resources such as a printer and a notebook-based server providing shared document folders. Figure 5 shows the network configuration. Figure 6 shows the portable case. It has been used at over 8 offsite seminars, conferences, and/or training events. Figure 7 shows an example of use as an Internet Café during breaks at a CIO conference.

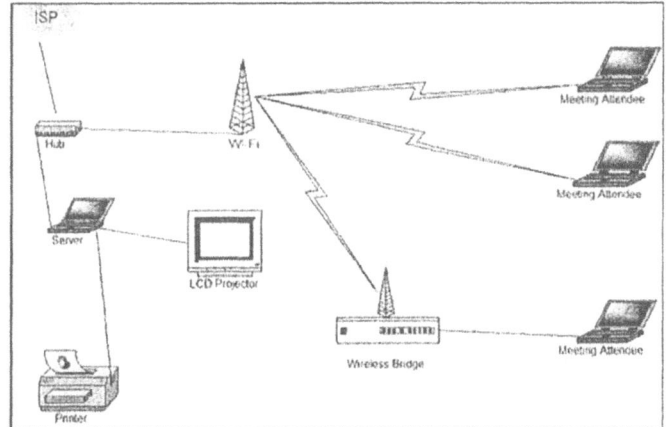

Figure 5. Instant Meeting Server WLAN and Shared Resources

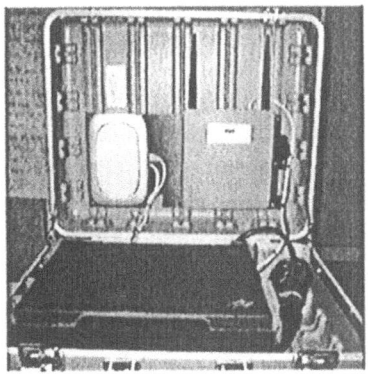

Figure 6. Instant Meeting Server Portable Suitcase

Figure 7. Internet Cafe using Instant Meeting Server

5.   LESSONS LEARNED

Through our research experiences, we have discovered that building IT Concept Cars help us to:
- Leverage longer term investments by harvesting and applying research from universities
- Increase our understanding of the desired end user experience and emerging usage models by applying design for usability principles and common processes for capturing user feedback across all concept cars
- Connect people to identify new ideas and synergies among IT concept cars through our weekly "community of practice" forum
- Tell our story through demos and exhibits, generating valuable feedback and more ideas

6.   CONCLUSION

Some IT concept cars will evolve into functional, productivity enhancing tools; others will inspire further innovation in areas that may not have otherwise been considered. Regardless of their ultimate fate, the process of conceiving and implementing these concept cars energizes the creative spirit within Intel, enabling the innovation that will drive IT into a new paradigm: delivering proactive solutions that anticipate needs before they arise.

SOFTWARE PATENTS: INNOVATION OR LITIGATION?

Linda Levine[1] and Kurt M. Saunders[2]

[1]*Software Engineering Institute, Carnegie Mellon University, USA;* [2]*College of Business and Economics, California State University, USA*

Abstract: The Directive on the Patentability of Computer-Implemented Inventions recently approved by the European Parliament may have significant implications for the software industry, public policy and patent protection. In this paper, we summarize the scope of patent protection in the European Union, the United States, and Japan. We examine the patentability of computer software under E.U. and U.S. patent law and also consider two instances of software patenting and their effects. We provide an overview of the Directive and finally assess the legal, economic, and public policy implications for software developers and users.

Key words: software patents, scope of patent protection, patentability of computer-implemented inventions

1. INTRODUCTION

The Directive on the Patentability of Computer-Implemented Inventions recently approved by the European Parliament may have long range implications for the software industry, public policy and patent protection for years to come. Patents can be, at once, a spur and a roadblock to innovation. At the same time, the world is moving toward greater economic integration, driving current debate on the harmonization of intellectual property rights.

In this paper, we begin by summarizing the scope of patent protection in the European Union, the United States, and Japan. In doing so, we examine the patentability of computer software as inventions allowed under E.U. and U.S. patent law. The value of software patents and the U.S. experience with patent liberalization for 20 years is considered. Two instances of software patenting and their effects are discussed. We then provide an overview of the proposed Directive recently approved by the European Parliament. The approval also included several amendments. Finally, we assess the legal, economic, and public policy implications of the Directive for software developers and users.

We do not attempt a comprehensive examination of the utility and rationale for patent protection. This is beyond the scope of our paper. Ours is also not an empirical research study; rather, our investigation is more contemporary and topical, looking at the conditions surrounding software patentability as a key barrier and enabler for innovation and competition.

2.   PATENT LAW AND COMPUTER SOFTWARE IN THE EUROPEAN UNION

The European Patent Convention (EPC) serves as the basis for a harmonized system of patent protection for all members of the European Union. European patents have the same effect as patents granted by each nation under its own national patent laws. Article 52(1) of the EPC states: "European patents shall be granted for any inventions which are susceptible of industrial application, which are new and which involve an inventive step." Article 52(2)(c) of the EPC specifically excludes "methods for ... doing business, and programs for computers" from the definition of inventions eligible for patent protection.

Although Article 52(2)(c) appears to completely exclude computer programs and business methods from patentability, Article 52(3) qualifies this exclusion by stating that the exclusion only applies to the extent an invention "relates to such subject matter or activities as such." This suggests that while computer programs standing alone are unpatentable, the presence of a computer program as one component of a larger invention does not prevent the invention from qualifying for a patent. The European Patent Office (EPO) has likewise taken this position in its *Guidelines of the European Patent Office*, which explains: "A computer program claimed by itself or as a record on a carrier, is not patentable irrespective of its content. ... If however the subject-matter as claimed makes a technical contribution to the known art, patentability should not be denied merely on the ground that a computer program is involved."

The leading case of the EPO Technical Board of Appeal involving software patentability is *Computer-related Invention/VICOM* (1986), which involved a program for digitally processing images. The Board held that the program was patentable since it was related to a technical process leading to a change in the physical images and was not merely a mathematical algorithm that manipulated numbers to calculate a purely numerical result. As to computer-implemented business methods, the key decision was *Pension Benefit Systems* (2001), where the application claimed a method for controlling a pension benefits program. The Technical Board of Appeal affirmed that the method was unpatentable, stating that "methods only involving economic concepts and practices of doing business are not [patentable] inventions."

3. THE SCOPE OF PROTECTION FOR SOFTWARE IN THE UNITED STATES AND JAPAN

Under section 101 of the U.S. Patent Act, subject matter that may be patented includes any "process, machine, manufacture, or composition of matter, or any ... useful improvement thereof." In addition, the invention must be useful, novel, and nonobvious to qualify for a patent. To be novel and nonobvious, the invention must be new in that it is not part of the field of existing technology or "prior art," and it must represent an inventive next step beyond the prior art, rather than an obvious variation.

Before the 1980s, it was generally accepted in the U.S. that software was not patentable. However, in 1981 the U.S. Supreme Court decided the case of *Diamond v. Diehr*, which involved a computer program, applied to a process for curing rubber, and expressly held for the first time that computer software was patentable so long as the claimed invention was not merely a procedure for solving a mathematical formula.

In 1998, the U.S. Patent & Trademark Office (USPTO) issued its *Examination Guidelines for Computer-Related Inventions* for its examiners to use in evaluating the patentability of software. Essentially, the USPTO Guidelines make clear that any type of software or computer-implemented invention is eligible for patent protection if the other tests for patentability – utility, novelty, and nonobviousness – are also met. The Japanese Patent Office (JPO) has adopted a similar position in its *Guidelines for Computer Software Related Inventions*. Under the JPO Guidelines, computer programs are patentable in Japan as long as they possess a high degree of technical creativity and utility.

Also in 1998, the U.S. Court of Appeals for the Federal Circuit, in the case of *State Street Bank & Trust Co. v. Signature Financial Group, Inc.* (1998), upheld as valid a patent directed to a computer-implemented business method designed to perform financial calculations and data-processing for mutual fund investments. This case was important because prior to this decision, it was widely believed that business methods and systems were not patentable. The *State Street* case made clear that business methods were to be evaluated in the same manner as any other type of process. In 2000, the JPO followed suit and revised its Guidelines to allow for the patenting of computer-implemented business methods when there is clear "involvement of inventive step."

Since 1976, the USPTO has granted a steadily increasing number of patents for software-related inventions (see Appendix). Presently, 15% of all patents granted in the U.S. are software patents and the growth in software patents accounts for over 25% of the total growth in the number of patents issued between 1976 and 2001. Moreover, one recent study has reported that: "Overall, software patents are more likely to be obtained by larger firms, established firms, U.S. firms, and firms in manufacturing (and IBM); they are less likely to be obtained by individuals, small firms, newly public firms, foreign firms, and software publishers" (Bessen & Hunt, 2003, p.9).

Bessen and Hunt found that the correlation between R&D and patenting in the U.S. over time has been significantly *negative.* In other words, as software patenting rates have risen, R&D investment in sectors using information technologies has declined. This is not to say that other confounding factors may not have contributed to this phenomenon. We comment more extensively on the consequences of patent liberalization in the Reassessment section.

4.   THE COMPLEX REALITY OF SOFTWARE PATENTS: TWO EXAMPLES

In this section, we examine two instances of software patenting and their effects. One of the patents led to litigation and one led to settlement. Both illustrate some of the unforeseen consequences that follow a decision to obtain a patent on software and then to enforce it.

4.1   "One Click" Shopping

The recent case of *Amazon.Com, Inc. v. Barnesandnoble.Com, Inc.*, No. 00-1109 (Fed. Cir. 2001) demonstrates the real-world implications of patenting software and business methods. The case involved U.S. Patent No. 5,960,411 ("the '411 patent"), which was issued to the inventor/programmer on September 28, 1999, and later assigned to Amazon. On October 21, 1999, Amazon sued BarnesandNoble.com (BN) alleging infringement of the patent.

Amazon's patent was directed to a method and system for "single action" ordering of items in a client/server environment such as the Internet, known as the "One-Click" shopping model. The '411 patent described an approach in which a consumer could purchase items via an electronic network using only a "single action," such as the click of a computer mouse button. Amazon developed the patent to deal with what it considered to be the frustrations presented by the "shopping cart model" for online purchasing. This method is described in the following excerpt from the '411 patent:

1. A method of placing an order for an item comprising:
under control of a client system, displaying information identifying the item; and
in response to only a single action being performed, sending a request to order the item along with an identifier of a purchaser of the item to a server system;
under control of a single-action ordering component of the server system, receiving the request;
retrieving additional information previously stored for the purchaser identified by the identifier in the received request; and
generating an order to purchase the requested item for the purchaser identified by the identifier in the received request using the retrieved additional information; and
fulfilling the generated order to complete purchase of the item whereby the item is ordered without using a shopping cart ordering model.
2. The method of claim 1 wherein the displaying of information includes displaying information indicating the single action.

Ultimately, the court ruled that Amazon demonstrated that it would likely succeed in its infringement suit. However, the court also observed that BN had raised substantial questions as to the validity of the '411 patent based on obviousness. As such, the court returned the case to the district court for further review of the patent and the alleged infringement by BN.

4.2   Graphical Interchange Format (GIF)

A second case concerns the patent award and alleged infringement pertaining to the LZW encoding method that underlies the .GIF document format, which is used widely for graphic images. This infringement suit was unsuccessful and eventually withdrawn.

GIF images are compressed to reduce file size, using a technique called LZW (after Lempel-Ziv-Welch). The technique was initially described by Welch in *IEEE Computer*, June 1984. Unisys holds a patent on the technique described in the article, but the article describing the algorithm made no mention of this. The LZW procedure quickly became a popular technique for data compression because it provided economical, high performance, adaptable, and reversible data compression. Likewise, GIF became a standard in its field. Apparently, neither CompuServe, the programmer who designed GIF, nor the software community were aware of the patent. Claim 1 of the patent reads as follows:

> 1. In a data compression and data decompression system, compression apparatus for compressing a stream of data character signals into a compressed stream of code signals, said compression apparatus comprising
> storage means for storing strings of data character signals encountered in said stream of data character signals, said stored strings having code signals associated therewith, respectively,
> means for searching said stream of data character signals by comparing said stream to said stored strings to determine the longest match therewith,
> means for inserting into said storage means, for storage therein, an extended string comprising said longest match with said stream of data character signals extended by the next data character signal following said longest match,
> means for assigning a code signal corresponding to said stored extended string, and
> means for providing the code signal associated with said longest match so as to provide said compressed stream of code signals.

CompuServe released GIF as a free and open specification in 1987. GIF soon became a world standard and played an important role in the Internet community. In December 1994, CompuServe and Unisys suddenly announced that developers would have to pay a license fee in order to use this technology that had been patented by Unisys. This caused immediate confusion.

The original licensing agreement text which had upset so many was soon followed by clarifications from CompuServe and Unisys. Unisys faced a delicate challenge and risked alienating a large segment of the software community. While having the right to file suit against LZW users, Unisys has been accommodating and fair. Most likely, the success of LZW and its widespread use caught Unisys off guard. It is difficult to understand how else Unisys could first allow a large number of developers to use LZW for for free for years and then, after the establishment of de facto standards based on LZW, abruptly change its attitude.

5. DIRECTIVE ON THE PATENTABILITY OF COMPUTER- IMPLEMENTED INVENTIONS

The primary requirement of the Directive is that, "in order to be patentable, an invention that is implemented through the execution of software or on a computer, or similar apparatus has to make a contribution in a technical field that is not obvious to a person of normal skill in that field." A computer-implemented invention is defined as "any invention implemented on a computer or similar apparatus which is realized by a computer program." (Proposal for a Directive, 2002, p.13) Under the current situation, inventors have two avenues for obtaining protection for inventions. They may (1) apply for patents at the European Patent Office (EPO) under the auspices of the European Patent Convention, or (2) seek patents through the national patent offices in the Member States. Practically speaking, this scenario has resulted in differences of interpretation and claims that "there is no unifying structure with binding effect on national courts." Thus, the Directive represents a call for harmonization. (www.europa.eu.int/comm/internal_market/en/indprop/comp/0232.htm)

The formulation of the Directive has fueled a heated debate on the limits of what is patentable—and the benefits of software patenting—both currently and for the future. Detractors argue that the problems of strategic patenting are only now gaining attention in the U.S. and they warn that "software patents damage innovation by raising costs and uncertainties in assembling the many components needed for complex computer programs and constraining the speed and effectiveness of innovation" (Open letter, 2003). Clearly, however, some action is inevitable and required: "over the years, national courts have decided that there is no reason why a patent should not be granted for a machine programmed to carry out some technical function, or a technical process carried out using a computer or similar machine." As a result, since the EPC came into force in 1978, more than 30,000 software patents have been granted and a body of case law has been built up by the EPO and Member States' courts.

Those supporting the Directive, including Arlene McCarthy, JURI's rapporteur and the European Parliament member responsible for the draft legislation, maintain that nothing will be made patentable which is not already patentable. Moreover, there are "links between the patentability of computer-related inventions and the growth of IT industries in the United States" (Broersma, August 27, 2003). Opponents to the Directive have responded that economic research does not support such links between software patents and business growth. They argue that patents are harmful in casting "in concrete the so-powerful oligopolies that naturally emerge in information-based industries" and that in the "United States, where such patents are allowed, large corporations such as IBM routinely stockpile patents to be used against competitors--usually to the detriment of smaller companies" (Broersma, April 28 & August 28, 2003). Others have long argued that patents can be used to delay or stifle innovation through the use of patent pools, patent thickets, exclusive licensing, and other abusive patent enforcement tactics (Rivette & Kline, 1999).

The authors of the Critique perceive significant uncertainties surround the scope, reach, and effect of software patents and limits on the ability of small companies to grow in competition with large companies. The Fall of 2003 saw demonstrations, protests, a call for action petition, and an open letter and critique written by a high ranking group of scientists. On September 15, 2003, the European Parliament voted to approve the Directive. However, this approval also included several amendments. The amendments outlaw patents for algorithms, make it impossible to register business-method patents, and restrict the definition of the sort of software that should be eligible for a patent. Supporters of software patents argue that the amendments could deprive existing patent holders of their ability to enforce their property rights and that such a law may violate international intellectual property agreements. Experts predict that the European Commission is likely to be displeased with the adjustments and that ministers of the 15 national governments of the union are likely to challenge the amendments to the Directive.

6.   THE VALUE OF SOFTWARE PATENTS

Those who argue in favor of software patents identify five types of public benefits that can be realized:

Encouraging inventive efforts through the promise of economic rewards to inventors. Patents for new software are likely to encourage increased discovery and development of new types of useful software. Patent rewards can overcome "free rider" risks. Free riding slows development of new software because these efforts involve considerable financial outlays that may not produce accompanying returns if others can freely use the resulting product (Gruner, 2000, pp. 999-1007).

Promoting public disclosures of useful inventions through issued patents. In exchange for the public disclosure of an invention in a published patent, an inventor is given the reward of exclusive rights in the invention for 20 years. Disclosure makes the availability of the invention, under sale or license, known to those who may have a use for it. Disclosure encourages others to consider making improvements or other substitutes. Most important, public disclosure can avoid unnecessary duplication of efforts in creating something that already exists (Gruner, 2000, pp. 1007-12).

Providing incentives for product refinement and commercialization. Substantial post-invention refinement may be needed to begin mass production, distribution, and marketing of a new invention. Absent patent rights, a potential developer may be reluctant to be the first to take on the production problems and the marketing costs for a new product. Patent protection may be needed to convince investors to back start-up companies in the development and marketing of new software. Subsequent developers' costs will be lower if they can gain some benefit from the first mover's product introduction efforts (Gruner, 2000, pp. 1012-20).

Encouraging Efficient Prospecting for Applications and Improvements. Patents may encourage software developers to search for further applications of the patented software in the same field in which the software was originally developed, or promote cross-domain searching for new applications in other fields. (Gruner, 2000, pp.1019-23)

Limiting duplicative efforts to discover, perfect, and improve patented inventions, thereby maximizing society's net gain from each patented invention. Software patents may prevent wasteful investments. Following a software developer's patenting of a particular advance, others working to develop similar or less effective software will be averted from further wasteful or expensive efforts. (Gruner, 2000, pp.1023-28)

Finally, we might ask whether alternative forms of intellectual property protection would provide the same public benefits. Those who argue that patents are preferable often compare them to copyright or trade secret protection. As to whether copyright protection is preferable to patent protection, we must remember that patents protect functional and utilitarian features of software that cannot be protected by copyright, which protects only descriptive or expressive aspects (Gruner, 2000, p. 994) In addition, unlike trade secret protection, patents provide rights that prevent reverse engineering and reuse of software (Gruner, 2000, p.995)

7.   A REASSESSMENT

How is the European Union to proceed? While software patents may stymie innovation especially in small companies, and while the economic benefit is not resolved, some kind of harmonization appears necessary. Is the backlash alarmist given that "nothing will be made patentable which is not already patentable" or are the detractors wise to observe that this formalization of existing practice is dangerous?

We find limitations to the persuasiveness of the current arguments on both sides. Those who support the Directive warn of possible future conflicts in interpretation and application of patent law. But there is no hard evidence to corroborate this concern. Nor is there sufficient evidence and justification for whether the solution—the Directive as it is currently formulated—will remedy the problem. Those opposed to the Directive have taken an extreme, even Draconian position, arguing that this remedy will have dire consequences. They claim the Directive will spawn new problems for R&D and competition, framing slogans such as: "patent inflation is not a victimless crime" and "software patents kill innovation."

Some have argued that patents on software may actually slow innovation by making it more costly and difficult to advance or build on existing information technology. This position draws on the "tragedy of the anticommons", which theorizes that an over-assignment of property rights for a privately-held resource leads to under-utilization of that resource. Preventing an anti-commons tragedy "requires overcoming transaction costs, strategic behaviors, and cognitive biases of participants, with success more likely within close-knit communities than among hostile strangers" (Heller & Eisenberg, 1998, p. 280).

On the other hand, an example of where the cautionary reaction may be well founded concerns the patentability of pure business methods. Much controversy has surrounded this issue in the U.S. as to its effects on software development. In the EU, these will not be patentable; however, "some inventions involving business methods could fall within the definition of computer-implemented inventions. These inventions would be dealt with in accordance with the Directive, and in particular patents would only be granted for inventions that made a technical contribution" (Proposal FAQ p.5). Critics rightfully anticipate a natural progression of decisions that will follow, leading to more patents for business methods.

Critics have continued to argue for reforms to the patent system. For example, Gleick (2000) claims "that in the US, the patent office has grown entangled in philosophical confusion of its own making; it has become a ferocious generator of litigation; and many technologists believe that it has begun to choke the very innovation it was meant to nourish" (Gleick, 2000).

One compromise solution would be to limit the term for software patents. Product life cycles in software are very short: approximately 3 years. What is seen as an invention at inception, can by virtue of adoption be mainstream within six months. On the other hand, a patent protects an invention for 20 years, which is clearly disproportionate. Likewise, Jeff Bezos of Amazon has suggested that a patent term of 3-5 years is more realistic and fair.

Moreover, the Directive has exposed ideological differences between two competing paradigms: Open Source versus pro-intellectual property rights.

The Directive has alienated the Open Source community and the Euro Linux alliance—over 200 commercial software publishers and European non profit associations with the goal of promoting and protecting the use of Open Standards, Open Competition, and Open Source Software.

Regardless of the stated positions, the world is moving toward globalization and greater integration of trade, commerce, and intellectual property rights: "Finding the right balance will not be easy. Patents can be a spur to innovation, but they can also be an obstacle, and the great advantage of digital technology was supposed to be its very malleability. Moreover, there is another headache. The harder it is to patent computer-related inventions in Europe, the wider will be the legal gap with America" (Economist, 2003). In future, the U.S., EU, Japan, China, and the developing world will need to compromise and take steps in the direction of one another to find common ground.

Much remains to be understood about software patentability. Future directions for research might consider a range of empirical and economic issues. Specifically, we have yet to understand the relationship between the number of software patents awarded versus the number of patents litigated. We might discover, in fact, that most software patents are not the cause or source of litigation.

REFERENCES

Battilana, Michael C. (2003). The GIF Controversy: A Software Developer's Perspective. http://cloanto.com/users/mcb/19950127giflzw.html

Bessen, J.& Hunt, R.M. (2003). An empirical look at software patents, (*MIT and the Federal Reserve Bank of Philadelphia: working paper*, 2003, pp.7, 9, 31) http://www.researchoninnovation.org/swpat.pdf

Broersma, Matthew (2003). Developers gather to protest patents. *CNET News.com*, Aug. 28, 2003. http://news.com.com/2102-1008_3-5069279.html?tag=ni_print

Broersma, Matthew. (2003). EU rapped over software plan. *CNET News.com*, Aug. 27, 2003. http://news.com.com/2102-1008_3-5068842.html?tag=ni_print

Broersma, Matthew (2003). Patent battle to culminate in Brussels. *CNET News.com*, August 26, 2003. http://news.com.com/2100-1012_3-5068007.html

Broersma, Matthew (2003). Software patent vote delayed. *CNET News.com*, June 30, 2003. http://news.com.com/2100-1012_3-1022181.html?tag=rn

Broersma, Matthew. (2003). Scientists protest EU software patents. *CNET News.com*, April 28, 2003. http://news.com.com/2102-1012_3-998547.html?tag=ni_print

Burgunder, Lee. (2003). *Legal Aspects of Managing Technology.* Mason, Ohio: West Legal Studies in Business, pp.173-203.

Computer-related Invention/VICOM, T 208/84, 1987 OJ EPO 14 (1986).

A Critique of the Rapporteur's Explanatory Statement accompanying the JURI Report to the European Parliament on the proposed Directive on the Patentability of Computer-Implemented Inventions. (August 2003). Paul A. David, Oxford Internet Institute and

Stanford University; Bronwyn H. Hall, University of California, Berkeley and Scuola Sant'anna Superiore Pisa; Brian Kahin, University of Michigan; W. Edward Steinmueller, Science and Technology Policy Research Unit, University of Sussex, August 2003 http://www.researchineurope.org/policy/critique.htm

Diamond v. Diehr, 450 U.S. 175 (1981).

The Economist. (2003). A clicking bomb: An explosive row over how to protect intellectual property in Europe. September 4, 2003. http://www.economist.com/business/displayStory.cfm?story_id=2043416

Examination Guidelines for Computer-Related Inventions, (1998) U.S. Patent & Trademark Office. http://www.uspto.gov

Final Report: The Results of the European Commission Consultation Exercise on the Patentability of Computer Implemented Inventions (PbT Consultants). http://www.europa.eu.int/comm/internal_market/en/indprop/comp/studyintro.htm

Dinwoodie, Graeme B., Hennessey, William O., & Perlmutter, Shira. (2002). *International and comparative patent law*. Newark, New Jersey: LexisNexis Group, pp. 89-115.

Gleick, James. (2000). Patently absurd. *New York Times Magazine*, March 12, 2000.http://www.nytimes.com/library/magazine/home/20000312mag-patents.html

Gratton, Eloise, (2003). Should patent protection be considered for computer-related innovations? *Computer Law Review & Technology Journal 7*, 223-253.

Gruner, Richard S.(2000). Better living through software: Promoting information processing advances through patent incentives, *St. John's Law Review 74*, 977-1068. *Guidelines for computer-software related inventions*, (2000).Japanese Patent Office. http://www.jpo.go.jp

Heller, Michael A. (1998) The tragedy of the anticommons: Property in the transition from Marx to markets, *Harvard Law Review 111*, 621-688.

Heller, Michael, A. & Eisenberg, Rebecca S. (1998) Can patents deter innovation? The anticommons in Biomedical Research, Science 280 (5364), 698-701.

Japanese Patent Law, articles 2 & 29. http://www.jpo.go.jp

League of Programming Freedom. http://www.lpf.ai.mit.edu/

An Open Letter to the European Parliament Concerning the Proposed Directive on the Patentability of Computer-Implemented Inventions (25 August 2003). Birgitte Andersen, Birkbeck, University of London; Paul A. David, Oxford Internet Institute and Stanford University; Lee N. Davis, Copenhagen Business School; Giovanni Dosi, Scuola Sant'anna Superiore; David Encaoua, Université Paris I; Dominique Foray, IMRI Université Dauphine; Alfonso Gambardella, Scuola Sant'anna Superiore; Aldo Geuna, SPRU, University of Sussex; Bronwyn H. Hall, University of California, Berkeley and Scuola Sant'anna Superiore; Dietmar Harhoff, Ludwig-Maxmiliens Universitaet; Peter Holmes, SEI, University of Sussex; Luc Soete, MERIT, University of Maastricht; W. Edward Steinmueller, SPRU, University of Sussex. http://www.researchineurope.org/policy/patentdirltr.htm#_ftn1

Pension Benefit Systems, 2001 OJ EPO 413 (2001).

Proposal for a Directive of the European Parliament and of the Council on the Patentability of Computer-Implemented Inventions. Brussels, 20.02.2002. COM(2002)92 final. 2002/0047 (COD)

Proposal for a Directive of the European Parliament and of the Council on the Patentability of Computer-Implemented Inventions-frequently asked questions. (FAQ) http://www.europa.eu.int/comm/internal_market/en/indprop/comp/02-32.htm

Rivette, Kevin & Kline, David. (1999*). Rembrandts in the attic: Unlocking the hidden value of patents*. Cambridge, Mass: Harvard Business School Publishing.

Software Patent Institute - http://www.spi.org/primintr.htm,

State Street Bank & Trust Co. v. Signature Financial Group, Inc., 149 F.3d 1368 (Fed. Cir. 1998).
Tocups, Nora M., & O'Connell, Robert J. (1997). Patent protection for computer software. *Computer Law 14*(11), 18-23.
United States Code, title 35, sections 101, 102, 103, & 271.

APPENDIX

Table 1A. Number of Software Patents Granted in the U.S.[*]

Year	Software Patents	Aharonian Estimates	Other Utility Patents	Software/Total
1976	766	100	69,460	1.1%
1977	885	100	64,384	1.4%
1978	902	150	65,200	1.4%
1979	800	200	48,054	1.6%
1980	1,080	250	60,739	1.7%
1981	1,281	300	64,490	1.9%
1982	1,404	300	56,484	2.4%
1983	1,444	350	55,416	2.5%
1984	1,941	400	65,259	2.9%
1985	2,460	500	69,201	3.4%
1986	2,666	600	68,194	3.8%
1987	3,549	800	79,403	4.3%
1988	3,507	800	74,417	4.5%
1989	5,002	1,600	90,535	5.2%
1990	4,738	1,300	85,626	5.2%
1991	5,401	1,500	91,112	5.6%
1992	5,938	1,624	91,506	6.1%
1993	6,902	2,400	91,440	7.0%
1994	8,183	4,569	93,493	8.0%
1995	9,186	6,142	92,233	9.1%
1996	11,664	9,000	97,981	10.6%
1997	12,810	13,000	99,173	11.4%
1998	20,411	17,500	127,108	13.8%
1999	21,770	21,000	131,716	14.2%
2000	23,141	--	134,454	14.7%
2001	25,973	--	140,185	15.6%

[*] Source: J. Bessen & R. M. Hunt, "An Empirical Look at Software Patents," p. 31 (MIT and the Federal Reserve Bank of Philadelphia: working paper, 2003) http://www.researchoninnovation.org/swpat.pdf

PART IV

INNOVATION ADOPTION

TEMPORAL DISCLOSEDNESS OF INNOVATIONS
Understanding Innovation Trajectories in Information Infrastructures

Edoardo Jacucci
Department of Informatics, University of Oslo, Norway

Abstract: The paper addresses the research question of how to analyze and evaluate innovations in complex information infrastructures. Recent studies of innovation processes involving standard-based networking technologies have called for new theoretical insights than the ones provided by traditional studies on Diffusion Of Innovations. Understanding this type of innovations becomes particularly difficult in the context of complex information infrastructures. Based on the empirical material from a case study of an IS innovation in a Norwegian hospital, the paper provides insights on temporal aspects of innovations. The paper makes two key contributions to the ongoing discussion. It introduces the concept of *temporal disclosedness*, which describes the tendency of innovations to disclose along the past, present, and future trajectory of the infrastructure, breaking traditional analytical and temporal boundaries. Secondly, with the help of this concept, it shows how some of the current theoretical approaches may be adequate to study the *disclosedness* of innovations, but their use is currently limited and hence should be extended.

Key words: Innovation, Time, Information Infrastructure

1. INTRODUCTION

This paper is concerned with the conceptualization of innovation in information systems (IS) (Kwon & Zmud 1987; Cooper & Zmud 1990; Baskerville & Pries-Heje 1997; McMaster *et al.* 1997). Innovation here refers to the process of implementation of a new idea in an organizational setting (Van de Ven 1986). In particular, the paper addresses the question of how innovations implemented in complex infrastructures of interconnected IS should be analyzed and evaluated (Hanseth *et al.* 1996; Damsgaard & Lyytinen 1997). Recent discussions have raised the need to find new theoretical approaches to understand the complexity of such innovation processes (McMaster *et al.* 1997; Lyytinen & Damsgaard 2001). Examples of the complexity can be found in the difficulty in evaluating the success or failure of an infrastructural innovation; for instance, the difficulty to temporally and spatially isolate it from the surrounding context. When implemented, the type of innovations at case become largely dependent from the existing context, are often subject to frequent re-planning, and in some cases never seem to end. As a response, this paper aims to contribute to the ongoing discussion by focusing the attention on the study of temporal aspects of innovations. It investigates how innovations relate to the context before, during, and after its complete or incomplete implementation. The paper also discusses how approaches to the study of innovations have thus far come short in addressing in particular the influence of innovations on the future trajectory of the surrounding infrastructure.

Analyzing a case study on an innovation implemented in a large Norwegian hospital, the paper illustrates how understanding these temporal aspects is paramount to the study of the innovation. More specifically, the case deals with the implementation process of an Electronic Patient Record (EPR) system in a Norwegian hospital. The case shows how, in order to understand the dynamics of the innovation, the analysis must include the study of the context (the surrounding infrastructure) in the time before and during the implementation. Additionally, the analysis must investigate the influence of the implemented innovation on the generation of future innovations.

Based on the empirical evidence of the case, and by introducing the concept of *temporal disclosedness*, the paper aims to illustrate how innovations in complex information infrastructures tend to break analytical, temporal and spatial boundaries, and *disclose* along the entire trajectory of the infrastructure. Hence, the paper suggests that evaluations of innovations in complex information infrastructures take into consideration this particular property by extending their analysis beyond apparent temporal and spatial boundaries. The analysis of the case presented in the paper provides a practical example.

The paper is structured as follows. First, a reasoned review of recent literature on the study of innovations in IS is provided. Secondly, the methodology of the study conducted is illustrated. Thirdly, the case description will provide the empirical evidence for the subsequent discussion. Finally, conclusions are drawn summarizing the main message of the paper.

2. THEORETICAL FRAMEWORK

This paper aims to contribute to the ongoing discussion on how to conceptualize innovation in the context of Information Systems (IS) implementation (Kwon & Zmud 1987; Cooper & Zmud 1990; Baskerville & Pries-Heje 1997; McMaster *et al.* 1997). In particular, similarly to the work by Lyytinen & Damsgaard (2001) it is concerned with innovations involving complex standard-based networked technologies in large organizations. Additionally, it adopts the perspective of technology as suggested by Hanseth *et al.* (1996) whereby technology is not seen as isolated from its social-technical context. Accordingly, IS innovation can not be considered as the implementation of isolated pieces of technology: it is rather a process of continuous negotiation and re-appropriation of the existing installed base (Hanseth *et al.* 1996; Monteiro & Hepsø 1998).

The paper will try to focus the attention of the current discussion on aspects of temporality in innovations by addressing the following research question: how does the innovation relate to the past, present, and future of its lifecycle and to the surrounding infrastructure in which it is implemented?

To address this question, we will first identify two main streams of theoretical works in the field. Then we will briefly discuss their temporal conceptualization of the innovation. Finally, we will provide the background and motivation to elaborate the theoretical contribution in the discussion section.

2.1 A review of research in IS innovation

Adapting a framework proposed by Wolfe (1994) in her study of organizational innovations, the field of IS innovation may be categorized in the two research streams of *stage-model* and *process-model* innovation process theories. Accordingly, while more traditional Diffusion Of Innovations (DOI) studies seem to address the question of "What is the pattern of diffusion of an innovation through a population of potential adopter organizations?", process theory approaches (stage-model and process-model) tend to address the question of "What are the processes organizations go through in implementing innovations?". In turn, each of the two research streams reinterprets this question by either focusing on the identification of implementation stages (stage-model research) or by analyzing the chain of events resulting in the innovation implementation (process-model research) (ibid.).

The review will first provide an overview of selected works in the stage-model research stream. Following this, the process-model stream will be presented. It is worth noting how this research in the IS field, although not purely DOI research according to Wolfe (1994), has traditionally been heavily influenced by the theoretical frameworks of DOI research (Rogers 1995). Examples of such studies are Cooper & Zmud (1990), Fichman & Kemerer (1999), and Gallivan (2001). In each of these interesting works the researchers seem, on the one side, embrace the DOI framework, and yet on the other side struggle with its clear limitations when applied in the context of complex IS. As a result, in one case the model consistently describes the adoption and infusion processes only if the innovation has a low complexity (Cooper & Zmud 1990). In another case, the struggle resulted in the need to refine the too generic stage model by Rogers (1995) into a more sophisticated one involving *primary* and *secondary* adoption (or *voluntary* and *mandatory*) (Gallivan 2001). In a third case, the researchers strived to go beyond the appearance (or *illusion*) of diffusion of an IS innovation, by measuring the often elusive and invisible gap between *acquisition* of an innovation and its factual *deployment* (Fichman & Kremer 1999).

Despite the clear attempt to overcome theoretical shortages of the DOI framework, these kinds of studies tend to remain within that framework. Yet, they differ from more classical DOI studies in their focus on the post-adoption phase of the innovation. In particular they tend, in line with DOI theories, to model the innovation in distinct (though possibly overlapping) stages.

A second set of studies on IS innovation is the *process-model* innovation process theory research stream (Wolfe 1994). These studies often recognize the limitation of the DOI framework and adopt alternative frameworks with the clear intention to dive into the messiness of the innovation process without the urge to model it in distinct stages (Lyytinen & Damsgaard 2001; McMaster *et al.* 1997; Attewell 1992).

For instance, Swanson & Ramiller (1997) introduce the notion of *organizing vision* in IS innovation providing an institutional view of how a "[...] collective, cognitive view of new technologies enables success in information systems innovation [...]" (ibid.). Other works indirectly contribute to the conceptualization of IS innovation as a mutual adaptation of technology and organization (Leonard-Barton 1988) or as sense-making of IT adaptation (Weick 1995; Henfridsson 1999).

Other researchers have focused on the role of organizational learning (Attewell 1992), or of *formative contexts* (Ciborra & Lanzara 1994). In these studies *innovation* is defined as the learning process of the organization in a particular context and in this sense independent of the technology. Ciborra also interprets the appropriation of a Groupware application in an organization as a process of *taking care* of the innovation (Ciborra 1996). This latter conception pushes the study of IS innovation in the philosophy of existentialism (Heidegger 1962).

A more recent stream of research developed the concept of "interactive innovation" (Rothwell 1994; Newell *et al.* 2000; Newell *et al.* 1998). This concept contributes to the conceptualization of innovations as the emergent process of knowledge exchange among inter- and intra- organizational networks. From this perspective, innovations are seen as the "[...] interaction of structural characteristics of organizations and the actions of individuals" (Newell *et al.* 1998).

Finally, other studies on IS innovation have adopted the conceptual framework of Actor Network Theory (ANT) (Latour 1987; McMaster *et al.* 1997; Monteiro & Hepsø 1998). For instance, Aanestad & Hanseth (2000) conceptualize the implementation of an open network technology as the cultivation of a "hybrid collectif" of humans and non-humans, technologies and non-technologies.

These studies share the interest of understanding the process of innovation as it unfolds, and not as the completion of a stage process. Accordingly, drifts, deviations, and complexities are not feared, rather are acknowledged and included in the analysis and theorization.

Specifically, the contributions by Lyytinen & Damsgaard (2001), and Monteiro & Hepsø (1998) regarding the diffusion of complex infrastructural innovations are particularly relevant for the case discussed in this paper. These two works open the discussion of IS innovation to address aspects of complexity: innovations are considered embedded in their organizational context where the existing socio-technical network has a great influence on the unfolding of their process. In this context, phenomena such as path-dependence (Arthur 1994; David 1985), lock-in (Arthur 1989), network effects (Antonelli 1992), and drift (Ciborra 2000) due to the existing *installed base*, tend to influence the trajectory of the innovation (Dosi 1982; Hanseth *et al.* 1996). In this sense, aspects of complexity are inherent to the infrastructural nature of IS innovation.

As the case study presented further in the paper will suggest, this is an area of discussion which needs to be expanded and which requires deeper understanding of the innovation phenomena.

2.2 On the current temporal conceptualization of innovation

The aim of this section is to unfold how the two broad research streams, stage-model and process-model, deal with the temporal aspects of innovation. Specifically, we mean to unveil the discourse about temporality in the two streams focusing on the three temporal analytical categories of *past, present,* and *future.*

The *past* category refers to the past heritage of the context where the innovation is implemented. The first research stream tends to overlook or underestimate the role of the past in the conceptualization of innovations when analyzing complex IS. The past, or the initial situation where the innovation needs to be adopted, is conceptualized as a stage which needs to be *unfreezed*, changed, and *refreezed* to implement the innovation (Kwon & Zmud 1987; Lewin 1952). It is therefore taken for granted that there is a beginning or an initiation of the innovation, which is a particular point in time where the innovation starts with a clear cut from previous events.

On the other hand, the second research stream often adopts theoretical frameworks which explicitly take into consideration the influence of the past in the implementation effort of the innovation. For example, studies on organizational learning using concepts such as double-loop learning (Argyris & Schön 1996) or formative contexts (Ciborra & Lanzara 1994), underline the relevance of existing cognitive structures and knowledge in the process of innovation. Other studies in this stream adopt ANT to take into account the role of the past showing how strength of inscriptions, black boxing, and irreversibility contribute to the shaping and establishment of socio-technical networks (Latour 1987; McMaster *et al.* 1997; Monteiro & Hepsø 1998).

As for the relation of the innovation with the *present*, the question remains as to the extent to which the implementation is regarded as realization of a "[...] 'black box' of readily available technological possibilities [...]" or as an emergent and open process or trajectory (Dosi 1982). According to the first research stream, the innovation unfolds in stages which have a logical order. Independent from the sophistication in quantity, quality, or temporal arrangement of the stages, the assumption of the stage-model is the possibility of reaching a *closure* of the innovation where predicted benefits are achieved.

In the second research stream the implementation process is often seen as emergent (Leonard-Barton 1988), context-dependent (Ciborra & Lanzara 1994), continuously negotiated (Monteiro & Hepsø 1998), or interactive (Robertson *et al.* 1996). Thus the range of possibilities of the innovation is continuously re-invented, re-negotiated, re-contextualized and re-appropriated undermining the closure of the implementation according to predicted benefits, and opening for potentially new ones.

Finally, the relation of the innovation with the *future* concerns the effects of the innovation on the future development of the surrounding context. In our particular case, it regards the role of the innovation in influencing the future trajectory of the infrastructure. Studies based on DOI limit their temporal scope to their unit of analysis: the particular innovation process. Hence no account of evaluation of the innovation on the future development of the context in which it is implemented is provided. Studies in the second research stream also seem to come short in addressing this temporal dimension. That is, the analysis and evaluation of the innovation is also in this case focused on the "event" of the implementation rather than on the influence of such implementation on future innovations.

On this point, the information infrastructure (II) perspective as proposed by Hanseth (*et al.* 1996) provides useful insights. By conceptualizing the complex system of interrelated IS in an organization as an II, the critical role of the installed base in the development of the existing socio-technical network is put forward. This conceptualization acknowledges the role of the past (e.g. of path dependence) in the current shaping of the II. This implicitly points at the importance of considering the innovation process itself as a process contributing to the shaping of the new installed base influencing future innovation. It is consequently relevant to consider the present innovation in the light of future possible developments of the II.

In the case and discussion sections of this paper, we will illustrate how the temporal dimension of the future is crucial to the understanding and evaluation of the innovation discussed. We will then elaborate in more detail a temporal conceptualization of the innovation in II by introducing the concept of *temporal disclosedness*.

3.   METHODOLOGY

The research reported in this paper is grounded in the interpretive approach to case study in IS (Klein and Myers, 1999; Walsham 1993, 1995). The research was guided by the following general question: "What are the complexities involved in the implementation of a standard-based technology in the setting of a hospital information infrastructure?". The fieldwork has been conducted over a period of over two years, from October 2001 to the present. Ethnographically inspired methods were employed for data collection, including 32 interviews with doctors, nurses, secretaries and IT managers, 8 instances of observations of daily work and users training sessions, documents analysis, and participation in several discussion meetings. The interviews were recorded on MiniDisc©, the most relevant (about 30%) have been transcribed and circulated among the research group. The others have been re-listened and summarized. The research group consisted of two Ph.D. students (including the author) a lecturer and a professor. After coding and discussions, the data has been analyzed by organizing it in process narratives (Langley 1999). A second round of analysis was used for theorization (e.g. production of this paper). The head of research of the IT department of the hospital joined regularly these meetings to update the research group on the project, and to suggest interesting area for further research.

Concerning the use of theories, the process of theorizing from the collected process data was inspired by the study of temporality in the existential philosophy of Heidegger (1962). In doing so we acknowledge the risk of appropriating a concept or perspective in a different theoretical milieu than the one in which it was conceived. We do in fact recognize that, if inspired by Heidegger, the study of temporality should not be detached from a study of the *state-of-mind* (*Befindlichkeit* in German) (Heidegger 1962: p.389; Ciborra 2002). Nevertheless, we find that our appropriation of the concept is adequate as we see the current paper as a step towards the breaking-down of misleading metaphysical categories.

4.   CASE DESCRIPTION

The case illustrates the nature and the context of an IS innovation implemented in a major Norwegian hospital (referred to as NorHospital). Specifically the innovation regards the implementation of a hospital wide clinical information system called Electronic Patient Record (EPR). The EPR represents a networking technology which aims to standardize and centralize the structuring, storing, and use of patient related clinical information in the entire hospital.

In order to provide the necessary historical and contextual perspective, the case describes three analytical stages of the development of the Information Infrastructure (II) in NorHospital; the second stage being the one when the EPR was conceived and implemented. The aim is to provide an account of the three different stages as three different technological paradigms (Dosi 1982).

We will now provide some contextual information regarding the hospital and the role of its IT department.

NorHospital is a large University hospital. It is the second largest hospital in Norway, with approximately 600 beds, 4000 employees and an annual budget of 2.5 billion NOK (around 360 million US Dollars). In 2002 more than 193'000 patients were treated.

The role and function of the IT department in NorHospital have considerably shifted from purely technical support in early 90s, to encompass "high-risk" clinical information systems development. At the beginning of 1990s the IT departments' staff was approximately 20 persons on a budget of approximately 10-15 MNOK. Currently over 80 people are running projects on a budget around 80 MNOK. Moreover, for the next four years (2003-2006), the IT department has set up a budget of 267 MNOK alone for development and implementation of clinical information systems. There are currently requests to raise the budget to as much as 500 MNOK. The budget for similar projects before 1995 was 0 MNOK. Moreover, in the last decade the scope of intervention of the IT department has changed from being a "technical problems solving" department to act as a service department, oriented towards the needs of their "customer": the different hospitals departments. Finally, after a health sector reorganization in Norway, the department started to address new "customers" outside the hospital. Indeed, during the last year, the department has been active in positioning itself as a regional actor in providing services related to clinical information systems.

4.1   LAN: From Mainframe to PC

We will now describe the first of the three analytical stages of the development of the infrastructure: the transition from a Mainframe to a LAN and PC infrastructure.

In early 1990s the information infrastructure at the hospital was primarily based on a few mainframe systems used for administrative purposes. Examples of applications were a Patient Administrative System, Human Resource management systems and a financial system. Accordingly, competencies in the IT department were limited to technical knowledge, and the type of service provided was rather routinized. At this time there was no involvement of the IT department in any activity related to clinical information.

Around mid 1990s the IT department installed a Local Area Network (LAN) and started to diffuse the first Personal Computers (PC). As a consequence, some clinical departments started to develop local systems. The systems were usually developed by doctors who were also amateur programmers and were usually serving local needs of organizing and storing data. Most of these systems have survived until today and represent an important and efficient part of several clinical departments' practices. They were often used as local Electronic Patient Record (EPR) systems, sometimes with statistical functionality for research purposes. The systems were developed entirely inside the departments without any support from the IT department, apart from providing the basic infrastructure. Also in this transition of the infrastructure from mainframe-terminal to LAN-PC the role of the IT department was merely one of technical support and maintenance.

4.2   EPR: Centralization and Tight Integration

In this section we will describe the second stage of the development of the infrastructure: the flourishing of clinical information systems and the implementation of the centralized EPR.

It is not until 1995 that the IT department faced the need to embrace new challenges in the uncharted area of clinical information systems. During that year a considerable amount of new projects were started, alongside the traditional technical support and maintenance activity. The new projects included the development of four new laboratory systems, a Radiology Information System (RIS), and a picture archive system for x-rays (PACS). Moreover, in the same year an EPR project was started with the aim of developing a hospital-wide centralized clinical information system. The aim of the project (which was done in cooperation with four other regional hospitals) was to develop a centralized version of an EPR substituting the locally developed EPR and integrating it with the other clinical systems. The project was supposed to last three years with the vendor delivering the final version of the EPR by the end of 1999 and the IT department implementing the EPR to reach 3500 potential users. As of now, all users have been reached, but the final version has not yet been delivered, while the version currently implemented is covering only about 30-40% of the requested functionality. Moreover, local EPR are still being used, and integration with clinical systems is partial or non-existent.

Reasons for the delay and the limited implementation may be found in the complexity of the design of a standardized and centralized EPR; in the complexity of the integration process with local clinical systems (complexity which involves political and power issues besides purely technical ones); and in an overall vendor-dependent strategy of reaching for perfection.

At the time the EPR project was started the IT department consisted of 19 people including two developers. None of these people had any formal background in clinical practice and thus were not ready to take on such risky projects. However, the IT department began to systematically acquire competencies and resources in order to manage the new type of projects. For the development of any new clinical information system, a project manager from the clinical department was appointed, and some end-users were involved part-time in the project.

With the EPR and other clinical systems, the IT department started to deal with projects with an increasing complexity. Some of them were rather straightforward, others required far more time and resources than planned (e.g. the development of PACS and of the EPR). All of them required innovative ways of working. One key element was interdisciplinarity and co-responsibility. As a consequence, an increasing number of people working in clinical departments moved to the IT department covering key positions in development projects. The scope of the IT department was expanding from supporting administrative systems to also developing and implementing clinical systems. At the same time, the core competence was shifting from purely technical knowledge to an interdisciplinary approach to the design, implementation and adaptation of clinical information systems.

With the EPR project the range of action and influence of the IT department definitely reached the width of the entire hospital organization and the depth of the complex clinical work-practices.

4.3 CSAM: Layering and Loose Integration

The development and implementation of the hospital wide EPR proved more difficult and complex than predicted. Despite the considerable change in staff, competence, and resources, the IT department of NorHospital seemed incapable of meeting some of the challenges that the vision of the EPR had inscribed in its conception. In particular, the EPR was seen as the central system meant to substitute all local systems. Furthermore, it was supposed to integrate all the information in one standardized format, while until then the information was handled, stored and structured according local standards. Finally the vision of the EPR was to provide a single solution to a variety of often contradicting requirements. For example, the EPR had to fulfill users' needs, while satisfying divergence of opinions in the different professions of the users (e.g. doctors and nurses), and disciplines (in almost each medical discipline there is an international standardization committee concerned with the structuring of their clinical information). Additionally it had to comply with national laws and regulations regarding information security, patient privacy, and the definition of the legal documentation of the patient treatment.

The idea of implementing the "one" standardized and centralized clinical system was slowly being substituted by the idea that the implementation of the EPR was the "Art of the impossible!" (in Norwegian: *Det umuliges kunst!*; cited from a presentation by a manager at the IT department). The understanding of the information infrastructure of the hospital as a puzzle of systems was leaving the stage to a more "systemic" view; that is a view of a complex lattice of interconnected systems, practices, power relations, and acquired knowledge.

At this point, the leadership at the IT department sentenced the idea of the holistic ERP to *death*, instead replacing it with a new vision focusing on *services*. The new vision was labeled Clinical Systems All Merged (CSAM) and may briefly be summarized as follows:

- Focus on information services instead of information containers (such as the EPR)
- CSAM is a process: focus on delivery of tangible benefits (e.g. better coordination between two departments) rather than of particular products (e.g. the EPR)
- Stepwise bottom-up approach leveraging existing resources rather than top-down long term planning
- Bottom-up approach responding to factual emergent needs of services
- Technological solution based on portal technology: layering and loose integration
- Increased control over direction of development, e.g. by
- Increased independency from vendors: vendors provider of tool-box technology instead of complete solution (as with EPR)

Along with a Norwegian proverb, the motto which facilitated the transition was: "The best is enemy of the good!" (in Norwegian: *Det beste er det godes fiende!*; citation from interview with a manager of CSAM). In this motto the EPR vision was seen as the ideal (i.e. the best) solution which proved impossible to reach. The CSAM philosophy, on the contrary, aimed at providing a good enough solution, although probably not the best, but achievable in a reasonably short time and with available resources.

5.   DISCUSSION

We will now proceed with the analysis of the case with the aim to highlight how the implementation of the EPR related to the past, present, and future of the trajectory of the information infrastructure. We will first provide an interpretation of the three stages as three technological paradigms (Dosi 1982). Then we will discuss the extent to which the reviewed theories on innovation can provide insights in the case, highlighting their structural limitations or their limited use. Finally we will introduce the concept of *temporal disclosedness* to underline the particular temporal characteristic of innovations in the context of information infrastructure, and will accordingly suggest the direction of research in similar cases.

5.1   Three paradigms and their relation

The following table summarizes the three analytical stages of the management of the Information Infrastructure at NorHospital (Table 1).

Each of the stages represents a different paradigm of strategy, management, and technology which shapes the trajectory of the infrastructure (Dosi 1982). In order to comprehend the dynamic of this trajectory, it is important to identify the influence of each paradigm on the next. Many of the elements are external to the context of the hospital (e.g. advances in networking technologies, the diffusion of the internet as technology and as a technological paradigm, the development of user interface technologies, laws and regulations, health reforms etc...), yet still there are several internal elements which have a great influence in the future trajectory of the infrastructure.

Table 1. Three paradigms in the management of the Information Infrastructure

Dimensions	LAN	EPR	CSAM
General Strategy (IT dpt.)	No big strategy	Long term focus: achieve complete integration and standardization	Short term focus: achieve benefits on focused and emergent needs using available resources
Competence (IT dpt.)	Technical	Technical + Clinical	Technical + Clinical
Business Model (IT dpt.)	Maintenance and Support	Delivery of solution (product focus)	Provision of information services (process focus)
Management Style (IT dpt.)	Routine management	Traditional project management: hierarchies, planning, heavy documentation	"Creative" management: emergent, informal, dynamic
Role of Users	Almost none	Ideally participation of the whole user base	User base involvement secondary: heavy use of acquired knowledge from EPR implementation
Main Role of Vendor	Provider of infrastructural components (cables, computers etc...)	Provider of Complete Solution (high dependency on vendor)	Provider of infrastructural components (portal technology, database systems etc...) (lower dependency on vendor)

First, the radical change from the LAN to the EPR paradigm was mostly influenced by external factors. Yet, one determinant internal aspect of the LAN paradigm was the great potential of the establishment of a Local Area Network. This implementation was accomplished by the IT department still within the vision of it being a technical support and maintenance provider, thus without any idea of enabling the flourishing of local (and later interconnected) clinical systems.

Thus, the EPR paradigm exploited possibilities provided by the LAN paradigm. At the same time, the diffusion of PCs inside the hospital generated the flourishing of a myriad of local EPR and other clinical systems, which at the end hugely enriched the installed base of the information infrastructure, heavily influencing the course of the EPR implementation. In this very step, it is clear how on the one hand the EPR paradigm built on the previous one, yet on the other hand, the same infrastructure which allowed the EPR dream to be formalized, hampered the course of its implementation.

So far the influence of the past (the LAN paradigm and the infrastructure which it generated) and the dynamic of the present of the implementation of the EPR (the way it drifted away from plans producing delays and a long list of compromises) should be clear. We will now briefly analyze how the implementation of the EPR, and the paradigm in which it was conceived, heavily influenced the generation of the new paradigm for CSAM.

Similarly to the EPR, the CSAM vision was the by-product of an external shift of the technological paradigm. At that time the IT industry was strongly focused on the integration of existing systems. Accordingly, often the proposed technical solutions were a portal architecture, where the existing infrastructure of systems is loosely integrated with a layering strategy. Yet, the CSAM vision is not solely shaped by this new technical hype but it is largely the evolution of the EPR paradigm. In fact, each of the dimensions in the CSAM paradigm is the outcome of the experience accumulated with the EPR implementation. Indeed, the EPR had been the first attempt by the IT department to provide clinical systems with a hospital-wide scope. The people behind the CSAM vision were the same people who had been struggling for eight years in a particular management and technological paradigm (EPR) which they then thought was inappropriate. Most importantly, the new paradigm was not simply a change in a new direction with a fresh start. It is the opposite: it is the evolution in a new direction with the inertia and enrichment of the huge amount of knowledge and competence acquired in the previous paradigm. Examples of the paradigmatic shift enabled by this learning process are the change from a long-term to a shorter term strategy; or the evolution from a product to a process focus. Even clearer is the capitalization of knowledge about users and work practices from the EPR implementation, used to reorganize the role of users in the CSAM process (see Table 1).

From this perspective, the EPR innovation was both the (partial) realization of an infrastructural technology, and a huge learning process. The learning process resulted in the tangible confidence the IT department now has in addressing services for clinical information systems. It is also tangible in the installed base of (partly) standardized and centralized routines running on the EPR. This installed base represents a very important launching pad for CSAM like the LAN technology was for the EPR. Finally, but not least important, the implementation of the EPR created over the years a strong network of trust between the clinical departments and the IT department. Also in this case, the "installed base" of trust relationships represents an important starting point for the accomplishment of the CSAM vision.

5.2   Limits of current studies

We will now briefly discuss the literature reviewed in the theory section, and evaluate the usefulness of the proposed theoretical frameworks. Subsequently, we will point out how current theorizations tend to overlook the influence of an innovation on future innovations. We will conclude this exercize by pointing at some of the mentioned frameworks which could be used to open the analysis of innovation in the direction we are suggesting.

If we try to apply theories pertinent to the first research stream to the case, it will immediately become clear that almost each assumption and hypothesis made in those theories is hardly met. Not to say that such theories are not relevant: on the contrary. The mentioned theories simply are not adequate to the study of the innovation case presented in this paper. For instance, the model proposed by Cooper & Zmud (1990) would be flawed by the excessive complexity of the technology (here intended as socio-technical network). The study of assimilation gaps by Fichman & Kemerer (1999) would seem pointless as the deviation of *deployment* from the envisioned *acquisition* of the innovation would seem as a natural property of the trajectory of the infrastructure. Finally, any such theory based on a stage model would systematically fall short in describing the complexity of the innovation process. For one, in the model no account is given to the role of the installed base in shaping the innovation (if not for the negative connotations of something that needs to be changed). Nor does it take into consideration the fact that the innovation may remain open. According to each of these models the cycle of the innovation has to come to a final stage. This is certainly not the case of the innovation presented in this paper. While only about 40% of the EPR is implemented and operational, the remaining 60% is still under design, development, and testing.

If we turn to the second research-stream on IS innovation processes, we find at least two approaches which may be useful in the study of our case: studies on organizational learning (Attewell 1992; Ciborra & Lanzara 1994; Argyris & Schön 1996; Ciborra 1996) and studies adopting the theoretical framework of ANT (Latour 1987; McMaster *et al.* 1997; Monteiro & Hepsø 1998; Aanestad & Hanseth 2000). As illustrated previously in the paper, the change of paradigms may be interpreted as a learning process. As the mentioned theories on learning suggest, innovation as learning does not happen only as plain learning, or linear learning. In order to learn and innovate, people need to change their cognitive framework, formative context, or, finally, perform double-loop learning. While showing that the heritage of the past (in terms of e.g. formative context) influences the present nature of the innovation, such approaches tend to limit the study by identifying the innovation with the change to a different specific formative context (or set of values, or cognitive frame). For instance, the EPR paradigm entailed a new way of thinking, a new way of cooperating between departments, and a new way of looking at the technology. Nevertheless, the new formative context of the "organizing vision" (Swanson & Ramiller 1997) of the EPR was not the only one generated by the lengthy implementation process. The new way of thinking in CSAM (which we want to identify as yet another formative context) is also an instance of the EPR innovation.

Similarly, studies using ANT are effective at describing the influence from existing network and the emergence from present implementation as complex and dynamic processes of enrolment, inscription and translation. They tend however to end the analysis at the point when the network is stabilized. In the context of an infrastructure, socio-technical networks (such as "hybrid collectifs") hardly stabilize and are hardly isolated. As a consequence, their emergent and interconnected nature continuously influences the surrounding context making it extremely difficult to draw the boundaries between one innovation and the next. We sustain that the cited literature comes short in highlighting this aspect.

In conclusion, we submit that theories from the first research stream have a structural limitation which does not allow them to be used outside the given spatial and temporal boundaries. Whereas regarding the second research stream, theories on learning and studies using ANT seem to have the adequate methodological and analytical framework to investigate beyond those boundaries, but their application seems to be intentionally limited.

An exception to the proposed theories is the concept of *path creation* (Garud & Karnøe 2001). Path creation is a concept developed exactly to address the importance of being aware that present innovations will have a path dependent influence on the future trajectory. While this intuition is extremely relevant, its application to the study of innovation in IS seems to be lacking and should thus be developed.

5.3   Temporal disclosedness of innovations

In this paragraph, in light of the collected and analyzed empirical evidence, we will introduce the concept of *temporal disclosedness* of innovations in information infrastructures (Heidegger 1962). The aim of introducing this concept is to highlight a peculiar characteristic of innovations occurring in complex infrastructures of interconnected IS: as illustrated in the case, once implemented, it becomes extremely difficult to draw temporal and spatial boundaries around the innovation.

If implementation of the innovation usually means the realization of predetermined possibilities in time, *temporal disclosedness* of the innovation means that the implementation is dealing with possibilities beyond the traditional time and space occupied by the implementation process; i.e. before the implementation starts, as the implementation unfolds, and beyond the completion (or incompletion) of the implementation.

With regard to the past and the present, the concept of *temporal disclosure* is near to other concepts highlighted in the theoretical section (from path dependency and installed base, to improvisation, interaction, and emergence). The concept proposed here is particularly relevant in that it also extends the perspective on the future evolution trajectory of the infrastructure.

5.3.1   Disclosure of the past

Temporal disclosure of the past refers to how phenomena such as path dependency, lock-in, and formative contexts of the existing installed base provide pre-existing possibilities and inertia with which the innovation has to deal. Disclosure means that the innovation is already-there in the possibilities allowed or precluded by the installed base which wait to be disclosed.

5.3.2   Disclosure of the present

Temporal disclosure of the present refers to how the complexity of the infrastructure affects the implementation as it unfolds: it drifts, it becomes emergent, interactive, renegotiated, and reinvented. Thus, compared to a view on implementation as *awating* of the realization of predetermined possibilities, implementation becomes also *anticipation* of new possibilities through improvisation, tinkering, and bricolage (Heidegger 1962; Ciborra 1999).

5.3.3   Disclosure of the future

As the influence of the EPR paradigm in the conception of CSAM shows, the innovation becomes part of the installed base shaping and triggering new innovations. Temporal disclosure of the future means that the new possibilities are not only regarding the implementation of the current innovation, rather they may affect the conception of new and future innovations.

5.3.4   Temporal disclosedness of innovations

Temporal disclosedness of innovations in information infrastructures means that the lifecycle of an infrastructural innovation is not closed in time boundaries defining its possibilities; it is rather *disclosed* along the entire past and future trajectory

Based on this characteristic, we propose that the analysis of future influence of the innovation on the trajectory of the infrastructure is not a matter of choice, rather a methodological imperative. This implies that the evaluation of an innovation in an infrastructure should also consider the influence of the innovation at case on the implementation of later innovations. For instance, the evaluation of the EPR should include an analysis of the influence of the EPR on the generation and implementation of CSAM. We also suggest that theories from the process-model research stream may provide adequate analytical and conceptual tools (particularly theories on learning and ANT). Yet, limiting the analysis to the particular apparent lifecycle of an innovation may be useful but it risks giving the wrong impression that innovations in such complex settings can be analytically isolated in time and space. In our case, evaluating the EPR innovation considering or not considering its influence on CSAM would bring different results. At first it could appear as a failure. At second analysis it is possible to observe the tangible (and unpredicted) benefits brought to the CSAM innovation.

More provokingly, when studying innovations in information infrastructure, one should not be fooled by the apparent suitability of spatial and temporal categories of the innovation. To understand the particular innovation, one is forced to understand also the surrounding context in which it is embedded and analyze it in an historical perspective including past and future.

6.   CONCLUDING REMARKS

In this paper, we have addressed the question of how to conceptualize temporal aspects of IS innovations occurring in complex infrastructures of interconnected information systems. Using the framework of Information Infrastructure (Hanseth *et al.* 1996), we have adopted a view on information technology in organizations which acknowledges the complex dynamics of its trajectory. We have then reviewed current literature on the study of innovation processes in the field of IS, highlighting different perspectives and shortcomings in the analysis of temporal aspects of innovations. In particular, we have illustrated how there is a general lack of study on the influence of the implemented innovation on the future trajectory of the infrastructure. With the help of a case study based on an IS innovation implemented in a hospital, we have shown how it is paramount to study such influence. In order to sustain this argument and based on the evidence of the case, we have introduced the concept of *temporal disclosedness* of the innovation. This concept is useful to understand the limits of current conceptualizations of innovation processes in complex infrastructures. The limits, which we suggest should be overcome, refer to the tendency of keeping a temporally closed view on the innovation. This view, we sustain, veils the nature of the innovation phenomenon, which, on the contrary, tends to become temporally disclosed along the entire trajectory of the infrastructure.

REFERENCES

Aanestad M., Hanseth O.. (2000), "Implementing open network technologies in complex work practices. A case from telemedicine". In *Proceedings from IFIP WG 8.2 International Conference. The Social and Organizational Perspective on Research and Practice in Information Technology*, June 10-12, 2000, Aalborg, Denmark

Antonelli, C., (1992), "The economic theory of information networks". In *The Economics of Information Networks*, ed. C. Antonelli, North-Holland, 1992.

Argyris C., Schön D.A., (1996), *Organizational Learning II*, Addison-Wesley, Reading, Mass., 1996.

Arthur, W.B., (1989), "Competing technologies, increasing returns, and lock in by historical events", *Economic Journal* 99: 116-131, 1989.

Arthur W.B., (1994), *Increasing Returns and Path Dependency in the Economy*, The University of Michigan Press.

Attewell P., (1992), "Technological Diffusion and Organizational Learning: The Case of Business Computing", *Organization Science*, Vol. 3, Issue 1 (Feb., 1992), pp. 1-19.

Baskerville R., Pries-Heje J., (1997), "IT diffusion and innovation models: the conceptual domains", in *Facilitating Technology Transfer through Partnership: learning from practice and research*, Proc. IFIP TC8 WG8.6 Conference, Tom McMaster Enid Mumford E.Burton Swanson Brian Warboys and David Wastell (eds.), Kluwer Academic Publisher.

Ciborra C.U., Lanzara G.F., (1994), "Formative Contexts and Information Technology: Understanding the Dynamics of Innovation in Organizations", *Accounting Management and Information Technology*, Vol. 4, No. 2, pp. 61-86, 1994.

Ciborra C.U., (1996), "What does Groupware Mean for the Organizations Hosting it?", in *Groupware and TeamWork: Invisible Aid or Technical Hindrance?*, Claudio U. Ciborra ed., John Wiley & Sons, pp. 1-19.

Ciborra, C., (1999), "Notes on Improvisation and Time in Organizations", *Accounting, Management and Information Technologies* 9 (1999).

Ciborra, C., Braa, K., Cordella, A., Dahlbom, B., Failla, A., Hanseth, O., Hepsø, V., Ljungberg, J., Monteiro, E., Simon, K., (2000), *From Control to Drift. The Dynamics of Corporate Information Infrastructures*, Oxford University Press, 2000.

Ciborra C. U., (2002), *The Labyrinths of Information – Challenging the Wisdom of Systems*, Oxford University Press

Cooper R. B., Zmud R.W., (1990), "Information Technology Implementation Research: A Technological Diffusion Approach", *Management Science*, Vol. 36, No. 2 (Feb. 1990), pp.123-39.

Damsgaard J., Lyytinen K., (1997), "Hong Kong's EDI bandwagon: derailed or on the right track?", in *Facilitating Technology Transfer through Partnership: learning from practice and research*, Proc. IFIP TC8 WG8.6 Conference, Tom McMaster Enid Mumford E.Burton Swanson Brian Warboys and David Wastell (eds.), Kluwer Academic Publisher.

David P., (1985), "Clio and the economics of QWERTY", *American Economic Review*, Vol. 75 (1985), Issue 2 (May), pp. 332-37.

Dosi G., (1982), "Technological paradigms and technological trajectories: a suggested interpretation of the determinants and directions of technical change", *Research Policy*, Vol. 11 (1982), pp. 147-62.

Fichman R.G., Kemerer C.F., (1999), "The Illusory Diffusion of Innovation: An Examination of Assimilation Gaps", *Information Systems Research*, Vol. 10, No. 3, September 1999, pp. 255-75.

Gallivan M.J., (2001), "Organizational Adoption and Assimilation of Complex Technological Innovations: Development and Application of a New Framework", *The DATA BASE for Advances in Information Systems*, Summer 2001, Vol. 32, No. 3, pp. 51-85.

Garud R., Karnøe P., (2001), *Path Dependence and Creation*, Lawrence Erlbaum Associates Publishers

Hanseth O., Monteiro E., Hatling M., (1996), "Developing information infrastructure: The tension between standardization and flexibility". *Science, Technology and Human Values*. Vol. 21 No. 4, Fall 1996, 407-426.

Heidegger M., (1962), *Being and Time*, Blackwell Publisher Ltd.

Henfridsson O., (1999), *IT-adaptation as sensemaking: inventing new meanings for technology in organizations*, Ph.D. Thesis, Department of Informatics, Umeå University, Sweden.

Klein, H.K., and Myers, M.D., (1999), "A set of principles for conducting and evaluating interpretive field studies in information systems", *MIS Quarterly*, vol.23, no.1, 1999, pp.67-93.

Kwon T.H., Zmud R.W., (1987), "Unifying the fragmented models of information systems implementation", in *Critical Issues in Information Systems Research*, R.J. Boland and R.A. Hirschheim eds., John Wiley & Sons, pp.227-51.

Langley A., (1999), "Strategies for theorizing from process data", *Academy of Management Review*, 1999, Vol. 24, No. 4, 691-710.

Latour B., (1987), *Science in Action*, Harvard.

Leonard-Barton D., (1988), "Implementation as mutual adaptation of technology and organization", *Research Policy*, Vol. 17 (1988), pp. 251-67.

Lewin K., (1952), "Groups Decision and Social Change", in *Readings in Social Psychology*, Newcombe and Hartley (Eds.), Henry Holt, New York, pp. 459-73.

Lyytinen K., Damsgaard J., (2001), "What's wrong with the Diffusion Of Innovation Theory?", in *Diffusing Software Product and Process Innovations*, Proc. IFIP TC8 WG8.6 Conference, Ardis M.A., Marcolin B.L. (eds.), Kluwer Academic Publisher.

McMaster T., Vidgen R.T., Wastell D.G., (1997), "Technology Transfer – diffusion or translation?", in *Facilitating Technology Transfer through Partnership: learning from practice and research*, Proc. IFIP TC8 WG8.6 Conference, Tom McMaster Enid Mumford E.Burton Swanson Brian Warboys and David Wastell (eds.), Kluwer Academic Publisher.

Monteiro E., Hepsø V., (1998), "Diffusion of infrastructure: mobilisation and improvisation", In *Information systems: current issues and future challenges* , Proc. IFIP WG 8.2 & 8.6, T. Larsen L. Levine (eds.), IFIP 1998, pp. 255 - 274.

Newell, S., Swan, J.A., Robertson M., (1998). "A cross-national comparison of the adoption of BPR: An interactive perspective". *31st International Conference on Systems Sciences*, IEEE Computer Society Press.

Newell, S., Swan, J.A., Galliers R.D.. (2000). "A knowledge-focused perspective on the diffusion and adoption of complex information technologies: The BPR example." *Information Systems Journal* 10: pp.239-259.

Robertson, M., Swan, J., Newell S. (1996). "The role of networks in the diffusion of technological innovation." *Journal of Management Studies* 33(3): pp.333-359.

Rogers E.M., (1995), *Diffusion of Innovations*, 4th Edition, Free Press.

Rothwell, R. (1994) Towards the fifth generation innovation process. *International Marketing Review*, 11, 7-31.

Swanson E.B., Ramiller N.C., (1997), "The Organizing Vision in Information Systems Innovation", *Organization Science*, Vol. 8, Issue 5 (Sep.-Oct., 1997), pp. 45874.

Van de Ven A.H., (1986), "Central Problems in the Management of Innovation", *Management Science*, Vol. 32, No. 5, Organization Design (May, 1986), pp. 590-607.

Walsham, G., (1993), *Interpreting Information Systems in Organizations*, Wiley, 1993.

Walsham, G., (1995), "Interpretive case study in IS research: nature and method", *European Journal of Information Systems*, Vol.4, 1995, pp.74-81.

Weick K.E., (1995), *Sensemaking in Organizations*, Thousand Oaks: Sage.

Wolfe R A., (1994), "Organizational Innovation: Review, Critique and Suggested Research Directions", *Journal of Management Studies*, 31:3, May 1994, pp. 405-31.

HOW IS AN IT INNOVATION ASSIMILATED

E. Burton Swanson
Anderson School, UCLA, USA

Abstract: The concept of organizational assimilation of information technology (IT) innovations is under-explored in the research literature. Here we rethink the concept, focusing on assimilation in use, in particular. Taking an organizational learning perspective, we propose that experimentation in use serves as the assimilative engine, driving the innovation's interpretation and routinization, leading in turn to its eventual conceptual sublimation and taken-for-grantedness. From this model, several new opportunities for research are identified.

1.   INTRODUCTION

Innovating with information technology (IT) has in the last decade become a popular research subject (see, e.g. Fichman, 2000; Gallivan, 2001). Studies have been particularly inspired by the broader literature on innovation diffusion, with its focus on early versus late adoption (Rogers, 1995). Among practitioners, firms that adopt an IT innovation earlier than others are characteristically viewed as the more "innovative," earning admiration where they are successful. Research has understandably sought to explain what enables firms to become innovators and early adopters of IT (see, e.g., Cooper and Zmud, 1990; Grover and Goslar, 1993; Grover, et al, 1997; Kauffman, et al, 2000).

But while important, research focused on the IT adoption decision can be myopic. In innovating with IT, mere adoption does not get it done. More specifically, IT implementation is characteristically problematic (Lucas, 1981; Markus, 1983; Swanson, 1988; Klein and Sorra, 1996). Accordingly, research has also sought to assess what enables firms to be successful in their IT implementations (see, e.g., Leonard-Barton and Deschamps, 1988; Orlikowski, 1993; Fichman and Kemerer, 1997). Whether earlier or later adopters will find implementation to be easier has been raised as an interesting research question, for instance (Swanson, 2003). And so the research literatures on IT innovation and implementation, each with their separate histories, have also been cobbled together (Kwon and Zmud, 1987).

Beyond implementation, the IT innovation literature has yet to achieve a coherent focus, even though important research has been done (see, e.g. DeSanctis and Poole, 1994; Tyre and Orlikowski, 1994; Orlikowski, 1996; Orlikowski, 2000; Majchrzak, et al, 2000). Successful implementation has remained the primary dependent variable of research interest, although Zmud and colleagues have identified additional stages to the process that commence with IT innovation in *use*, specifically, those of adaptation, acceptance, routinization, and infusion (Kwon and Zmud, 1987; Cooper and Zmud, 1990; Zmud and Apple, 1992; Saga and Zmud, 1994). More recently, researchers have addressed the organizational *assimilation* of IT innovations, a concept given broad and various interpretations, as we shall see. In principal, one would think that assimilation of IT innovations would be understood to result substantially from their use, but this notion has been little engaged, notwithstanding substantial evidence of assimilative problems subsequent to implementation with innovations such as ERP (Ross, 1999; Markus and Tannis, 2000).[1]

In the present paper, we reconsider the question of how an IT innovation comes to be organizationally assimilated. Taking an organizational learning perspective (Levitt and March, 1988; Huber, 1991; Crossan, et al, 1999; Robey, et al, 2000), we examine IT innovation assimilation in use, in particular. We begin in the next section with a brief review of the key research literature. We then introduce a framework within which we identify the "learning mechanism" underlying IT innovation assimilation. We follow with an extended discussion. Our broad purpose is to provide a new focus for research into IT innovation beyond implementation, in the context of use, rather than prior to use, where past research has mostly focused.

2.   LITERATURE DEFINITIONS

The concept of organizational assimilation of an IT innovation is under-explored in the research literature, even while it has attained some

---

[1] Ross (1999) observes that business performance often dips when ERP is first implemented. The performance improvements promised by ERP are usually achieved only with considerable time and effort.

prominence. Table 1 summarizes articles that have featured the concept (in their titles). From these, we can identify several strands of interpretation.

Table 1. Organizational Assimilation of Innovations as Defined in Literature

Article	Assimilation Definition	Remarks
Meyer and Goes (1988)	"Assimilation is defined here as an organizational process that (1) is set in motion when individual members first hear of an innovation's development, (2) can lead to the acquisition of the innovation, and (3) sometimes comes to fruition in the innovation's full acceptance, utilization and institutionalization." (p. 897)	Decision-making stages in assimilation of medical innovations include: knowledge-awareness (apprehension, consideration, discussion), evaluation-choice (acquisition proposal, medical-fiscal evaluation, political-strategic evaluation), adoption-implementation (trial, acceptance, expansion). The assimilation of 12 innovations by 25 hospitals is studied.
Agarwal, Tanniru, and Wilemon (1997)	"…adoption connotes the development of the 'first' successful system using a new information processing technology, while diffusion is concerned with a transfer of this success to other relevant applications, i.e., the spread of the innovation through the target system. The term assimilation is used to refer collectively to both the adoption and diffusion phases of innovation." (p. 347)	"Generic" assimilation strategies include: support, advocacy, total commitment. Experiences of several organizations are described.
Fichman and Kemerer (1997)	"Following Meyer and Goes (1988), assimilation is defined as the process spanning from an organization's first awareness of an innovation to, potentially, acquisition and widespread deployment." (p. 1346)	Describes the assimilation of object-oriented programming languages (OOPLs) by 608 IT organizations. Employs a six-stage model of software process innovation: awareness, interest, evaluation/trial, commitment, limited deployment, general deployment. Theory draws from Attewell (1992).
Armstrong and Sambamurthy (1999)	"…we define IT assimilation as the effective application of IT in supporting, shaping, and enabling firms' business strategies and value-chain activities." (p. 306)	Surveys business executives from 153 companies as to their self-rated performance in applying IT to logistics, marketing, and business

Article	Assimilation Definition	Remarks
		strategies. IT infusion and routinization are presumed, rather than directly assessed.
Fichman and Kemerer (1999)	"For this study we define the assimilation gap as the difference between the pattern of cumulative acquisitions and cumulative deployments of an innovation across a population of potential adopters. Although this definition is made in reference to two particular events- acquisition and deployment- in principle, any two assimilation events could be used to define an assimilation gap..." (p. 258)	Follows Fichman and Kemerer (1997). Compares assimilation gaps for three software process innovations: relational database management systems (RDBs), general purpose fourth generation languages (4GLs), and computer aided software engineering tools (CASE), finding a pronounced gap for CASE.
Fichman (2000)	"Diffusion is the process by which a technology spreads across a population of organizations, while assimilation refers to the process within organizations stretching from initial awareness of the innovation, to potentially, formal adoption and full scale deployment." (p. ?)	Provides an overview of basic concepts, theories and research concerned with IT innovation diffusion and assimilation. See too Fichman and Kemerer (1997, 1999).
Gallivan (2001)	"Once secondary use occurs, it is meaningful to consider the organization's assimilation stage. (This) describes how deeply the innovation penetrates the adopting unit (e.g., the company, division, or workgroup)." (p. 62)	Substantial review. Drawing from Cooper and Zmud (1990), assimilation includes: initiation, adoption, adaptation, acceptance, routinization, and infusion (extended use, integrative use, and emergent use).
Purvis, Sambamurthy, and Zmud (2001)	"Assimilation is defined as the extent to which the use of the technology diffuses across the organizational projects or work processes and becomes routinized in the activities of those projects and processes." (p. 121)	Surveys organizational assimilation of CASE tools. Examines effects of management championship and knowledge embeddedness within CASE platform. Routinization not directly assessed.
Cho and Kim (2001-2002)	"Assimilation of innovation is the process spanning from an organization's first awareness of an innovation to its potential acquisition and widespread deployment. The process consists of awareness, interest, evaluation, trial, commitment, and finally, deployment of the new technology." (p. 133)	Relies upon Fichman and Kemerer (1997). Surveys object-oriented programming language assimilation in 220 Korean firms.

The leading interpretation stems from the research of Meyer and Goes (1988), and may be characterized as *extensive* over the organization's innovation life cycle. Assimilation begins with first awareness, well before adoption, while it "comes to fruition" with "full acceptance, utilization and institutionalization." The work of Fichman and Kemerer (1997, 1999) follows this interpretation, although it takes a more *abbreviated* view, with assimilation culminating in the innovation's "full scale deployment" (Fichman, 2000). Absent here is any notion of assimilation in use. Cho and Kim (2001-2002) follow the Fichman interpretation.

Agarwal, Tanniru, and Wilemon (1997) take a different view, offering a *strategic* interpretation of assimilation as purposeful organizational action. Here the focus is on "transfer," or facilitating adoption and diffusion among sub-units within the firm. The innovation's assimilation is essentially "pushed" by management, and leadership is a key research focus[2] Again, there is little notion of assimilation in use.

Armstrong and Sambamurthy (1999) offer what might be termed a *value-added* perspective of assimilation. Emphasis is placed on achieving business value from the innovation. According to the authors, this requires that the "implicit functionality" of the IT innovation be "assimilated within the ongoing actions of individuals and teams." Here, assimilation in use is clearly suggested. However, it is arguably also conflated with achieving successful business outcomes.

Purvis, Sambamurthy, and Zmud (2001) more directly address assimilation in use. Their definition incorporates both the spread of use (internal "diffusion") within the enterprise, as well as routinization of this use. However, their study of CASE tool assimilation focuses more on the former than the latter. They define assimilation operationally as "the proportion of an organization's systems development projects where the technology is being used to support systems development work activities." (p. 121)

Gallivan (2001) attempts a theoretical synthesis. Organizational assimilation is conceived as stemming substantially from individual ("secondary") use. Assimilative stages go beyond initiation and adoption ("early stages") to include adaptation, acceptance, routinization, and infusion ("later stages"), as suggested by Cooper and Zmud (1990). The notion of *infusion* (Cooper and Zmud, 1990) refers to the elaborated use of an

---

[2] See also Martinsons and Schindler (1995) who argue, "Inspirational leadership and a needs-driven organizational vision are crucial for the effective assimilation of a new or emerging technology" (from the abstract). For additional perspectives on IT transfer, see McMaster, et al (1997).

innovation, whereby the innovation is increasingly embedded in the organization's work and managerial systems (Zmud and Apple, 1992). But Gallivan's detailed model does not differentiate analytically among the later assimilative stages.[3]

In sum, while the notion of assimilating an IT innovation has been featured in the literature, interpretations have varied. While organizational learning is arguably a theme in these interpretations, it has received little explication. Most importantly, assimilation *in use* has received surprisingly little attention. Here we lack integrative theory and studies of the assimilative and learning mechanism. In the next section, we take up the question of how through learning an IT innovation comes to be assimilated in its use.

3.   RETHINKING THE CONCEPT

The term "assimilate," from its definitional roots, suggests that something is taken in and absorbed by a system, such that it is no longer distinct from it (i.e. dissimilar to it).[4] We shall want to rethink the concept of assimilating IT in these terms. The "something" assimilated will be the IT innovation. The "system" absorbing it will be the organization and its work systems. Our premise will be that the innovation process is purposeful. The firm's work systems are to be renewed by the IT innovation, such that they better serve the enterprise. Once assimilated, the IT innovation is no longer distinct as such from the work systems. It becomes "taken for granted" in practice. However, the renewed work systems are likely to be (indeed, are usually intended to be) discernibly altered. In the extreme, the work systems may be *transformed* (Orlikowski, 1996; Robey and Sahay, 1996).[5]

The process by which the IT innovation is absorbed by an organization's work systems may be characterized as one of *organizational learning*, in particular, *by doing* (Levitt and March, 1988). Basically, the organization

---

[3] Cooper and Zmud's (1990) rationale for these stages (adapted from Kwon and Zmud, 1987) is founded on Lewin's (1952) classic change model. Thus interpreted, assimilation would presumably conclude with the "refreezing" of changed organizational processes.

[4] See, e.g., Webster's 3rd New Int. Dictionary. The term "assimilate" is widely applied, e.g. to the incorporation or conversion of nutrients by plants and animals, to the adaptation of adjacent sounds in spoken languages, and to the absorption of one social culture by another.

[5] We thus allow for innovations that may be either incremental or radical in their effects (Dewar and Dutton, 1986; Orlikowski, 1993). Whatever old competencies are to be displaced, we presume that new competencies are intended to be superior.

innovates so as to attain new capabilities, embodied in new *routines* (Nelson and Winter, 1982). Such capabilities can't simply be bought in the marketplace. Rather, each firm has some *absorptive capacity* for new capabilities, which is largely a function of its prior related knowledge (Cohen and Levinthal, 1990). Many IT innovations are made popular by *organizing visions* that describe what the innovation is all about and what it is good for (Swanson and Ramiller, 1997), but while widespread adoption may be motivated by these visions, implementation and organizational assimilation can be highly problematic (Attewell, 1992). Here each firm in achieving its new capabilities must inevitably rely heavily upon its own resources, even while it may draw from the expertise of intermediaries such as vendors and consultants.

But how in such circumstances will learning actually occur? Beyond implementation, how are IT innovations assimilated *in use*, in particular, such that new capabilities are achieved?[6] Drawing from a long tradition suggesting that *attention* is fundamental to organizational behavior (Simon, 1947; Cyert and March, 1963; Nelson and Winter, 1982; Ocasio, 1997), we propose that an analysis of work-situated *attention to the innovation* provides insights as shown in Figure 1. Here a two-dimensional framework is suggested. On one dimension we distinguish between attention to the innovative concept (in coordinative communication) and attention to the innovative practice (in performance of the task). That is, we distinguish between attention to saying and attention to doing where innovative work is concerned.[7] We are inspired here by Weick's (1995) work on organizational sensemaking, whereby firms and their decision makers are theorized to learn by making retrospective sense of their collective actions (see too Weick and Roberts, 1993). In the same vein, we argue that attention to the innovative concept is motivated largely by attention to the innovative practice, where organizational *learning with use* is concerned. In achieving new

---

[6] Robey, et al (2002), provide an illustrative study focusing on ERP implementation, as distinct from use. They report, "Assimilation was a challenge not only for users, but also for core team members and other stakeholders such as customers. Before users could use an ERP system effectively, they needed to learn the business processes that were revised following system implementation. Firms addressed the need for users to learn new systems by providing *formal training* for users and by taking an *incremental approach* to systems implementation." (p. 32) In the present paper, by focusing on use, we seek to get *beyond* such an implementation perspective.

[7] Both communication and task performance are regarded to be forms of action. Because task performance is typically a social activity, it is often accompanied by, indeed, often requires, communication. While certain of this communication may be integral to "doing", other will serve the broader role of organizational sensemaking (Weick, 1995) and management of the coordinative task apart from its direct performance.

organizational capabilities, what the firm knows how to *do* is what is of central discursive concern.

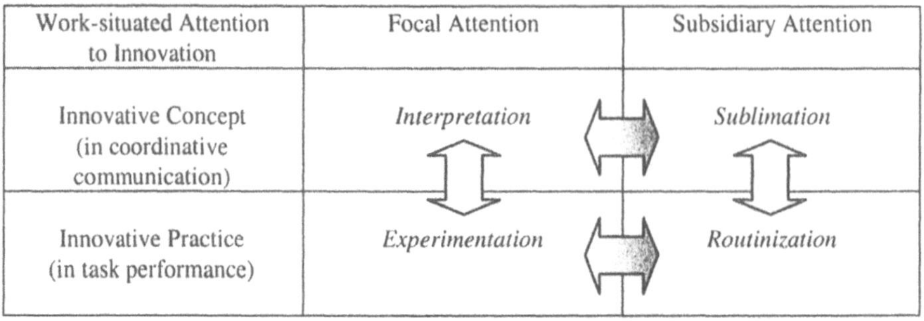

Work-situated Attention to Innovation	Focal Attention	Subsidiary Attention
Innovative Concept (in coordinative communication)	*Interpretation*	*Sublimation*
Innovative Practice (in task performance)	*Experimentation*	*Routinization*

Figure 1. How are IT Innovations Assimilated in Use?

Notes: Double-headed arrows indicate dual relationships. Shaded arrow-heads indicate primary assimilative direction over time, with subsidiary attention gradually displacing focal attention.

On the framework's other dimension, we distinguish between focal attention and subsidiary attention, as suggested originally by Polanyi (1958).[8] By focal attention, we mean conscious attention as commonly understood.[9] By subsidiary attention, we mean attention, conscious or not, to certain *means* for achieving focal attentive *ends*. Such means are characteristically attended to as extensions to the purposeful individual, as Polanyi (1958, p. 59) beautifully describes:

"Our subsidiary awareness of tools and probes can be regarded ...as the act of making them form a part of our own body. The way we use a hammer or a blind man uses his stick, shows in fact that in both cases we shift outwards the points at which we make contact with the things that we observe as objects outside ourselves. While we rely on a tool or probe, these are not handled as external objects. We may test the tool for its effectiveness or the probe for its suitability, e.g. in discerning the hidden details of a cavity, but the tool and the probe can never lie in the field of these operations; they remain necessarily on our side of it, forming part of ourselves, the operating persons. We pour ourselves out

[8] Polanyi (1958) uses the terms "focal awareness" and "subsidiary awareness" in introducing the basic theoretical distinction (p. 55). However, he also refers in subsequent discussion to "focal attention" and "attending subsidiarally." We will presume that in interchanging these terms, he means that awareness of a kind requires equivalent attention.

[9] Ocasio (1997, p. 187) quotes James (1890, pp. 403-404): "Everyone knows what attention is. It is the taking possession by the mind, in clear and vivid form, of one out of what seem several simultaneously possible objects or trains of thought. Focalization, concentration of consciousness are of its essence. It implies withdrawal from some things in order to deal effectively with others ..."

into them and assimilate them as parts of our own existence. We accept them existentially by dwelling in them."

Extrapolating from this notion, we shall be interested here in the focal and subsidiary attention given to an innovation by a purposeful organization. Following Occasio (1997), we define organizational attention to be the socially structured pattern of attention by the organization's decision makers. Such attention is collectively focused, contextually situated, and structurally distributed according to the firm's rules, resources, and social relationships (Occasio, 1997, p. 188). Where organizational innovation is concerned, we argue that focal and subsidiary attention are directed both to the innovative concept and to the innovative practice.

From an organizational learning perspective, four processes are suggested to comprise assimilation in use, as mapped in the two-dimensional framework. In *experimentation*, focal attention is given to carrying out the innovative practice, to confirm and discover what works, as part of the practice itself. Trial-and-error experimentation is well understood to be fundamental to learning by direct experience (Levitt and March, 1988; Cyert and March, 1963). In *interpretation*, focal attention is given to the innovative concept, which is shaped, elaborated, and communicated in coordinating the organization's work. As with experimentation, interpretation of experience is also known to be important to learning, as "people in organizations form interpretations of events and come to classify outcomes as good or bad" (Levitt and March, 1988, p. 323, citing Thompson, 1967). In *routinization*, subsidiary attention displaces focal attention to the innovative technology in practice, now understood from experience to function acceptably (in the absence of new evidence to the contrary). The organization learns from its history, encoding its successful experiments into routines guiding future behavior (Levitt and March, 1988). The IT becomes a tool attended to subsidiarily; as an object it becomes "invisible."[10] Finally, in what we will term *sublimation*, subsidiary attention displaces focal attention to the innovative concept, now increasingly taken-for-granted and largely subsumed within (and inseparable from) the broader everyday discourse.[11] Decision makers speak in terms of the innovation, but largely without being

---

[10] Weiser (1991, p. 94), in speculating on "ubiquitous computing," remarks, "The most profound technologies are those that disappear. They weave themselves into the fabric of everyday life until they are indistinguishable from it."

[11] The root notion of sublimation involves the conversion of something base (here the innovative concept) into something of higher worth (here the notion of the broader enterprise).

aware of it. The organization's cognitive resources are freed to be directed to learning elsewhere, completing the assimilative process.

A basic dynamic to assimilation in use is therefore suggested (note the arrows in Figure 1). Learning in use is posited to arise most fundamentally through experimentation, undertaken out of necessity to perform the unroutinized task. Experimentation in turn motivates interpretation and provides the basis for eventual routinization. It drives the assimilative process; it serves as its engine. When the need for experimentation diminishes, so does the need for related interpretation. The need for routinization continues, but now to reinforce what has already been learned. Learning in use is "completed" with sublimation, absent further interpretation, but present the continually reinforced routine, every execution of which reinforces the institutional order (Barley and Tolbert, 1997). Through sublimation, the innovation becomes "taken for granted" in discourse. Notwithstanding this overall pattern, there are dual effects (feedback) among the processes as shown in Figure 1. In particular, ongoing experimentation may be guided by the interpretation of outcomes as well as by failures in routinization. The four processes may all be active at any one time, in particular, where the innovation has been implemented in stages, to provide for gradual infusion more easily assimilated over time (as illustrated by an ERP system implemented by stages according to functional modules and/or organizational sites, to control the depth and/or breadth of the system's infusion).

Broadly, then, we conceive assimilation in use over time to involve ongoing communicative sensemaking from the innovative practice (as indicated by the upward pointing arrow-heads in Figure 1), as well as the gradual displacement of focal attention by subsidiary attention to the innovation (as indicated by the shaded rightward pointing arrow-heads in Figure 1).

4.   DISCUSSION

In this section, we elaborate briefly on our learning model of IT assimilation in use. As a reminder, we assume that the firm has already adopted and implemented (at least in usable part) the IT innovation. The technology has been deployed. Users have been trained. A "switch" has been thrown and the firm is now "live" with its newly IT-supported work practices. Whatever assimilation has already taken place in the firm's adoption and implementation of the IT innovation (and there may have been extensive and elaborate preparations), the assimilative focus turns now to everyday hands-on use.[12]

We revisit each of the component processes identified in our model as shown in Figure 1. We discuss each in further detail and illustrate our points by drawing from the existing research literature.

4.1   Experimentation

As already mentioned, our model is anchored in a learning-by-doing perspective. The assimilative process (in use) originates with focal attention given to carrying out the innovative practice. We term this activity "experimentation," recognizing that initially upon implementation the organization can't be said to have effective knowledge of the new practice, i.e., knowledge confirmed through experience.[13] Until the firm executes the practice and discovers that it functions satisfactorily, every task action constitutes a kind of experiment. It calls for focal attention to following a particular new procedure, for instance.   Or, it requires conscious improvisation in the absence of such a procedure. Essentially, every task performance is ungoverned by an organizational routine (Pentland and Reuter, 1994).   It is an experiment that can go well as intended (the outcomes will after all be "for keeps"), or badly, but, either way, the firm learns from it.[14]

In the long run, we note, experimentation will lead to *adaptation* of the innovation to the enterprise (Leonard-Barton, 1988; Tyre and Orlikowski, 1994).   That is, the learned work practice as a set of routines will come to be specific to the organization in certain important aspects.   No two firms will undertake the same sequence of experiments.   They will take different *experimental paths*.   Regardless of their implementation approaches, no two

---

[12] Again, implementation activities may not be complete even as everyday use begins. Implementation may be ongoing, in particular, where it has been staged, or where the IT undergoes subsequent improvement (see, e.g., Kling and Iacono, 1984), even in "maintenance" (see, e.g., Hirt and Swanson, 2001). Too, should things go very badly in use, even the adoption decision may come to be reconsidered.   In sum, assimilation in use does not imply that other innovation activities have ended (Swanson, 2003).

[13] Obviously, much may be learned about the IT innovation prior to assimilation in use, especially in the course of implementation, where more formal testing and experimentation may be undertaken, without risks to organizational outcomes. But, we suggest, no amount of such testing and training substitutes for the learning that inevitably takes place when the firm goes live with its system and organizational actions are "for keeps."

[14] Such learning can in part be superstitious; that is, the inferred connections between actions and outcomes can be mis-specified (Levitt and March, 1988).

firms will discover the identical functionality in the innovation. They will learn differently from their usage experiences. Thus, while two firms may become more similar to each other by adopting the same innovation (reflecting mimetic isomorphism as described by DiMaggio and Powell, 1983), in their assimilations of the innovation they are likely to again become different (although not so different as before).

Similarly, it is through experimentation that the innovation is likely to be *reinvented* (Rice and Rogers, 1980), for example, through the discovery of unanticipated uses. While some of these uses may be specific to the firm, others may prove to be widely imitable. Beyond such new "features," certain "work-arounds," initially devised to cope with problematic aspects of the innovation, may come to be essential to the learned work practice (Gasser, 1986). Broadly, such adaptations may occur entirely in the context of use, or they may feed back into continued development of the technology, suggesting the provisioning of new features. They will also motivate the innovation's interpretation among participants, to be discussed below.

Experimentation can also contribute to the innovation's *infusion*, where increasingly sophisticated use is made of the new IT, with new functionality often introduced in stages (Cooper and Zmud, 1990). Achieving this desired usage requires experimental initiatives, not just advanced training. But experimentation is also effortful and may wane prematurely, leaving the new IT under-explored and the firm's routines much the same as before. The more advanced features of the new IT may not be used at all and the firm may find itself simply "paving over the cow-paths," rather than purposefully changing its work practices. For the innovation's benefits to be achieved, the firm may need to push on the experimental frontier, so as to engage and leverage the full functionality offered by the new IT.

4.2   Interpretation

In interpretation, the firm gives focal attention to the innovative concept in its coordinative communications, a process that originates long before the firm puts the innovation to use. Indeed, the firm's initial encounter with an IT innovation is likely to be an interpretive one, as it attempts to grasp the associated organizing vision (Ramiller and Swanson, 2003). Similarly, in choosing to adopt the innovation, the firm typically articulates the know-why for its decision, making a further interpretive effort (Swanson, 2003). This know-why is further reconstructed and reinterpreted as the firm moves forward, first in implementation, then in use, where the innovative benefits are presumably to be found. The firm asks again, "Why are we doing this? Where are the benefits?"

Once the IT innovation is put to use, its interpretation is substantially driven by the experimentation described above. As the firm experiments, it interprets and evaluates outcomes. It organizes itself and initiates coordinative activities, both formal and informal, to bridge interpretive gaps. It establishes cross-functional teams and liason roles. It engages in coordinative meetings of all kinds. It undertakes progress reporting, policy setting, information dissemination, and propagandizing on behalf of the innovation. All of this serves the interpretive and coordinative effort and aids the firm in dealing with the uncertainties of its experimentation.

Broadly, the identification of successful outcomes also reshapes the notion of the innovation and what it is good for, in a way more specific to the firm. As a concept, the innovation is not reinvented through experimentation alone, as described above, but by the closely linked interpretive process. Thus, two firms that have made substantial use of the "same" innovation, having taken different experimental paths, are likely to arrive at somewhat different notions of what the innovation is in practice all about. In some cases, through professional interactions, these interpretations may come to influence the broader community's notion of the organizing vision.

For the firm itself, interpretation will also be important to the innovation's local *institutionalization* (Scott, 1995).[15] The prevailing institutional order will of course strongly condition this discourse. Even ostensibly radical interpretations may be couched so as to reproduce this order. Scripted patterns of interaction can therefore persist, notwithstanding change rhetoric suggesting otherwise. Still, in some instances, institutional change may be entertained. Organizational members may seek to make changes in their scripts (Barley and Tolbert, 1997). They may choose to interpret the innovation as altering the organization's status quo, and argue that a new context for action prevails. Thus, while the discourse itself reflects current institutional arrangements, it is also the occasion for renegotiating these arrangements. The firm that has implemented an ERP system, for instance, may struggle to achieve improved cross-functional coordination, breaking down the authority structure of its old functional "silos" (Davenport, 1998). Much talk, reflective of different contesting interests, may be expended toward this end.

---

[15] We understand institutionalization to take place simultaneously both within the enterprise, i.e. "locally," and more broadly among firms, i.e. "globally." The former is substantially shaped by the latter, especially to the extent the innovation is widely diffused (Tolbert and Zucker, 1983). Even so, local institutionalization will reflect firm-specific regulative, normative, and cognitive forces (Scott, 1995).

4.3   Routinization

As firms carry out task experiments as described above, they will gradually encode their successful performances into organizational routines. That is, they will gradually identify sets of possible performances governing organizational tasks, where individual actors know what "moves" they can make (Pentland and Reuter, 1994). The new IT will over time pose fewer puzzles in this respect. Attention to the IT innovation will be increasingly undifferentiated from attention to the routines and their moves; it will be increasingly subsidiary to the focal routines.

Routinization describes not only the initial encoding process for the production of new organizational routines; it describes too how organizational routines are maintained, namely through their repeated execution. What the firm knows how to do via its routines, it "remembers" only by executing the routines again and again. The firm's routines are much like their employees' skills; both routines and skills can become rusty and both are honed primarily in repeated execution (Nelson and Winter, 1982). Both are also associated with the building of tacit knowledge into the firm's capabilities.

Routinization will thus also be the basis for performance improvements in task execution, likely to be necessary to achieving the innovation's promise. With repeated executions, the firm comes up the "learning curve" associated with the innovation. It becomes more efficient and wastes less of its efforts. It achieves the desired "continuous improvement" in this regard. Routinization thus incorporates further learning (Feldman and Pentland, 2003). However many times a task has been executed, improvisation under the governing routine may be needed the next time around. Small experiments are likely to persist.[16]

Finally, routinization does not necessarily mean that task performances meet nominal standards (set by management or process engineers). Rather, it is likely to reflect a "stable accommodation between the requirements of organizational functioning and the motivations of all organization members" (Nelson and Winter, 1982, p. 108). In short, the firm's routines rest upon a political resolution of multiple interests. Moreover, every performance of the routine tends to re-affirm this resolution. Routinization thus supports the normative and regulative aspects of institutionalization (Scott, 1995).

---

[16] Small experiments under a governing routine are very different, we suggest, from experimentation in the absence of a governing routine.

4.4   Sublimation

As experimentation and novel outcomes wane, so too does the interpretation effort. Routinization takes over, and focal attention is displaced by subsidiary attention to the innovation. The firm now makes "sense" of the innovation by turning its terminology to other interpretive ends. The innovation is not so much absent as it is "sublimated" in the broader organizational discourse. Focal discursive attention is directed elsewhere, as the IT innovation becomes taken-for-granted, even while its terminology lingers, invoked in subsidiary roles. Much like the new IT becomes "invisible" as a tool in the innovative practice, it becomes "invisible" as an idea in coordinative communication. Organizational members "speak the language of ERP," for instance, in their local dialect, without focusing on ERP as such.[17]

The IT innovation is now *accepted* in that political resolution of multiple interests is reflected in the innovation's routinization as just described.[18] The innovation is itself no longer a focus of contending interests in the firm's discourse. To the extent that problems persist, they are likely to bubble up elsewhere. Thus, for example, where an IT innovation leads to certain deskilling, while this issue may be resolved in this context, it may surface in another guise elsewhere in the firm's labor relations.

Because organizational discourse no longer focuses on the IT innovation as such, the firm's verbal memory of it is likely to atrophy. At the same time, certain stories associated with the innovation may take hold and be passed along over the years, entering into the firm's *mythology*. The organizational lessons drawn from these stories may serve as reminders of the firm's assimilative struggles.

---

[17] To illustrate further, advocates for CRM (Customer Relationship Management) foresee the day when "CRM will take its place in mainstream American business and will be practiced as seriously as financial management." (Dver, 2003, p. 64) In this circumstance, the language of CRM would be identifiable, but would blend with that of management more broadly. In our terms, the CRM concept would be sublimated to the extent it permeated the broader discourse, without calling special attention to itself.

[18] The traditional literature on user acceptance of IT takes individual, voluntary use to be definitive of acceptance (Venkatesh, et al, 2003). We make no such assumption. Rather, we take usage of the IT as the context within which to evaluate acceptance.

5.   CONCLUSION

To summarize, we have argued in this paper that IT innovations are assimilated substantially in their use, beyond their implementation. The current literature on IT innovation assimilation has until now had relatively little to say about this. Here we have sought to open some new doors. Taking an organizational learning perspective, we have introduced a framework that suggests the particular mechanism by which an IT innovation is assimilated in use. In our discussion, we have sought to tie this notion back to related concepts and research results reported elsewhere in the literature.

Several opportunities for future research may be identified. These include, first of all, longitudinal case studies of IT assimilation in use within enterprises, to further explore the assimilation mechanism suggested here. Of particular interest would be studies that illuminate experimentation and its role in facilitating the innovation's routinization. We have argued here that experimentation is crucial to the establishment of routines, and thereby to assimilative outcomes, but it remains to subject this notion to careful empirical examination.

Of equal interest would be case studies that focus on the role of interpretation in the firm's assimilative process. Here we have argued that interpretation is fueled substantially by experimentation, but that with routinization, focal attention is displaced by subsidiary attention, enabling the innovation to become taken-for-granted in discourse. Again, this notion requires empirical examination.   Opportunities exist, in particular, for situated discourse studies (Phillips and Hardy, 2002) in tandem with observational studies of collective task performance.[19]

Comparative case studies of different firms' assimilation of the same IT innovation offer additional research opportunities. Here we note, firms may bring very different strategies, structures, and innovation histories to their assimilations. How do these affect assimilative outcomes? Ocasio (1997) suggests, for instance, that the purposeful organization of communications and procedures can channel decision makers' attention, with important consequences for a firm's learning. How might structural differences of this kind guide or otherwise affect innovation assimilation, in particular?

---

[19] See Rose and Kraemmergaard (2003) for a study of the discourse in implementing one ERP system. Interestingly, this study includes the period of the system's first use. The authors suggest that the term "implementation" may be inappropriate for large ERP projects; rather the experience is more like a never-ending journey.

Lastly, a few caveats. First, we note that the completion of the firm's assimilative process, as we have described it, should not be conflated or confused with the amount of innovation achieved. Consider, for instance, that two firms may take different assimilative paths with the same innovation, with one firm choosing to transform itself, while the other tailors the innovation more to existing practice. Here both firms may "complete" their assimilations, while achieving very different innovative outcomes.[20]

We note too that while organizational assimilation of IT innovations may be important to achieving desired outcomes, it is not in itself necessarily a good thing. Organizational learning with new IT may eventually result in the setting of *competency traps* (March, 1981; Levitt and March, 1988) that impede further innovation. Certain of these traps may in effect be technology traps, where the firm is disadvantageously "locked in" to a particular IT solution (Shapiro and Varian, 1999). Finally, the political resolution of different interests achieved through routinization, an essential aspect of assimilation, in our view, does not necessarily advance *social justice*. Thus, the different consequences of IT innovation assimilation are also an important subject of research and critical study.

ACKNOWLEDGEMENTS

I'm grateful to Yutaka Yamauchi and Arnaud Gorgeon for their helpful research assistance, and to Brian Pentland, Neil Ramiller and Ping Wang for their comments on an earlier draft. I also benefited from a panel discussion of IT assimilation with Rob Fichman and Bob Zmud at DIGIT 2003 in Seattle.

REFERENCES

Agarwal, R., Tanniru, M., and Wilemon, D., "Assimilating Information Technology Innovations: Strategies and Moderating Influences," *IEEE Transactions on Engineering Management*, 44, 4, 1997, 347-358.

---

[20] More broadly, firms may undertake the same innovation, beginning from very different positions, with respect to their capabilities. They may adopt the innovation for different reasons. They may take different approaches to implementation. Their assimilative paths may be different, and, of course, they may achieve very different outcomes. I'm grateful to Ping Wang for this important observation.

Armstrong, C. P., and Sambamurthy, V., "Information Technology Assimilation in Firms: The Influence of Senior Leadership and IT Infrastructure," *Information Systems Research*, 10, 4, 1999, 304-327.

Attewell, P., "Technology Diffusion and Organizational Learning: the Case of Business Computing," *Organization Science*, 3, 1, 1992, 1-19.

Barley, S. R., and Tolbert, P. S., "Institutionalization and Structuration: Studying the Links Between Action and Institution," *Organizational Studies*, 18, 1, 1997, 93-117.

Cho, I., and Kim, Y.-G., "Critical Factors for Assimilation of Object-Oriented Programming Languages," *Journal of Management Information Systems*, 18, 3, 2001-2002, 125-156.

Cohen, W. M., and Levinthal, D. A., "Absorptive Capacity: A New Perspective on Organizational Learning," *Administrative.Science Quarterly*, 35, 1990, 128-152.

Cooper, R. B., and Zmud, R. W., "Information Technology Implementation Research: A Technological Diffusion Approach," *Management Science*, 36, 2, 1990, 123-139.

Crossan, M. M., Lane, H. L., and White, R. E., "An Organizational Learning Framework: From Intuition to Institution," *Academy of Management Review*, 24, 3, 1999, 522-537.

Cyert, R. M., and March, J. G., *A Behavioral Theory of the Firm*, Englewood Cliffs, NJ: Prentice-Hall, 1963.

Davenport, T. H., "Putting the Enterprise into the Enterprise System," *Harvard Business Review*, July-August 1998, 121-131.

DeSanctis, G., and Poole, M. S., "Capturing the Complexity in Advanced Technology Use: Adaptive Structuration Theory," *Organization Science*, 5, 2, 1994, 121-147.

Dewar, R. D., and Dutton, J. E., "The Adoption of Radical and Incremental Changes: An Empirical Analysis," *Management Science*, 32, 11, 1986, 1422-1433.

DiMaggio, P. J., and Powell, W. W., "The Iron Cage Revisited: Institutional Isomorphism and Collective Rationality in Organizational Fields," *American Sociological Review*, 48, 1983, 147-160.

Dver, A., "Customer Relationship Management: The Good. The Bad. The Future," Special Advertising Section, *Business Week*, April 28, 2003, 52-64.

Feldman, M. S., and Pentland, B. T, "Re-theorizing Organizational Routines as a Source of Flexibility and Change," *Administrative Science Quarterly*, 48, 1, 2003, 94-118.

Fichman, R. G., "The Diffusion and Assimilation of Information Technology Innovations," in Zmud, R. (Ed.), *Framing the Domains of IT Management Research*, Pinnaflex Educational Resources, Cincinnati, OH, 2000.

Fichman, R. G., and Kemerer, C. F., "The Assimilation of Software Process Innovations: An Organizational Learning Perspective," *Management Science*, 43, 10, 1997, 1345-1363.

Fichman, R. G., and Kemerer, C. F., "The Illusory Diffusion of Innovation: An Examination of Assimilation Gaps," *Information Systems Research*, 10, 3, 1999, 255-275.

Gallivan, M., "Organizational Adoption and Assimilation of Complex Technological Innovations: Development and Application of a New Framework," *Data Base*, 32, 3, 2001, 51-85.

Gasser, L., "The Integration of Computing and Routine Work," *ACM Transactions on Office Information Systems*, 4, 3, 1986, 225-250.

Grover, V., Fiedler, K., and Teng, J., "Empirical Evidence on Swanson's Tri-Core Model of Information Systems Innovation," *Information Systems Research*, 8, 3, 1997, 273-287.

Grover, V., and Goslar, M., "The Initiation, Adoption and Implementation of Telecommunications Technologies in U.S. Organizations," *Journal of Management Information Systems*, 10, 1, 1993, 141-163.

Hirt, S. G., and Swanson, E. B., "Emergent Maintenance of ERP: New Roles and Relationships," *Journal of Software Maintenance and Evolution*, 13, 2001, 373-397.

Huber, G. P., "Organizational Learning: the Contributing Processes and the Literatures," *Organization Science*, 2, 1, 1991, 88-115.

James, W., *The Principles of Psychology*. New York: Holt, 1890.

Kauffman, R.J., McAndrews, J., and Wang, Y-M., "Opening the 'Black Box' of Network Externalities in Network Adoption," *Information Systems Research*, 11, 1, 2000, 61-82.

Klein, K. J., and Sorra, J. S., "The Challenge of Innovation Implementation," *Academy of Management Review*, 21, 4, 1996, 1055-1080.

Kling, R., and Iacono, S., "The Control of Information System Developments After Implementation," *Communications of the ACM*, 27, 1984, 1218-1226.

Kwon, T. H., and Zmud, R. W., "Unifying the Fragmented Models of Information Systems Innovation," in Boland, R. J., and Hirschheim, R. A. (Eds.), *Critical Issues in Information Systems Research*, New York: Wiley, 1987, 227-251.

Leonard-Barton, D., "Implementation as Mutual Adaptation of Technology and Organization," *Research Policy*, 17, 1988, 251-267.

Leonard-Barton, D., and Deschamps, I., "Managerial Influences in the Implementation of New Technology," *Management Science*, 34, 10, 1988, 1252-1265.

Levitt, B., and March, J. G., "Organizational Learning," *Annual Review of Sociology*, 14, 1988, 319-340.

Lewin, K., "Group Decision and Social Change," in Swanson, G. E., Newcomb, T. M., and Hartley, E. L. (Eds.), *Readings in Social Psychology*. 2nd Edition. Holt, New York, 1952, 459-473.

Lucas, H. C., Jr., *Implementation: The Key to Successful Information Systems*, New York: Columbia University Press, 1981.

March, J. G., "Footnotes to Organizational Change," *Administrative Science Quarterly*, 26, 1981, 563-577.

Markus, M.L., "Power, Politics, and MIS Implementation," *Communications of the ACM*, 26, 6, 1983, 430-445.

Markus, M. L., and Tannis, C., "The Enterprise Systems Experience- From Adoption to Success," in Zmud, R. (Ed.), *Framing the Domains of IT Management Research*, Pinnaflex Educational Resources, Cincinnati, OH, 2000, 173-207.

Majchrzak, A., Rice, R.E., Malhotra, A., King, N., and Ba, S., "Technology Adaptation: The Case of a Computer-Supported Inter-Organizational Virtual Team," *Management Information Systems Quarterly*, 24, 4, 2000, 569-600.

Martinsons, M. G., and Schindler, F. R., "Organizational Visions for Technology Assimilation: The Strategic Roads to Knowledge-based Systems Success," *IEEE Transactions on Engineering Management*, 42, 1, 1995, 9-18.

McMaster, T., Mumford, E., Swanson, E. B., Warboys, B., and Wastell, D. (Eds.), *Facilitating Technology Transfer through Partnership*, London: Chapman & Hall, 1997.

Meyer, A. D., and Goes, J. B., "Organizational Assimilation of Innovations: A Multilevel Contextual Analysis," *Academy of Management Journal*, 31, 4, 1988, 897-923.

Nelson, R. R., and Winter, S. G., *An Evolutionary Theory of Economic Change*, Cambridge, MA: Harvard University Press, 1982.

Ocasio, W., "Towards an Attention-based View of the Firm," *Strategic Management Journal*, 18 (Summer Special Issue), 1997, 187-206.

Orlikowski, W. J., "CASE Tools as Organizational Change: Investigating Incremental and Radical Changes in Systems Development," *MIS Quarterly*, 17, 3, 1993, 309-340.

Orlikowski, W. J., "Improvising Organizational Transformation Over Time: A Situated Change Perspective," *Information Systems Research*, 7, 1, 1996, 63-92.

Orlikowski, W. J., "Using Technology and Constituting Structures: A Practice Lens for Studying Technology in Organizations," *Organization Science*, 11, 4, 2000, 404-428.

Pentland, B. T., and Reuter, H. H., "Organizational Routines as Grammars of Action," *Administrative Science Quarterly*, 39, 1994, 484-510.

Phillips, N., and Hardy, C., *Discourse Analysis: Investigating Processes of Social Construction*. Thousand Oaks, CA: Sage Publications, 2002.

Polanyi, M., *Personal Knowledge*. Chicago: University of Chicago Press, 1958.

Purvis, R. L., Sambamurthy, V., and Zmud, R. W., "The Assimilation of Knowledge Platforms in Organizations: An Empirical Investigation," *Organization Science*, 12, 2, 2001, 117-135.

Ramiller, N. C., and Swanson, E. B., "Organizing Visions for IT and the IS Executive Response," *Journal of Management Information Systems*, 20, 1, 2003, 13-43.

Rice, R., and Rogers, E., "Reinvention in the Innovation Process," *Knowledge: Creation, Diffusion, Utilization*, 1, 4, 1980, 499-514.

Robey, D., Boudreau, M., and Rose, G. M., "Information Technology and Organizational Learning: A Review and Assessment of the Literature," *Accounting, Management and Information Technologies* 10, 2, 2000, 125-155.

Robey, D., Ross, J. W., and Boudreau, M.-C., "Learning to Implement Enterprise Systems: An Exploratory Study of the Dialectics of Change," *Journal of Management Information Systems*, 19, 1, 2002, 17-46.

Robey, D., and Sahay, S., "Transforming Work Through Information Technology: A Comparative Case Study of Geographic Information Systems in County Government," *Information Systems Research*, 7, 1, March 1996, 93-110.

Rogers, E. M., *Diffusion of Innovations*, 4th Edition, New York: Free Press, 1995.

Rose, J., and Kraemmergaard, P., "Dominant Technological Discourses in Action: Paradigmatic Shifts in Sense Making in the Implementation of an ERP System," in Wynn, E. H., Whitley, E. A., Myers, M. D., and DeGross, J. I. (Eds.), *Global and Organizational Discourse about Information Technology*, Dordrecht: Kluwer, 2003, 437-462.

Ross, J. W., "The ERP Revolution: Surviving versus Thriving," MIT Sloan School, CISR Working Paper #307, 1999.

Saga, V. L., and Zmud, R. W., "The Nature and Determinants of IT Acceptance, Routinization, and Diffusion," in Levine, L. (Ed.), *Diffusion, Transfer and Implementation of Information Technology*, North-Holland, 1994, 67-86.

Scott, W. R., *Institutions and Organizations*, Sage Publications, Thousand Oaks, CA, 1995.

Shapiro, C., and Varian, H. R., *Information Rules*. Cambridge, MA: Harvard Business School Press, 1999.

Simon, H. A., *Administrative Behavior*. Chicago, IL: Macmillan, 1947.

Swanson, E. B., *Information System Implementation*, Homewood, IL: Irwin, 1988.

Swanson, E. B., "Talking the IS Innovation Walk," in Wynn, E. H., Whitley, E. A., Myers, M. D., and DeGross, J. I. (Eds.), *Global and Organizational Discourse about Information Technology*, Dordrecht: Kluwer, 2003, 15-31.

Swanson, E. B., and Ramiller, N. C., "The Organizing Vision in Information Systems Innovation", *Organization Science*, 8, 1997, 458-474.

Thompson, J. D., *Organizations in Action*, New York: McGraw-Hill, 1967.

Tolbert, P. S., and Zucker, L. G., "Institutional Sources of Change in the Formal Structure of Organizations: The Diffusion of Civil Service Reform, 1880-1935," *Administrative Science Quarterly* 30, 1983, 22-39.

Tyre, M. and Orlikowski, W., "Windows of Opportunity: Temporal Patterns of Technology Adaptation in Organizations," *Organization Science*, 5, 1, 1994, 98-118.

Venkatesh, V., Morris, M. G., Davis, G. B., and Davis, F. D., "User Acceptance of Information Technology: Toward a Unified View," *Management Information System Quarterly*, 27, 3, 2003, 425-478.

Weick, K. E., *Sensemaking in Organizations*, Thousand Oaks, CA: Sage Publications, 1995.

Weick, K. E, and Roberts, K. H., "Collective Mind in Organizations: Heedful Interrelating on Flight Decks," *Administrative Science Quarterly*, 38, 1993, 357-381.

Weiser, M., "The Computer for the 21st Century," *Scientific American*, September 1991, 94-100.

Zmud, R. W., and Apple, L. E., "Measuring Technology Incorporation/Infusion," *Journal of Product Innovation Management*, 9, 1992, 148-155.

GROUPWARE INTEGRATION IN VIRTUAL LEARNING TEAMS
A qualitative analysis based on the TAM-model

Pernille Bjørn and Ada Scupola
Roskilde University, Denmark

Abstract: In this paper we apply Davis' Technology Acceptance Model (TAM) in a qualitative fashion to analyze and interpret the chronological sequence of events leading to the acceptance of the Groupware technology, BSCW, in a virtual learning team. The research question investigated is: What are the factors influencing the integration process of Groupware technology in virtual learning teams in part-time adult education? The data were gathered through an in-depth qualitative action research study of one virtual learning team doing problem-oriented project work within a master education program. We find that one important factor influencing the integration process of Groupware is: How the technology provides support for social perspective awareness. In the case investigated the technology BSCW supported social awareness, which influenced both the ease-of-use and the perceived-usefulness of the Groupware technology, thus being an important condition that influenced the positive outcome of the integration process.

Key words: Technology Acceptance Model, Social Awareness, Groupware, Master Education, Virtual Team, Learning, BSCW, CSCL

1. INTRODUCTION

Problem-oriented learning and university teaching based on a pedagogy whereby students collaborate in teams are a central part of newer educational research (Olesen and Jensen, 1999). At the same time there is a demand that it should be possible to take an education anywhere at any time. Close collaboration in distance setting *could* be a contradiction in terms, however new technology such as Groupware gives us the opportunity to do both. Geographical distributed project teams need technology support for collaborative activities. We define a virtual team as a team comprising geographically distributed participants who mainly mediate their collaborative activities through technology. The team collaboration can be supported by e-mail correspondence, phone meetings or by using more advanced Groupware technologies such as the BSCW (Basic Support for Cooperative Work). The foundation of these kinds of technologies is advanced file-management systems based on web-technologies (Bentley et al., 1997). We however need new innovations in this field to insure success with Groupware technologies.

Current research on Groupware states that appropriate support for the integration, implementation and continued use of the technology is crucial for success (Karsten, 1999; Orlikowski, 1992; Grudin, 1994 and Majchrzah et al., 2000). Drawing on these findings we conducted an action research project, where we intervened to facilitate a virtual learning team in integrating Groupware. Our action research approach was to support the integration by assisting the team to develop and implement coordination mechanisms (Schmidt and Simone, 1996). Here we describe how the virtual learning team went from using e-mail and phone to mediate their collaboration by integrating Groupware to support the distributed practice.

When analyzing the data we needed a model to examine why the integration succeed. We used Davis' Technology Acceptance Model (Davis, 1989) – the TAM-model – as a 'lens' or framework to describe and analyze the different factors that influenced the integration process at different periods of time. We chose to use the TAM-model for two main reasons. First the TAM-model is the most well-known, rigorously validated empirically and widely accepted model for examining technology acceptance (Legris et al., 2003; Adams et al., 1992; Davis, 1989; Davis et al., 1989), and secondly because we found that the model constructs Ease-of-use, Perceived-usefulness, Intentional-use and Actual-use provided a framework, which was useful when analyzing the integration process.

Groupware support for Problem Oriented Project work in distributed part-time education has been investigated in a range of studies (e.g. Dirckinck-Holmfelt and Sorensen, 1999; Cheesman and Heilesen, 1999; Bjørn, 2003). However none of these studies has used the TAM-model to investigate the Groupware integration process. This paper does so by reflecting and analyzing the integration of Groupware technology in a virtual learning team by using the TAM-model in a qualitative and interpretative way. The overall research question investigated is: What are the factors influencing the integration process of Groupware technology in virtual learning teams in part-time adult education?

We found that besides the importance of coordination mechanisms another aspect of Groupware technology influenced the integration in a positive way: social perspective awareness. We define social perspective awareness as background information on belief and knowledge of team members, similar to the information one can gather when working around in the physical office. Groupware technology affected the social setting in the virtual team by providing support for an unarticulated need in the team for social perspective awareness. So besides supporting the need for coordination and document handling, which we expected, the Groupware technology also supported the team's need for perspective awareness. This factor influenced the integration of this technology in a positive way. We use the TAM-model to illustrate this point while showing that social perspective awareness influences both the construct Ease-of-use and the construct Perceived-usefulness in the model.

The paper is structured as follows. In the next section, the TAM model and its application in a variety of contexts are presented. Then, the concept of social perspective awareness and its relevance to virtual teams is discussed. Following this, the research setting, research method and data analysis is presented. In the empirical part of the paper we analyze the integration of Groupware technology over four key checkpoints, followed by a discussion of the research results. Finally, the conclusions, limitations and implications of the study for further research using the TAM model in integration of Groupware in virtual learning teams are discussed.

2.   THE CONCEPTUAL BASE

2.1   Technology Acceptance Model (TAM)

Davis (1989) synthesized the findings of a range of diverse research streams to propose the Technology Acceptance Model (TAM), which identifies a number of constructs relevant to technology acceptance (First publish in Davis' dissertation in 1986). These constructs fall into two broad categories, *ease of use (EoU)* and *perceived usefulness (PU)*. Davis suggests a chain of causality between these categories: greater EoU leads to higher PU, which in turn leads to more usage of technology (see Fig. 1).

In the original TAM-model two more constructs were present in addition to those in fig. 1: *External variables* and *Attitude*. The external variables influenced both EoU and PU, while Attitude was influenced by both EoU and PU affecting the construct Intentional-use. We have chosen not to include the External-variables and Attitude in our model, drawing on the research of Legris et al. (2003) and Gefen et al. (2003). In fact, Legris et al. (2003) found that in most cases researchers have mainly considered EoU and PU and their effects on Intentional-use.

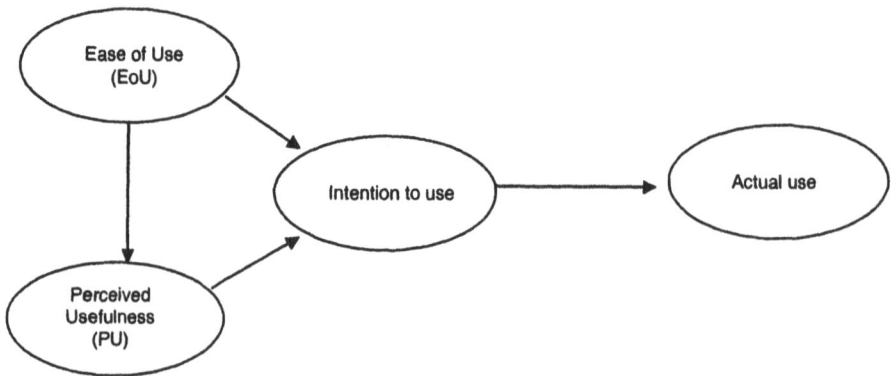

Figure 1. Davis' Technology Acceptance Model (TAM), (Davis, 1989)

The TAM-model has been used in a range of studies. Kwon and Chidambaram (2000) use and test the TAM model to examine patterns of cellular phone adoption and usage in an urban setting. The results of the study confirm that users' perceptions and especially perceived EoU, are significantly associated with the motivation to use cellular phones. Lederer et al. (2000) applies TAM in relation to work-related tasks in the context of the World Wide Web. They find full support for the TAM model and demonstrate that ease of understanding and ease of finding predict EoU, and that information quality predicts usefulness for revisited sites. Yager (1999) uses the TAM constructs to address the perceptions of currently available and yet-to-be-released IT support mechanisms among virtual and face-to-face (non-virtual) teams. The study shows that virtual team members reported greater EoU and PU of the IT support mechanisms than non-virtual team members together with more Intention-to-use. Pavlou (2001) extends the TAM model to incorporate the constructs of risk and trust in consumer intention to adopt electronic commerce. The TAM model has also been extended with variables such as control, intrinsic motivation and emotion (Venkatesh, 2000), and has been used in the marketing field to explain online consumer behavior (e.g. Koufaris, 2002). Social influence has been found important in technology acceptance. Venkatesh and Davis (2000) develop and test a modified version of the original TAM, which explains PU and usage intentions in terms of social influence (subjective norm, voluntariness, and image) and cognitive instrumental processes (job relevance, output quality, and result demonstrability). This study was however not done in concern to virtual teams or Groupware technology.

Originally the TAM-model was developed to study the integration and acceptance of an IT-system in an individual setting, and the different constructs of the model were measured at a given point in time by using quantitative data collected mainly through surveys. Three main points differentiate earlier use of TAM from the use in this paper. First our goal is not to measure the TAM's constructs at a given point in time, but instead we want to describe how the different constructs of the TAM-model *change* during the integration and acceptance process. Secondly we use the TAM-model in a *qualitative study*, whereas it is normally used in a quantitative fashion. Others have also used the TAM-model in a qualitative approach e.g. Neville and Fitzgerald (2002). Finally, while the TAM-model is typically concerned with an individual's acceptance of technology, we use it to investigate *a group's* acceptance of technology.

2.2 Virtual Teams and Social Awareness

Research on the integration of technology in virtual teams is important, because virtual teams in general have more malleable structures due to the typically 'not-yet-organized' and more informal organization (Majchrzak et al. 2000). In this way we can learn from virtual teams when interested in change management. Also research on virtual team is important to understand and support the practice of global organizations. Some research has taken up the challenge of investigating virtual teams by focusing on managerial aspects such as the role of the project manager (e.g. Kayworth and Leidner, 2002; Paré and Dubé, 1999; Piccoli and Ives, 2000), while other research has taken an empirical approach to exploring the process of virtual teams (e.g. Maznevski and Chudoba, 2000). Most of the research has however been conducted with emphasis on special aspect of virtual teams such as culture or trust (e.g. Jarvenpaa et al., 1998; Massey et al., 2001; Alexander, 2002). No research has however addressed the importance of social perspective awareness in virtual teams.

Social and people-centered issues such as social awareness have been found important when researching group collaboration (Steinfield et al., 1999; Prinz, 1999, Tollman et al., 1996; Schmidt, 2002). Awareness as a concept has been categorized in different ways. Prinz (1999) identifies two types of awareness: social awareness and task-oriented awareness. In our study the type of awareness relevant to the integration of Groupware can be categorized as social awareness according to Prinz (1999: p. 2) definition: namely, to provide information similar to 'information received when walking along the office floor'. Others have also located the importance of social awareness saying that 'Awareness involves knowing who is "around", what activities are occurring, who is talking with whom; it provides a view of one another in the daily work environments (Dourish and Bly, 1992)".

Steinfield et al. (1999) proposes a number of awareness categories, including activity awareness, availability awareness, process awareness, environmental awareness, and perspective awareness. A full discussion of each of these categories is beyond the scope of this paper. However, the one most relevant to our study is that of *perspective awareness* which is defined as 'giving group members information helpful for making sense of others' actions, such as background on team members belief and knowledge' (Steinfield et al., 1999: p. 84). There is a difference between knowing who is around and background knowledge necessary to interpret others' actions. In our empirical data, the importance and need was on the background knowledge: perspective awareness.

For the purpose of this study, we complement therefore Prinz definition of social awareness with aspects of perspective awareness in giving group members information helpful for making sense of each others' actions so that the information received is similar to the one gathered when walking along the office floor'. Finally, the setting investigated in our study is an asynchronous setting, meaning that the team members do not work synchronously or at the same time, while being apart from each other. The need for 'knowing if people are available in a synchronous perspective' is thus not an issue. Therefore, our concept of social awareness differs, from Tollmar et al. (1996) and Dourish and Bly (1992), which both focuses on a synchronous context. The awareness concept used in this paper can be defined as asynchronous social perspective awareness – here just referred to as social awareness.

3.   THE RESEARCH CONTEXT

3.1   Research setting

The empirical investigation was conducted within a part-time Master program (the Master of Adult Education at Roskilde University in Denmark), where a virtual learning team was closely followed. The focus of the investigation was to facilitate and in this way locate important factors for success when integrating the Groupware technology BSCW in the virtual team.

The master of adult education is a three-year part-time university education for people active on the labour marked. To apply and being accepted the students need a bachelor degree and at least two years of job experience. The average participant is a woman between 40 and 50 years old working with health or education in the public sector. Because it is a part-time education each semester is stretched over a whole year from September to June, and each year the program requires the students to attend five weekend-long seminars on campus in September, November, January,

March and April. Over 50% of the program is based on project work in groups of two to four participants with an academic supervisor.

The team in focus consisted of three students in their mid-thirties: Emma, Thomas and Lisa. They all had families and were in full-time employment, so study-time was on weekends and evenings. The three team-members lived far apart, leaving little opportunity to physically meet, apart from the five seminars on campus. Due to different working hours the team primarily collaborated asynchronously. On this basis, we classify the group as a virtual team according to Steinfield et al. (1999) definition: 'any group characterized by having members in different locations' and use the terms virtual learning team and group interchangeably.

It could be argued that there is a difference between groups in educational and working contexts. However, following the Schmidt and Simone' (1996: p. 158) definition of cooperative work as 'constituted by the interdependencies of multiple actors who interact through changing the state of a common field of work', we believe the setting of the study represented a true and realistic work context. The multiple actors were Emma, Thomas and Lisa, the common field of work was their project, and the state was changed through discussion, reading, writing, revising documents – which in the end led to the final project report they turned in to the exam.

Our research focused on the coordination activities surrounding the production of an outline of a project report that the group was required to submit to the supervisor by a particular deadline. The research was conducted during the group's last year at the university. All group members had experience with project work in virtual teams mediated by e-mail and phone, but no experience in using Groupware to support their collaboration.

3.2 Technology

The Groupware system used in the investigation was Basic Support for Cooperative Work (BSCW, further details at bscw.gmd.de), one of the most well known CSCW (Computer-Supported Cooperative Work) systems in the academic world (Bentley et al., 1997).

The BSCW system is a web-based CSCW system, which supports file-management, asynchronous and synchronous dialogs, management of URLs, and calendar functions. The BSCW system also supports different awareness functions such as monitoring which documents, folders and notes are new, read, revised or moved. It is also possible to get automatically e-mail notifications, when different events occur within the system. The BSCW broad functionality and versatility, which allow users to adjust the conceptual structures as needed makes it a strong tool, when there is a requirement to collaborate and coordinate different tasks within a distributed group.

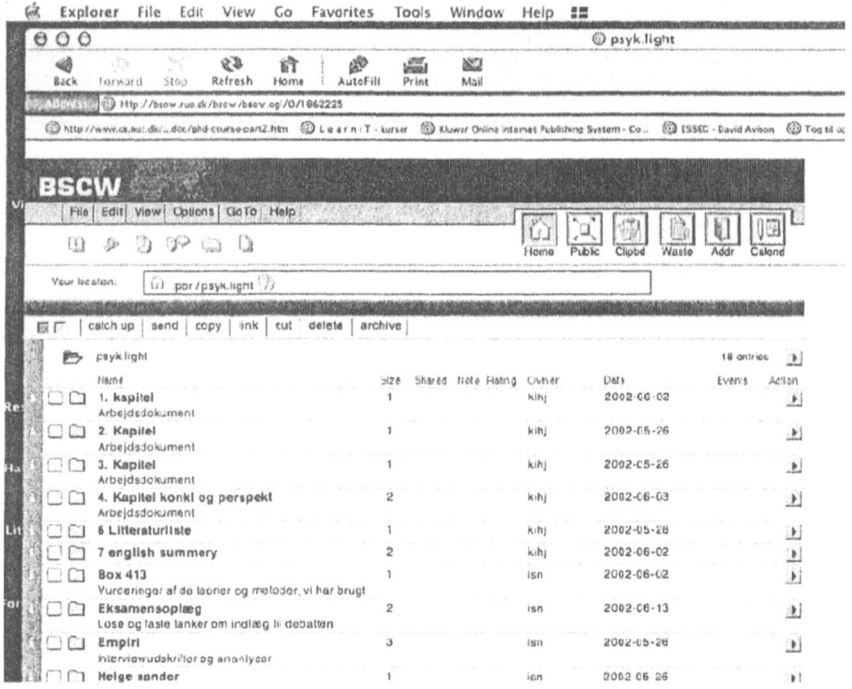

3.3   Research method

Research on Groupware integration and use based on 'experimental settings' with the sole purpose of evaluating Groupware has produced confusing and inconsistent results, because it is not possible to simulate real-life collaboration (Davison et al., 1998). Davison et al. (1998) proposes, instead, using action research to fully capture the complexity of Groupware use and collaboration. Using the action research approach in the Information Systems (IS) community is well known (e.g. Mathiassen 1998, 2002; Avison, Lau, Myers & Nielsen 1999; Vidgen and Braa, 1997; Braa and Vidgen, 2000; Donnellan, 2003). The IS action research approach combines theory and practice through change and reflection in a problematic real-life situation.

The triangle in Fig. 2 represents the unity of the three goals: to understand, to support and to improve. The arrows inside the triangle represent the distinct research activities through which the different goals are supported. Having the activities inside the triangle illustrates that each activity can benefit from the other activities. "First, our understanding is based on interpretations of practice. Second, to support practice we simplify and generalize these interpretations and engage in design of normative propositions or artifacts, e.g. guidelines, standards, methods, techniques, and tools. Third, we change and improve practices through different forms of social and technical intervention." (Mathiassen, 1998: p. 20). In this way different studies are placed in different locations of the triangle.

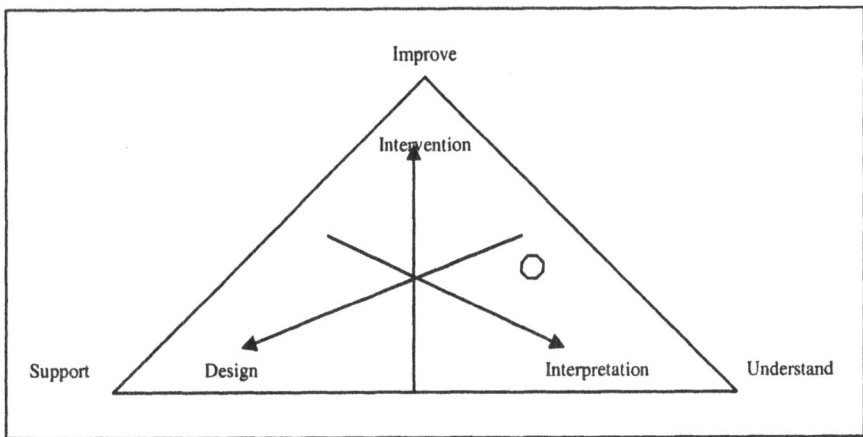

Figure 2. A Framework for Action Research (adapted from Mathiassen, 1998)

The study presented here was lead by the research question: What are the factors influencing the integration process of Groupware technology in virtual learning teams in part-time adult education? In pursuing this question we wanted to *understand* practice (the social practice and collaboration within virtual learning team) with the aim of *supporting* the practice (support of coordination and social aspect by Groupware) and then *intervene* with the practice (facilitate the integration of Groupware to support the group's needs). As predicted by the research question our emphasis is on the understanding of the practice in the virtual team (what are the factors...) in combination with a motivation for facilitating and improving the integration process. In this way we position ourselves in the triangle as closer to improve and understand than to support (see circle in fig. 2).

When doing action research it is crucial to be explicit about the role of the researcher. The role of the researcher in this study was to act as an outside facilitator and process-supervisor in the integration of Groupware in the virtual learning team. It was made explicit to the students that the integration of Groupware should support their collaboration, and if they did not find the Groupware useful they should state this. This approach made the students to be critical towards the technology. As a result, they clearly stated throughout the investigation if and when they were unhappy with the technology and what they would have liked to change. The researcher had no direct connection to the Master of Adult Education program, and was not one of the teachers within the Master education.

3.4   Research activities

The research took place over the period September 2001 to June 2002. In September the research project was presented to all the students attending the first on-campus seminar in the third year, and the students were asked if they were interested in participating. It was made clear to the students that the researcher not only wanted to make the technology available to the students, but also wanted to facilitate the use being a process-supervisor. The work of the researcher was then supporting the team in both the technical functions and in translating the teams work patterns into guidelines for adjusting the conceptual structures in BSCW. One team volunteered to participate to the research project.

During the whole research period, four intervention points in the process were analyzed: November 2001, January, March and April 2002. In November 2001 a workshop of six hours was held. The workshop activities were recorded for later analysis by using wall-graphs (Simonsen and Kensing, 1997) and tape-recording, in addition to a personal logbook written by the researcher just after the session. In January 2003 the researcher's intervention took the form of a focus group interview (Kvale, 1996). The interview lasted about 2 hours, was tape-recorded for later analysis combined with wall-graphs and rich-pictures (Checkland and Scholes, 1990).

The third researcher's intervention was another evaluation session held in March 2002. The researcher was not present due to external reasons. The team was, therefore, given a list of questions to discuss. The questions could be divided into two types. One type related to the evolution of the project itself and another aiming at finding out the role of BSCW in the collaboration. Examples of questions are: Which kind of document do you have at this time? How far are these? And are there documents not placed in BSCW? Try to describe what you have been doing in the past period and how BSCW or other kinds of technology (like phone and e-mail) have worked? The team recorded the conversation, and these data were later analyzed by the researchers. The researcher asked the team some clarifying questions through the BSCW system after listening to the tapes.

The last intervention in April was organized as one-hour focus group interview. The session was recorded on tape for later analysis. In between the different intervention sessions the researcher had observed and recorded the group's activities within BSCW and intervened if necessary or asked by the students. This recording was done in the personal logbook held by the researcher.

3.5   Data analysis

Each physical intervention with the team was tape-recorded and transcribed within 24 hours in a sequential order to ensure reliability of the data (Perakyla, 1997). Wall-graphs or rich-pictures were made during the sessions, containing rich-pictures and notes of the sessions. After each encounter two resumes of the session were made. One was done by the researcher and one was done by the team. The collected data were analyzed intuitively by the authors according to the theme of Groupware integration. The parts of the interview text relevant to this theme were then contextualized to the framework of the TAM-model and social awareness (Creswell, 1998; Walcott, 1994). The resumes and conclusions were presented to and discussed with the team in a following intervention session to ensure validity (Creswell, 1998).

4.   ANALYSIS OF THE INTEGRATION PROCESS OF THE BSCW GROUPWARE SYSTEM

In this section we use the TAM-model to analyze and describe the evolution of the integration process of BSCW in the virtual learning team over the four points in time described above – November, December, January and March. The analysis illustrates how BSCW was adopted and accepted by the virtual learning team, and how BSCW contributed to reduce coordination efforts, achieve a mutual understanding of each other and support collaboration, thus indirectly supporting also social perspective awareness among the members of the team.

4.1   November setting

Due to the geographic distance among Emma, Thomas and Lisa, the first workshop between the researcher and the team was conducted on a Friday evening between 6:00pm and midnight the day before a Master seminar on campus. The purpose of the workshop was for the researcher to support the team in the negotiation of both the content and the goal of the project, and to find out how the team could use the BSCW to support their collaboration.

The team negotiated the project and developed a first common understanding for the use of BSCW to support collaboration. The understanding was based on the team's earlier experience with project work, combined with examples of how to use BSCW efficiently to support the task at hand. This was then used to design the conceptual structures of BSCW e.g. which folders should be created, how they should be named, and more importantly, how the participants should use the different folders. The result of the November activity was a project contract mainly describing the overall plan for the project, and a designed BSCW workspace. The researcher also held a hands-on introduction to BSCW technical functionalities. When asked to reflect on how the distributed collaboration was perceived after this November workshop, a group member expressed it as follows:

> "I think we all had a feeling of being far more on track than the year before. It was a relief to have an overview of the project and process even though it might have been an illusion. (...) the hard thing about this part-time education is that you sometime lose feeling with the project and then something like this (BSCW) is extremely good to have." (Group member in January 2002)

Thus, the November introductory session on how the group could use the BSCW system to coordinate the work did induce a feeling that the BSCW system would be useful in supporting collaboration. In TAM's terms, the perceived usefulness (PU) was positive (depicted as + in Fig. 3). The group had gone through the different functions and constructed the folders agreed upon, however the question of how easy it was to use did not arise as an issue. Thus, the group's view on the EoU construct was 'missing' or neutral at this point in time (depicted as 0 in Fig. 3). Still the high PU made the Intention-to-use high, and the expectation of Actual-use was high. However, we did not know at that time whether the group would *actually use* the Groupware in the future (depicted as ? in Fig. 3).

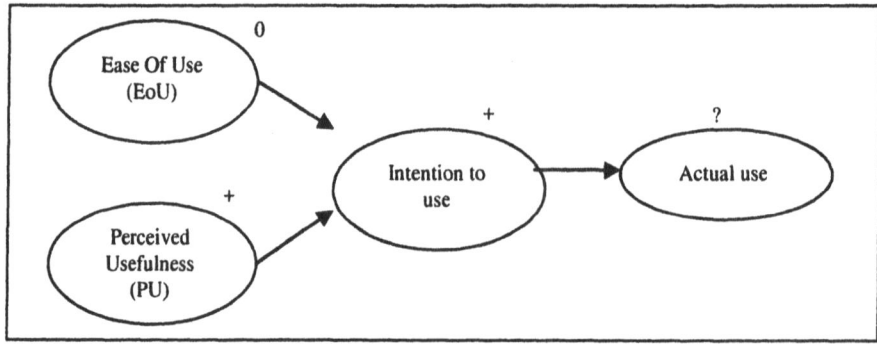

Figure 3. November setting; BSCW use in TAM.

4.2 December setting

The period from November to December was characterized by very low or almost no interaction within the group. Due to the members' daily work and family routine, the group did not have any kind of collaboration in this period. The members did logon to BSCW to 'see' if anything was happening and there were one or two small discussion-notes added but none were answered. Then between Christmas and New Year the group held a telephone-conference to "start up the communication again", as they put it. The main issue for the telephone meeting was to discuss and coordinate the production of an outline of the project to be sent to their supervisor before meeting him in January. In the telephone discussion about how to proceed to coordinate the document they decided *not* to use BSCW, but e-mail and telephone instead. When asked in January why they had decided not to use BSCW, a group member explained:

> "The thing with the BSCW is that when the working process is not continuous (...) then nothing happens (...) so it becomes like a stranger out there"
>
> (Group member in January)

In a part-time education program, where the participants use their free time to study, the process will never be continuous, and this makes it difficult to achieve sustained use of the Groupware technology. Analyzing the situation in December using the TAM model, the main issue emerging was the EoU (Fig. 4).

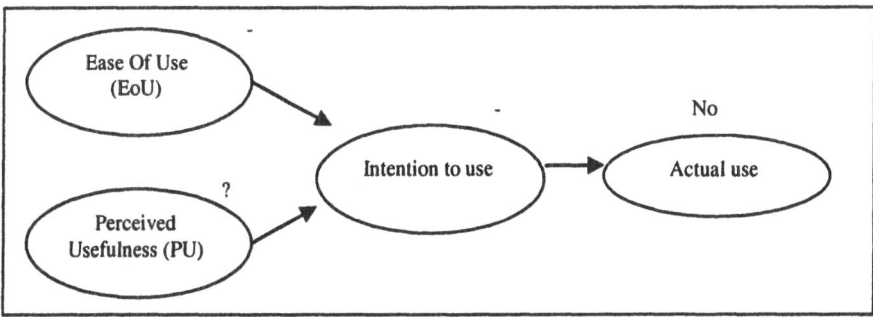

Figure 4. December setting: BSCW in TAM.

The high expectations and Intention-to-use BSCW present in November decreased in December because the focus in this period shifted from learning how to use the technology to creating the content and coordinating the project outline to be delivered to the supervisor. The perception that BSCW would be useful in the coordination process began to be questioned (depicted by ? in Fig. 4). The group started to question *whether* BSCW actually could help them reduce the complexity of coordination, and as the EoU factor started to be problematic (depicted by - in Fig. 4), the Intention-to-use was also reduced. The result was that the group did not actually use the groupware technology between November and December, but instead chose to rely on the more familiar e-mail and telephone. When we model the use of e-mail and telephone technology using the TAM model, the scenario in Figure. 5 emerges.

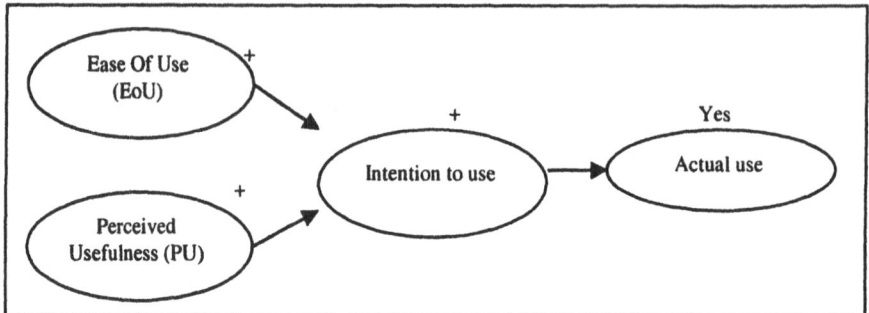

Figure 5. December setting: e-mail/telephone in TAM

The EoU concerning e-mail and telephone was positive. They were both familiar technologies and used previously by the group for coordination purposes. At the same time the expectations that these familiar technologies would support coordination were high due to earlier experience. So both the PU and Intention-to-use were high, leading to the Actual-use of e-mail and telephone to coordinate the production of the project outline. Thus, the Actual-use of these technologies was achieved.

4.3   January setting

In January a new workshop was held to evaluate the use of the Groupware system in the period from November to January. Here the group was encouraged to articulate the *actual* collaboration process as experienced from November to January. Knowing they had used e-mail and telephone to coordinate the outline, the main question was whether these "more traditional" technologies were successful in *coordinating* the document production. If they were, then the inevitable question would arise as to the need for the Groupware system. However, it transpired that the use of e-mail and telephone to support the project coordination had failed. The group did not actually realize this before the January meeting with their supervisor to whom they had already sent the document by e-mail two days before. They *thought* they had a common understanding of the content and the process by which the document had been produced, but it was not the case. The following discussion went on in the workshop:

Thomas: I think there is something missing here on the first few pages [pointing at a printed version of the document].

Emma: Is this not the last version you sent?

Thomas: No it is not.

Emma: The one you sent a couple of days ago?

Thomas: The one we mailed to Adam [the supervisor], the one we called version 4, the one Lisa had written on – unfortunately I don't have a printed version because my printer isn't working, but Lisa had put mine and hers together; it is about 6-7 pages long...(...)
(Group discussion, January)

This discussion continued and they got more and more frustrated about the situation. They did not have a printed last version, and at the end they decided to contact the supervisor to check if he had gotten the right version. They also started discussing the e-mail coordination process, and soon realized that they did not have a common understanding of the process underpinning the situation:

Emma: No I just had a thought, if I did get that e-mail I would have made a printed copy and taken it with me.

Thomas: Well, have you then got it or what? Sometimes I have trouble with my e-mail (...)"
(Group discussion in January)

By examining the situation using the TAM model, it becomes clear that the group had been incorrect in their December expectation regarding the perceived usefulness of using e-mail and telephone for coordinating the project outline and submission. The group now realized that the e-mail and telephone technology had not been adequate.

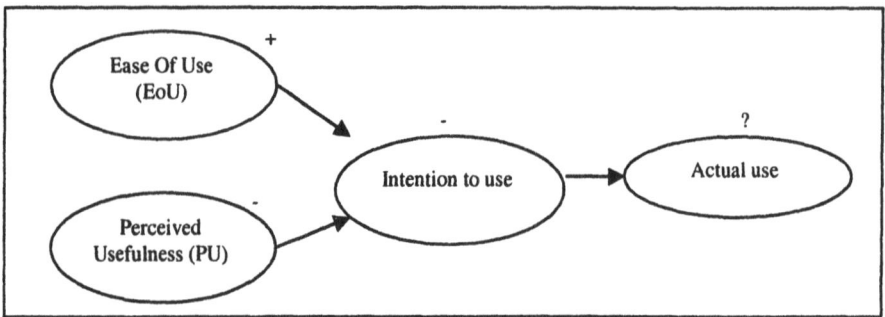

Figure 6. January setting: e-mail/telephone in TAM

As shown in Fig. 6, in January the PU of e-mail and telephone has been altered from positive to negative. The EoU was not changed, but the low PU affected the Intentional-use of e-mail for this kind of task. The low Intention-to-use made the Actual-use less likely. This experience affected, in turn, the use of BSCW. The PU of BSCW was restored also due to the need for 'something else.' Also, at the end of the workshop in January, a number of specific actions were taken by the researcher to help improve the EoU of the BSCW system. These actions were: 1) a new hands-on introductory session, 2) a written description of the functionalities of the BSCW system, 3) writing up three scenarios on how to use the BSCW for coordination, 4) turning on the BSCW direct notification function which would alert group members of relevant events occurring in the system, 5) setting up a discussion forum called weekly logbook, where group members could write comments about the project, together with personal information and other issues they wished to mention.

All these actions were meant to help the group overcome the troubles they had experienced using BSCW (low EoU). The PU for the BSCW system was high due to the coordination difficulties experienced with e-mail and telephone earlier. The EoU was also high as a consequence of the new training on the BSCW technical aspects. However, the main explanation for the positive increase in EoU was the January introductory hands-on session. One of the group members expressed the EoU this way:

> "Now we need to get past these [feelings towards BSCW] and say; ok it is not that difficult and instead realize that this [BSCW] actually makes it easier to get access to each other. I think that what we need is to commit to the system."

(Group member in January)

The expectation for using the BSCW for future coordination was high at the end of the January workshop. However, the group had not actually used BSCW yet.

4.4 March setting

The situation for the group had changed in two ways in March (Fig. 7). Firstly, they had managed to integrate the BSCW in their collaboration. This was clear by the large number of actions in the system, e.g. revising the structure of folders, uploading lots of documents and leaving notes in the weekly logbook. Secondly, the group collaboration had changed; they had developed a common understanding of how the project was taking shape, and also of the process of working on the project. The overall purpose of introducing BSCW was to reduce the effort required for coordination.

The group had managed to coordinate different documents while still keeping track of changes and versions. However, it emerged during the March workshop that the most interesting part of using BSCW was the weekly logbook. The weekly logbook had been originally established to encourage regular use of BSCW in the group to increase the EoU. However reflecting back on the situation in the March meeting, the group realized that they had expressed a need for 'something' not only related to the coordination of documents already in January.

> "(...) in the period [November to January] I needed to know how you were doing and so... or up to this seminar, how will we get the things we need to do done... I would have liked that kind of communication." (Group member in January)

The need being expressed here was awareness. In co-located teams the daily small interaction around the coffee-machine or water-cooler helps members to get a sense of each other. Lack of such information in virtual teams can affect group morale due to the possibility of misunderstandings and misinterpretations. When other group members are out-of-sight, this may be misinterpreted as inactive and unproductive. The group expressed a need for awareness related to the task they were doing, but also something more – the need for social awareness. In March a positive side effect of the integration of BSCW emerged. The use of BSCW and more specifically, the use of the weekly logbook had supported the need for social perspective awareness. The Groupware technology had affected the social setting of the virtual team.

> Lisa: I think this weekly logbook has been very good, because I have had a good sense of where you all have been – especially you Thomas, who have written all about your illness.

Emma: It makes it much more captivating to go in and read stuff like
           this too.

Thomas: Yes (...) because it gives you a good feeling of what is going
           on. (...) The constant response. It is especially good in these
           kinds of distance projects.

Lisa: Sometime you get hung up with work and lose contact with
           the project...

Emma: (...) you know you have it all in BSCW (...) it gives you a
           sense, psychologically, that there is a project forming."
           (Group discussion in March)

The use of the weekly logbook had given the group a sense of awareness
both in the task-oriented and social-oriented sense. The logbook content
provided extra information e.g. about interviews that had been conducted
and about members' health and family situation. The group members
pointed out how useful the first was for example in later analysis of the
transcriptions. The weekly logbook had been also a useful way to have
'casual social encounters' in an asynchronous way, simultaneously hosting
coordination activities such as planning and task location. One example was
the cancellation between January and March of a telephone meeting, which
had been planned during the workshop in January in favor of BSCW use.
When asked why, they explained that it was too expensive. The cost of a
telephone meeting had not been an issue between Christmas and New Year
because they needed it for coordination, but after BSCW integration in their
work, the perceived need for the telephone was reduced.

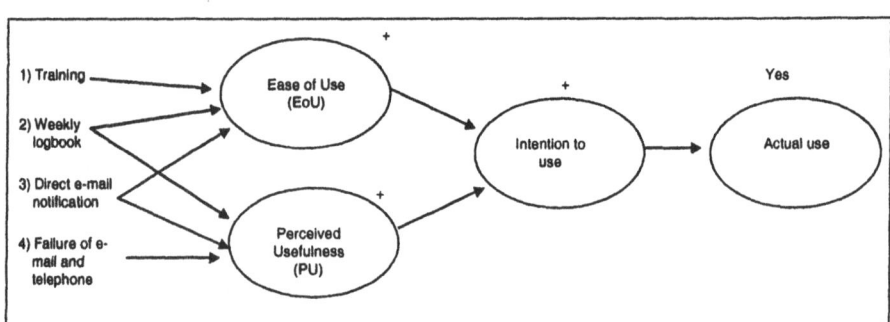

Figure 7. March setting: BSCW in TAM.

The weekly logbook in combination with the direct notification feature
supported social awareness, both in an active and a passive manner
(Steinfield et al., 1999). The passive manner was due to BSCW direct

notification feature, turned on in the January workshop. This meant that each time a member wrote, revised or moved objects within the system, an automatic e-mail was sent to all members informing them of the activity. Each member could therefore monitor when others had made a contribution in the weekly logbook. The active way was due to the fact that to actually read the content of the contributions, the members needed to logon to BSCW and actively click on the weekly logbook. As a result the weekly logbook had a huge impact on the EoU, because it caused the participants to use the system functionality regularly.

The weekly logbook facilitated spontaneous and informal interaction by being a free-form discussion forum, with no prescription as to usage or content. Still the participants needed to actively provide the awareness data (writing notes), which requires a deliberate and obtrusive strategy, as opposed to a situation where the data might be automatically generated. This aspect could have caused distraction and the related extra effort might cause resistance and non-acceptance of the technology. This was not the case. *If* the weekly logbook were perceived as a lot of extra work without relevance, the group would have perceived it as a distraction leading to decreased PU. It was very evident in March, instead, that the group had successfully integrated the BSCW technology to achieve the necessary collaboration.

5. DISCUSSION

So, what does our analysis using the TAM-model tell us about the integration process of Groupware in virtual teams in the Master of Adult Education? To summarize, in November we learned that highly PU did have a big impact on the Intentional-use and that the EoU was not even an issue. The data suggest that in the case of Groupware technology, high PU can supercede the importance of EoU, resulting in high *Intentional*-use. It was however not our goal to test, evaluate or change the TAM-model – and further elaboration in this direction would require more data.

However, we have learned in December that when it comes to the *Actual-use*, the EoU influence is vital. In fact, the conditions had changed in December due to little interaction and deadline pressure, which made EoU for BSCW more important than in November. This led to no Actual-use of BSCW, but to the use of the more familiar technologies instead: e-mail and telephone. Therefore we can conclude that the influence of both EoU and PU is present and important in the integration of Groupware technology in virtual learning teams.

During the January workshop the conditions for using BSCW to support the collaboration changed dramatically. It had become apparent to the group members that the December perceived usefulness of e-mail and telephone had not materialized. Instead, the new initiatives such as training, direct e-mail notification and weekly logbook had increased the EoU *and* PU of BSCW, even though the initial aim was only to increase EoU. The changed conditions of EoU and PU influenced the Intentional-use, and in March this led to Actual-use of BSCW. We had expected that the weekly logbook would increase only the EoU, instead we found out that the logbook also had increased the PU. Surprisingly the weekly logbook had supported a need for social interaction within the virtual learning team. Due to the large physical distance between the members, the group had a need to know about each other' intentions and actions, not only in relation to work but also in their social life. This need we have identified as social perspective awareness. The social awareness was supported by the weekly logbook within BSCW, and the use of the weekly logbook substantially contributed to BSCWs acceptance by the group. This in turn led to the integration of BSCW in the collaborative practice of the virtual team.

6.   CONCLUSIONS, LIMITATIONS AND SUGGESTIONS FOR FURTHER RESEARCH

The overall research question in this study was: What are the factors influencing the integration process of Groupware technology in virtual learning teams in part-time adult education? On the basis of an in-depth action research study (Mathiassen, 1998; 2002) conducted in a Problem Oriented Project Pedagogy Master Program at Roskilde University, Denmark, we got rich data describing the integration process of BSCW in a virtual learning team consisting of three students. We used the TAM-model to describe and analyze the data collected in four key checkpoints and found it useful in the description of successes and failures during the integration process of Groupware.

We can conclude that the TAM model can be used in a qualitative way to analyze and interpret technology acceptance. Furthermore in the specific setting of Groupware integration in virtual learning teams, we can conclude that both EoU and PU are important for the Intention-to-use and the Actual-use. We also found that social perspective awareness influenced the PU, which affected the Groupware integration in a positive way. Therefore we can conclude that one important factor influencing the integration process of Groupware technology in part-time adult education is: How the technology can provide support for social perspective awareness, which can in turn influence the PU of the technology.

The limitation of the study presented in this paper arises mainly in two aspects. Firstly the empirical work was collected with a virtual team consisting of students and thus it is somewhat questionable as to whether social awareness would have the same importance in virtual teams in business environments. However, even though it was not our goal to generalize our findings from an educational setting to a business setting, we expect social perspective awareness to be important within virtual teams in a business context as well, at least in relation to project coordination and document handling. We would propose that future research – qualitative as well as quantitative – should be done in business settings to test our results. The second limitation of the study lies in that, we on the one hand use the TAM-model in a total new setting (groupware, qualitative study, different goal) and on the other hand the study does provide some evidence to question the model's use in this setting (e.g. the causal relation between EoU and PU). Thus, we can conclude that the results of the study do not fully support the model, and this could be used to generate a new and revised TAM-model to be used in this setting. This is however beyond the goal of this paper. Here the main purpose was to use the TAM-model in a descriptive way when investigating the integration process, rather than trying to question its explanatory effect. To conclude we propose to use the TAM-model in qualitative studies in future research investigating Groupware integration in virtual learning teams. Also, we suggest that such investigation should focus on conditions influencing both EoU and PU in an integrated manner in the search for how to successfully integrate Groupware, and especially explore the role of social awareness in the integration of Groupware technology to support collaboration.

ACKNOWLEDGEMENTS

This paper is based on an earlier version co-authored by Pernille Bjørn, Brian Fitzgerald and Ada Scupola presented at IRIS 2003 conference in Haikko Manor, Finland. We would like to thank Brian Fitzgerald for reading and commenting on earlier drafts of this paper as well. Finally we also would like to thank the IFIP reviewers for profound and useful comments.

REFERENCES

Adams, D, Nelson, R. and Todd, P. (1992): Perceived usefulness, perceived ease of use and user acceptance of information technology: a replication, *MIS Quarterly*, (16, 2), July, 1992, pp. 227-247.

Alexander, P.M. (2002): Teamwork, Time, Trust and Information, in Proceedings of SAICSIT 2002. pp 65-74.

Avison, David; Francis Lau, Michael Myers and Peter Axel Nielsen: Action Research, in Communication of the ACM, January 1999, Vol. 42, No. 1

Bentley, R. T., T. Horstmann, and J. Trevor: The World Wide Web as enabling technology for CSCW: The case of BSCW, Computer Supported Cooperative Work: The Journal of Computer-Supported Cooperative Work, vol. 6, no. 2-3, Kluwer Academic Publishers, 1997, pp. 111-134

Bjørn, P. (2003): Re-Negotiating Protocols: A Way To Integrate GroupWare in Collaborative Learning Settings, Proceedings of European Conference in Information Systems (ECIS), Naples, 2003.

Bødker, K., Kensing, F. and Simonsen, J. (2003): Professional IT design – the foundation for sustainable IT usage, Forthcoming MIT Press.

Braa, K. and Vidgen, R. (2000): Research – From observation to intervention, Ch. 12 in Planet Internet. K. Braa, C. Sørensen, B. Dahlbom (eds.), Studentlitteratur, Lund, Sweden.

Carstensen, P. and Sørensen, C. (1996): From the social to the systematic: Mechanisms supporting coordination in design, Computer Supported Cooperative Work, The Journal of Collaborative Computing, vol. 5, no. 4, 1996, pp. 387-413.

Checkland, P. and Scholes, J. (1990): Soft Systems Methodology in Action, Chichester, West Sussex, UK, 1990.

Cheesman, Robin og Simon B. Heilesen: Supporting Problem-based Learning in Groups in a Net Enviroment, In Proceedings of the Computer Support for Collaborative Learning (CSCL) 1999 Conference, C. Hoadly & J. Roschelle (Eds.) Dec. 12-15, Standford University, Palo Alto, California. Mahwah, NJ: Lawrence Erlbaum Associates.

Creswell, J.W. (1998), *Qualitative Inquiry and Research Design*, Sage Publications.

Davis, F. (1989): Perceived usefulness, perceived ease of use and user acceptance of information technology, *MIS Quarterly*, (13, 3), September, 1989, pp. 319-340.

Davis, F., Bagozzi, R. and Warshaw, P. (1989) User acceptance of computer technology: comparison of two theoretical models, *Management Science*, (35, 6), August 1989, pp. 982-1003.

Davison, R., Qureshi, S., Vreede, G-J. d., Vogel, D. and Jones, N.: Group Support Systems through the Lens of Action Research: Cases in Organizations, Journal of Global IT Management, 3, 4, 6-23, 2000.

Dirckinck-Holmfeld, Lone og Elsebeth K. Sorensen: Distributed Computer Supported Collaborative Learning through shared Practice and Social Participation, In Proceedings of the Computer Support for Collaborative Learning (CSCL) 1999 Conference, C. Hoadly & J. Roschelle (Eds.) Dec. 12-15, Standford University, Palo Alto, California. Mahwah, NJ: Lawrence Erlbaum Associates.

Donnellan, B. (2003) Knowledge Management Systems Implementation to Support New Product Development, Unpublished Phd Dissertation, University of Limerick.

Dourish, Paul and Bly, Sara (1992): Portholes: Supporting Awareness in a Distributed Work Group, In P. Bauersfeld, J. Bennet and G. Lynch (eds.): CHI'92 Conference Proceedings: ACM Conference on human factors in computing systems, 3-7 May 1992, Montercy, California, New York: ACM press. pp. 541-577.

Gefen D., Karahanna, E. and D.W. Straub (2003): Trust and TAM in online shopping: An integrated model, MIS Quarterly 27, No. 1, pp 51-90.

Grinter, R. E. and Eldridge, M. A. (2001): y do tngrs luv 2 txt msg?, In Proceedings of the Seventh European Conference on Computer-Supported Cooperative Work, 16-20

September 2001, Bonn, Germany, pp. 219-238, Kluwer Academic Publishers, Netherlands, 2001

Grudin, J. (1994): Groupware and social dynamics: Eight challenges for developers, Communication of the ACM, Vol. 37, No. 1, 1994, pp. 92-105

Jarvenpaa, Sirkka L.; Knoll, Kathleen and Leidner, Dorothy E. (1998): Is Anybody Out There? Antecedents of Trust in Global Virtual Teams, Journal of Management Information Systems; Spring 1998; 14, 4; ABI/INFORM Global pg. 29

Karsten, H.: Collaboration and Collaborative Information Technologies: A Review of the Evidens, The DATA BASE for advances in Information Systems, Vol. 30, No. 2, 1999.

Kayworth T. R., and Leidner, D.E. (2002): Leadership Effectiveness in Global Virtual Teams, Journal of Management Information Systems, Winther 2001-2002, Vol. 18, No. 3, pp 7-40.

Koufaris, M. (2002): Applying the Technology Acceptance Model and Flow Theory to Online Consumer Behavior, Information Systems Research, Vol. 13, No. 2, pp. 205-223.

Kvale, S. (1996): InterViews: An introduction to qualitative research interviewing. Thousand Oaks, Ca.: Sage, 1996.

Kwon, H.S., Chidambaram, L. (2000): A Test of Technology Acceptance Model, The Case of Cellular Telephone Adoption, Proceedings of the 33rd Hawaii International Conference on System Sciences, IEEE Computer Society.

Lederer, A.L., Maupin, D.J., Sena, M.P., Zhuang, Y.(2000): The Technology Acceptance Model and the World Wide Web, Decision Support Systems, Vol. 29, No. 3, pp. 269-282.

Legris, P., Ingham, J. and Collerette P. (2003): Why do people use information technology? A critical review of the technology acceptance model, Information & Management 40 (2003), pp 191-204

Majchrazak, A., Rice, D.E., Malhotra, A., King, N.: Technology Adaptation: The case of a Computer-Supported Inter-organizational Virtual Team, MIS Quarterly, Vol. 24, No. 4, pp. 569-600, 2000.

Massey, A.P., Hung, Y.C., Montoya-Weiss, M., and Ramesh, V. (2001): When Culture and Style Aren't About Clothes: Perceptions of Task-Technology "Fit" in Global Virtual Teams, Proceedings of the 2001 – International ACM SIGGROUP Conference on Supporting Group Wokr – vol. 2001.

Mathiassen, Lars: Collaborative Practice Research, in Scandinavian Journal of Information Systems – Vol. 14, 2002; pp. 57-76.

Mathiassen, Lars: Reflective Systems Development, in Scandinavian Journal of Information Systems - Vol. 10, No. 1&2, 1998; pp. 67-134.

Maznevski, M.L. and Chudoba, K.M. (2000): Bridging Space Over Time: Global Virtual Team Dynamics and Effectiveness, Organization Science, Vol. 11, no. 5, September-October 2000, pp. 473-492.

Neville, K. and Fitzgerald, B. (2002): An Innovative Training Model for an Organization Embracing Technology, *Journal of IT Education*, Vol.1, No. 3.

Olesen, H.S. and Jensen, J.H. (1999): Project Studies – A late modern university reform?, Roskilde University Press, 1999.

Orlikowski, Wanda J.: Learning from NOTES: Organizational Issues in Groupware Implementation, in J. Turner and R. Kraut (eds.): CSCW '92. Proceedings of the Conference on Computer-Supported Cooperative Work, Toronto, Canada, 31. October - 4. November 1992, ACM Press, New York, 1992, pp. 362-369.

Paré, G. and Dubé, L. (1999): Virtual Teams: An exploratory study of key challenges and strategies. Proceedings of the 20th International conference on Information Systems, January 1999.

Pavlou, P.A. (2001) Consumer Intentions to Adopt Electronic Commerce-Incorporating Trust and Risk in the Technology Acceptance Model, Paper Presented at DIGIT Workshop, December, http://www.mis.temple.edu/digit, accessed the 4th February, 2003.

Perakyla, A. (1997), Reliability and Validity in Research Based on Transcripts, in *Qualitative Research*, Ed. Silverman D., Sage Publications, pp. 201-220.

Piccoli, G. and Ives, B. (2000): Virtual Teams: Managerial Behavior Control's impact on Team Effectiveness, International Conference on Information Systems. Proceedings of the 21 ICIS, Brisbane, Queensland, Australia, pp. 575-580.

Prinz, W. (1999): NESSIE: An Awareness Environment for Cooperative Settings, GMD-FIT, Germany, Proceedings of the Sixth European conference on Computer supported cooperative work, 1999, Copenhagen, Denmark, Kluwer Academic Publishers Norwell, MA, USA, 1999

Schmidt, K. (2002): The Problem with 'Awareness' – Introductory Remarks on 'Awareness in CSCW', Computer Supported Cooperative Work. *The Journal of Collaborative Computing*, vol. 11, 2002, pp. 285-298.

Schmidt, K. and Simone, C. (1996): Coordination mechanisms: Towards a conceptual foundation of CSCW system design, Computer Supported Cooperative Work. *The Journal of Collaborative Computing*, vol. 5, no. 2-3, 1996, pp. 155-200

Simonsen, J. and Kensing, F. (1997): Using Ethnography in Contextual design, Communication of the ACM, Volume 40, Issue 7, pp. 82-88.

Steinfield, C., Jang, C. and Pfaff, B. (1999): Supporting Virtual Team Collaboration: The TeamSCOPE System, Group 99, Phoenix Arizona USA, ACM 1999 - 1-58113-065-1/99/11

Tollmar, K., Sandor, O., and Schomer, A. (1996): Supporting social awareness @Work: Design and experience, in Proceedings of CSCW '96, Cambridge MA, November 1996, ACM Press, pp. 298-307

Venkatesh, V. (2000): Determinants of Perceived Ease of Use: Integrating Control, Intrinsic Motivation, and Emotion Into The Technology Acceptance Model, *Information Systems Research*, Vol. 11, Issue 4, p. 342-366.

Venkatesh, V., Davis F. (2000): A Theoretical Extension of The Technology Acceptance Model: Four Longitudinal Field Studies, *Management Science*, Vol. 46, No. 2, pp. 186-205.

Vidgen, R. and Braa, K. (1997): Balancing Interpretation and Intervention in Information System Research: The Action Case Approach. In Lee, A.S., Liebenau, J., and DeGross, J.I. (Eds.) (1997): Information Systems and Qualitative Research, London: Chapman & Hall.

Walcott, H.F. (1994), *Transforming Qualitative Data: Description, Analysis and Interpretation,* Thousand Oaks, Sage.

Yager, S.E. (1999): Using Information Technology in a Virtual Work World: Characteristics of Collaborative Workers, SIGCPR '99, New Orleans, LA, USA. Copyright ACM.

IOS ADOPTION IN DENMARK
Explanatory aspects of organizational, environmental and technological attributes

Helle Zinner Henriksen
Department of Informatics, Copenhagen Business School, Denmark

Abstract: Adoption and especially non-adoption of technologies, which are believed to yield operational and strategic benefits for businesses, are not always easy to comprehend. In this study we have surveyed businesses in the Danish steel and machinery industry to get an insight into what motivates or de-motivates adoption of IOS exemplified by EDI. The reason for choosing this particular industry is that it has been a target for massive information campaigns concerning the potential benefits of EDI adoption. In accordance with the Tornatzky & Fleischer (1990) model for organizational adoption fifteen opinion data items related to organizational, environmental, and technological attributes were operationalized and analyzed. The study suggests that environmental and organizational attributes rather than technological attributes are the main determining forces for adoption of EDI.

Key words: IOS adoption, quantitative measurement, impacts of organizational attributes, environmental attributes, and technological attributes.

1. INTRODUCTION

Why are some organizations adopting a technological innovation that is announced to yield both operational and strategic benefits while others hesitate, or decide not to adopt for the time being? This question is highly relevant especially in the case of interorganizational information systems (IOS) exemplified by EDI (electronic data interchange), due to the great attention this particular technology had received throughout the 1990s from academia, practitioners, and governments all over the world. Surprisingly few Danish organizations have adopted EDI. If their technical capabilities and their high degree of computerization are considered, then the organizations' reluctance to adopt EDI appears to be even more irrational and incomprehensible. The phenomenon of organizations not adopting IT regardless of their claimed opportunities to do so is well known (Harrison et al., 1997). What is missing so far are factors explaining the reason for this situation.

Small companies dominate the Danish business sectors. About two third of the approximately 50,000 companies within the industrial sector have less than ten employees. National and international industry and trade associations have created a number of awareness campaigns and have focused on creating beneficial conditions for the SMEs (small and medium sized enterprises) in order to get them to adopt IT, especially IOS such as EDI. The aim is to help the companies reduce or eliminate work routines and to support them in a market characterized by increased competition. The technological development has led to an increase in quality and functionality and a decrease in cost of hardware and software (Harrison et al., 1997). The traditional technological barriers for organizational adoption of IOS such as EDI might therefore not play the same dominant role as it did earlier. This new situation makes it highly relevant to examine explanatory factors for EDI adoption among SMEs, which traditionally have less resources allocated to IS compared to larger companies (Lai & Guynes, 1997).

In order to find an explanation for the puzzle of the limited EDI adoption and diffusion among Danish businesses a survey was conducted. The survey addressed SMEs in the Danish steel and machinery industry. The reason for choosing this sector was that business associations had targeted information campaigns towards this sector prior to the data collection and therefore the survey aimed at analyzing the patent priorities of the responders with respect to adoption of EDI.

The remainder of this paper is organized as follows. The next section conceptualizes adoption. The following section presents the applied research method. Thereafter, the research model is outlined. This section is followed by a description of how the research model is operationalized. The next section deals with the statistical analysis and the results. The penultimate section discusses the results from the statistical analysis. The paper ends with some concluding remarks concerning the implications of the results.

2.   THEORY: ADOPTION OF INNOVATIONS

There are two dominant views on adoption. Adoption can be seen as having or not having an innovation (Tornatzky & Fleischer, 1990), or it can be seen as using the innovation versus not having it (Rogers, 1995). Adoption is according to Rogers "... a decision to make full use of an innovation as the best course of action available and rejection is a decision not to adopt an innovation." According to this view the line between adoption and use of the innovation is non-existing. Rogers argues, that the adoption process is a process through which a decision-making unit passes. The steps are: gaining knowledge of an innovation, forming an attitude toward the innovation, making a decision to adopt or reject, implementing the new idea, and finally to confirmation of the adoption decision. In this course of events the adoption process is considered to be merely a mental exercise until implementation takes place.

In this paper the core understanding of the term adoption is "having versus not having" (Tornatzky & Fleischer, 1990) rather than "not having versus using" (Rogers, 1995). Consequently, measures on effects of the adoption of the innovation are not performed. The important point in this study is that some dividing line is crossed where the participants decide to invest resources necessary to accommodate the effort to change (Kwon & Zmud, 1987).

3.   RESEARCH METHOD

The method selected for the collection of survey data was a postal questionnaire. During Spring of 2000 the questionnaires were mailed to the managerial unit in the organizations involved. The questionnaire was mailed by two industry and trade associations and the cover letter carried the letterhead of the respective association. It was expected that data collection using this procedure would be perceived as more authoritative than usual questionnaires received in organizations, and that this procedure accordingly would result in a better response-rate.

The questionnaire was sent to the management of 917 manufactures and wholesalers in the steel and machinery sector in Denmark. It is part of the internal policy of the two involved associations not to burden their members with too much mail. Therefore, a second mailing was not allowed. A total of 252 responses were received, out of which 247 could be included in the analysis.

The 247 returned valid questionnaires equal a response rate of 27.4 percent. The response rate approximately equals similar other studies related to the adoption of IOS. For example 25.7 percent (Chau, 2001) and 27 percent (Masters et al., 1992). Some other studies focusing on both adopters and non-adopters of information systems have generally reached response rates that are somewhat lower. For example 18.4 percent (Lai & Guynes 1997) and 19 percent (Crum et al., 1996). Compared to other studies, which have included both adopters and non-adopters, the response rate of the present study is satisfactorily, especially when the collection procedure with only a single mailing is taken into consideration.

A number of similar studies have applied various multivariate parametric data analysis methods. These methods include factor analytic techniques such as principal component analysis, which has been applied by Chau (2001) and Premkumar and Ramamurthy (1995) and factor analysis, which has been applied by Chau and Tam (1997) and Grover (1993). The strengths of factor analytic techniques are their usefulness for summarizing items and data reduction, thereby possibly uncovering "latent" dimensions that hopefully describe data in a smaller number of concepts compared to the original variables.

Both the principal component method and the method of exploratory factor analysis are based on a matrix of Pearson correlation coefficients and data should therefore satisfy the assumptions for these statistical methods (Hatcher & Stepanski, 1994). Therefore, all relevant Likert-scale items from the survey of the steel and machinery industry were tested for normality. The results of the Shapiro-Wilks test showed p-values ranging from 0.0979 to < 0.0001. It is possible that some transformations of data could have overcome this non-normal distribution of data. However, it can be argued that Likert-scales are merely manifestations of ordered categories (Siegel & Castellan, 1988) and therefore the requirement of at least an interval-scale for the Pearson correlation coefficient is not met. Based on the above-mentioned considerations and findings it was found prudent to focus on non-parametric methods of analysis.

The main objective of the statistical analysis is to uncover the patent priorities of the responders with respect to adoption of EDI. In this respect patent priority refers to the manifest, face value expressions of the responders. This approach is different from searching for latent factors or structures as normally done in factor analytic techniques.

Firstly, Cronbach coefficient alpha was estimated. Cronbach coefficient alpha is used as a measure of reliability in relation to the operationalization of the constructs from the Tornatzky and Fleischer model (1990). The objective of these reliability estimations is to investigate the strength of the three groupings of items related to the organizational context, the environmental context, and the technological context given adoption level.

The next two steps in the quantitative analysis of data are exploratory searches for items related to the three adoption levels. Two steps are used to identify the factors motivating or de-motivating adoption of EDI. Fischer's Exact two-sided test on two-by-two tables is used to identify those items that are strongly related to each one of the three levels of adoption. Fischer's Exact test is chosen because many cell-counts in the two-by-two tables are rather small. Data was then analyzed by applying graphical models using the software application DIGRAM. This is done, because it is found prudent to analyze the relationships between all the items taken together and the respective levels of adoption. Contrary to the two-way analysis using Fischer's Exact test the analysis performed in DIGRAM is a multivariate analysis technique. One of the strengths of graphical modeling is the ability to analyze causal structures. In this particular case the causal structure between the items and a given level of adoption.

Logistic regression analysis is the final step in the search for patent priorities. The independent, explanatory items for logistic regression analysis are the items that were identified either by the two-way tables using Fischer's Exact tests and/ or through the exploratory analysis performed in DIGRAM.

Similar to other adoption studies (Moore & Benbasat, 1991) multi-item indicators were used for the opinion data items concerning the motivation for adoption. Seven-point Likert scales ranging from "fully agree" to "strongly disagree" were applied. Due to the limited number of responders it was necessary to reduce the seven-point scales to binary scales. These binary scales were constructed to reflect agreement and disagreement with the adoption item in question. The Likert scale points 1 to 3 defined agreement whereas the points 4 to 7 defined disagreement. Point four on the seven-point Likert scale in question is considered to indicate a neutral response. It was decided to include the neutral response with points 5 to 7, because a response value of 1 to 3 reflects definite agreement with the item, whereas response values 4 to 7 indicate neutral to non-agreement.

In order to analyze possible differences between those responders that had adopted EDI, those who were in the process of adopting EDI, and those that found it irrelevant to consider EDI adoption for the time being, the responders were divided into three sub-groups: adopters, planners, and non-adopters. Adopters and planners were asked the very same questions. For non-adopters all questions were phrased negatively.[1] In the subsequent analysis of data all opinion data item values were "reversed" for non-adopters.

Generally, one of the first tasks when conducting questionnaire research is to assess the reliability of the constructs. If the scales are not reliable there is no point in performing additional analyses. An analysis of Cronbach's coefficient alphas was performed based on the operationalization of the three constructs: Organizational context, environmental context, and technological context given adoption status.

---

[1] An example of negative phrasing is: "None of our business partners are using EDI" versus the positive phrasing for adopters: "All our business partners are using EDI."

Table 1. Cronbach's coefficient alpha for the three constructs

Construct	Adopters	Planners	Non-adopters
Organizational context	0.76	0.72	0.82
Environmental context	0.70	0.74	0.82
Technological context	0.52	0.34	0.49

Generally, the lower acceptable limit for summed scales is considered to be 0.70 (Nunnally, 1978). The constructs for organizational context and environmental context are of an acceptable reliability level independent of adoption status. On the other hand, the operationalization of technological context is below the generally acceptable reliability level independent of adoption status.

4.   RESEARCH MODEL

Tornatzky and Fleischer (1990) suggested that three explanatory contexts influence the process by which innovations are adopted in organizations. These three contexts are: the organizational context, the environmental context, and the technological context.

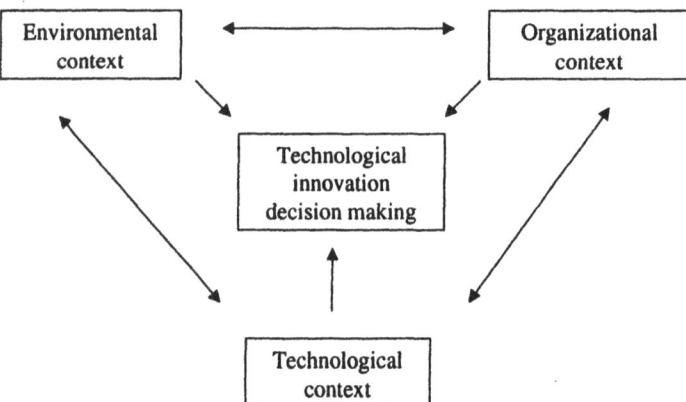

Figure 1. The Tornatzky and Fleischer (1990) model for adoption

This overall categorization in Figure 1 were operationalized in the further study of the Danish steel and machinery industry.

5.   OPERATIONALIZATION OF THE RESEARCH MODEL

Fifteen propositions are constructed based on the three categories presented in Figure 1. The fifteen propositions are based on secondary innovation attributes and the statistical analysis is therefore based on qualitative statements from informants.

5.1 Nature of the opinion data items

The operationalized adoption-decision variables in relation to EDI are mainly related to secondary innovation attributes (Downs & Mohr, 1976). Downs and Mohr distinguished between primary and secondary innovation attributes. Primary attributes are viewed as inherent in the innovation and invariant across settings and organizations e.g. company size, which can be measured fairly objectively. Secondary attributes are defined as perceptually based on subjective characteristics for example complexity and relative advantage. The perception of secondary attributes is assumed to be influenced by characteristics of both the particular setting as well as actors involved in adoption of a particular innovation. It is therefore acknowledged that the measures applied in the survey are subjective in the sense that they are perceived and interpreted in the mind of the responder.

The major distinction between the operationalization of the organizational context and the environmental context is whether the opinion data item is within or beyond the control of the organization. Control should in this context be understood as the organizations' (perceived) capability of influencing the adoption process and the possible outcome of adoption. That is for example the case for issues related to improvement of work environment or the organizations' capability of calculating the potential benefits of adoption. The opinion data items related to the environmental context are on the other hand facilitated or constrained by interaction with business partners. Examples include different degrees of power exerted on the organization and competitive forces.

5.2 The organizational context

The organizational context comprises attributes related to the organization. Tornatzky and Fleischer suggest that the organization provides a rich source of structures and processes that constrain or facilitate the adoption of innovations. These structures and processes can be formal or informal.

Generally profitability has been found to be a motivator for adoption of IS (Attewell, 1992). One rationale for adopting EDI is an expectation of increased efficiency due to improvement of intraorganizational and interorganizational routines (Timmers, 2000). Among the operational performance themes identified in the IS literature were: General performance improvements (Clark & Stoddard, 1996), accurate exchange of business information (Srinivasan et al., 1994), and benefits related to integration of EDI (Massetti & Zmud, 1996; Premkumar et al., 1994; Truman, 2000). Direct savings are rarely reported in EDI studies (Cox & Ghoneim 1996; O'Callaghan & Turner, 1995). Indirect savings are on the other hand often explored. These savings can be related to reduction in the workforce due to less re-keying and a decreased need for manual storing of documents, lower inventory costs, and shortened duration of transactions (O'Callaghan & Turner, 1995). The following proposition was formulated in order to examine whether the possible savings related to EDI have influenced a company's motivation for EDI adoption:

Proposition O1: Prospects of future savings motivate EDI adoption.

Tornatzky and Fleischer (1990) focus on human resources in relation to the organizational context. IS studies have included issues such as adequate education (Kurnia & Johnston, 2000) and employees' IS knowledge (Thong, 1999). The EDI literature has to a limited extent focused on issues related to work environment and human resources. Swatman and Swatman (1992) point out that adoption of EDI may lead to organizational restructuring involving staff retraining due to changing work functions. Especially training of all relevant employees has been found to be one of the major determinants for SMEs gaining benefits from EDI adoption (O'Callaghan & Turner 1995; Raymond & Bergeron, 1996). In order to investigate the importance of work environment and human resources in relation to motivation for EDI adoption three propositions were formulated.

Proposition O2 is related to, whether or not it is perceived that adoption of EDI will create better work conditions for employees leading to more independent job functions for employees. O'Callaghan and Turner (1995) characterized this situation as "Freeing professionals from administrative tasks."

Proposition O2: The notion that EDI will create a better work environment is a motivator for EDI adoption.

Proposition O3 is related to whether or not adoption of EDI will lead to training and education of employees. Re-training due to changes in staff functions as described by Swatman and Swatman (1992) was the underlying assumption for this proposition.

Proposition O3: The notion that EDI will benefit the development and utilization of human resources is a motivator for EDI adoption.

The theoretical inspiration for Proposition O4 can be found in the often-claimed benefit of EDI related to the elimination of redundant re-keying and elimination of manual reconciliation (Arunachalam, 1995; O'Callaghan & Turner 1995). One improvement of the work environment is seen in relation to elimination of trivial work and routines.

Proposition O4: The notion that EDI will eliminate trivial work is a motivator for EDI adoption.

Most EDI studies have focused on commodities and standardized products such as aircraft parts (Choudhury et al., 1998), hospital supplies (Steinfield et al., 1995), and office supplies (Jelassi & Figon, 1994). Research has especially shown a high level of implementation of EDI in the automotive industry (Tuunainen, 1998) and in the grocery sector (Andersen et al., 2000). Commodities and standardized products characterize both these sectors. Even though EDI is useful for exchanging business information regardless of whether the item is a commodity or something highly specific, the EDI literature and practice have so far mainly concentrated on commodities. To investigate whether or not the type of business activities influences the motivation for EDI adoption the following proposition was formulated:

Proposition O5: The decision-makers' awareness that the company's business activities are well suited for EDI is a motivator for EDI adoption.

One study (Iacovou et al., 1995) directly referred to organizational readiness for EDI. The study related organizational readiness to the level of financial and technological resources. Lai and Guynes (1997) referred to employee's positive attitude to organizational change. One aspect, which has been seen as a parameter for organizational readiness for EDI adoption, is related to whether the adopters are EDI initiators or followers (Swatman & Swatman 1992). Companies that are persuaded or directly forced to adopt EDI are not well prepared for EDI and as a consequence might not immediately reap the full benefits of EDI - if ever. In order to investigate the importance of organizational readiness proposition O6 was formulated:

Proposition O6: The notion that companies consider themselves to be well prepared for EDI are more likely to adopt EDI.

5.3   The environmental context

The environmental context is the arena in which a firm conducts its business. Included in this context are the organization's competitors, its access to resources supplied by others, and its dealings with government. All these elements can influence the way an organization views the need for adoption of innovations. Tornatzky and Fleischer suggest that two aspects of the external environment are key determinants of innovative activity: The competitive characteristics of its industry and the existence of a relevant technology support infrastructure.

Recalling that the environmental context is the arena in which the organization conducts its business (Tornatzky & Fleischer, 1990) the improvement of strategic performance is an important issue in relation to

IOS adoption. Three IS studies, which have included the environmental context in their survey instrument, have focused on competition (Kurnia & Johnston, 2000; Thong, 1999; Grover, 1993). A number of studies have found that adoption of EDI could lead to competitive advantages such as improved competitiveness (Chatfield & Bjorn-Andersen, 1997), new business opportunities (Jelassi & Figon, 1994), and changes in interfirm processes and politics (Lee et al., 1999). To investigate whether competitiveness influences the motivation for EDI adoption Proposition E1 was formulated.

Proposition E1 is directly related to possible improved competitiveness due to EDI adoption. The proposition rests on some of the "traditional" assumptions related to EDI. These include a possible increase in the company's competitiveness due to e.g. improved customer service, shorter lead times, and more timely information about transaction status (Arunachalam 1995; O'Callaghan & Turner 1995; Dearing, 1990).

Proposition E1: The prospect of improving the company's competitiveness motivates EDI adoption.

Proposition E2 is related to the strategic alliances between business partners for the purpose of maintaining a competitive edge (Chau & Tam 1997).

Proposition E2: The prospect of increasing the company's market share is a motivator for EDI adoption.

The items related to interorganizational forces contain issues related to power and pressure. This perspective when seen as a driver for IS adoption refers to the obligation of a firm to adopt an innovation in order to keep a customer or supplier content (Hart & Saunders, 1998). Hart and Saunders (1998) explored the different ways of exerting power in relation to business partners. They distinguished between persuasive and coercive power. Iacovou et al. (1995) distinguished between competitive pressures and imposition by trading partners. Bergeron and Raymond (1992) included the benefits from strategic repositioning of the firm due to implementation of EDI in their survey. Pressure related to imposition of business partners (Iacovou et al., 1995) was operationalized in Proposition E3 which is related to the knowledge that EDI is being used amongst business partners.

Proposition E3: The knowledge that several business partners already use EDI is a motivator for EDI adoption.

Proposition E4 is related to the situation where the company is subject to persuasive power (Hart & Saunders, 1998). The company is not directly forced to adopt EDI but business partners may take steps such as informing about EDI benefits and offering assistance in relation to the adoption and implementation process.

Proposition E4: The fact that EDI has been recommended by others is a motivator for EDI adoption.

Proposition E5 is related to direct pressure from business partners. Pressure can take different dimensions ranging from promises to threats (Iacovou et al., 1995). Promises include rewards such as rebates due to EDI usage and threats include negative sanctions such as suspension of the partnership.

Proposition E5: The fact that the company is put under pressure to use EDI is a motivator for adoption of EDI.

5.4   The technological context

The technological context comprises both the internal and external technologies relevant for the firm. It includes current practices and equipment internal to the firm as well as the pool of available technologies external to the firm. Decisions to adopt a technology depend on what is available, as well as how the available technology fits the firm's current technology. Tornatzky and Fleischer call attention to the fact that not all innovations are relevant to all industries.

One inevitable problem in relation to an examination of the three contexts in one questionnaire is the wide scope of the three contexts. It is difficult or nearly impossible to find a single person in an organization that can provide answers to all questions related to the organization. In order to avoid a situation where the manager was unable to answer issues related to the technological context, the questions were formulated in general terms focusing on the managerial aspects of technology.

Other things being equal, the more costly an innovation is, the less likely it is to be adopted, but once it is adopted and adapted, the large investment may strongly motivate implementation of the innovation (Cooper & Zmud, 1990). As for the cost of technical solutions, it can be seen as the direct price of purchase or in relation to both direct costs and expenses resulting from education and training of employees (Raymond & Bergeron 1996).

Proposition T1 is related to how managers perceive the importance of the technical level of EDI. Instead of specifically investigating issues related to standards, means for transportation, or prospect of integration this opinion data item was kept in very general terms.

Proposition T1: A satisfactory technical level of IT solutions is a motivator for EDI adoption.

Proposition T2 is directly related to price. A theoretical reference to T2 is the relative advantage of EDI adoption (Rogers, 1995) or cost-benefit considerations.

Proposition T2: A satisfactory price level of IT solutions is a motivator for EDI adoption.

None of the reviewed studies concerning EDI or adoption of IOS specifically included issues related to the threat of "technological marginalization" due to reluctance to adopt a technological innovation. The reviews were rich in examples of "economic marginalization" in the sense that non-adoption for example could lead to weakened competitive advantages. The adoption and diffusion theory is rich in examples related to the issue of technological marginalization. The S-curve often used to explain diffusion of technology (Attewell, 1992) conceptualizes the technological marginalization in the sense that it postulates that adopters can be divided into five groups: Innovators, early adopters, early majority, late majority, and laggards (Rogers, 1995). The last two opinion data items related to the technological context can from a theoretical point of view be related to the issue of managerial fashion innovation (Abrahamson, 1996). The fads and fashion perspective is related to the situation where managers actively search for innovations that can upgrade their businesses technologically (Newell et al., 2000).

The first perspective is related to the situation where the potential adopter is in a neutral position towards the innovation per se. However, mere knowledge that not having the innovation might exclude the company from being up-front might serve as a motivator for adoption. The threat of being a laggard (Rogers, 1995) with respect to adoption was formulated in Proposition T3.

Proposition T3: A feeling of being left behind with respect to EDI is a motivator for EDI adoption.

Proposition T4 is directly related to the fads and fashion phenomenon presented by Abrahamson (1996). It should be noted that though researchers do not perceive EDI as new and interesting this might not be the case for practitioners. Innovation is a relative term, which is conditioned by the perception of the potential adopter (Zaltman et al., 1973).

Proposition T4: The notion that EDI is new and interesting is a motivator for EDI adoption.

6.   STATISTICAL ANALYSIS AND RESULTS

Fischer's Exact tests were performed on all the original opinion data items transformed to binary variables related to the three adoption levels: Adopters, planners, and non-adopters.

Based on the results from Fischer's Exact test it was found that six items are strongly related to adoption. Four items were found to be of importance for planners, and nine items were of importance for non-adopters. The Fischer's Exact tests are based on analysis of two-by-two tables. These relationships are all two-dimensional. However, to uncover influences from multi-way dimensions and to identify possible causal structures these same opinion data items were included in a multivariate analysis based on graphical models. This was done, because it was found relevant to analyze the relationships between all items taken together versus the three levels of adoption.

The exploratory multivariate analysis suggests that four opinion data items that have a causal relationship with adoption. For planners there were causal relationships with three opinion data items. Finally, for non-adopters five causal relationships were found.

Table 2. Opinion data items identified for inclusion in binary logistic regression

Propo sition	Fischer's Exact			Graphical modelling			Items for inclusion		
	ADO	PLA	NON	ADO	PLA	NON	ADO	PLA	NON
T1		*	*						
T2		**	**					+	+
O1	***		***	←		←	+		+
O5	**		***			←	+		+
O2				←			+		
O3									
O4	**		***				+		+
O6	***		***				+		+
E1		***	***					+	+
E2		***	***	←	←	←	+	+	+
E3	***		***	←		←	+		+
E5									
T3	***						+		
T4					←			+	
E4		**			←	←		+	+

Legend: ADO = adopter, PLA = planner, NON = non-adopter, * = p <= 0.050, ** = p <= 0.010, *** = p <= 0.001, ← = p <=0.05, + = item for inclusion in the binary logistic regression analysis.

6.1 Binary logistic regression analysis

Binary logistic regression analyses were performed in order to estimate the explanatory power and strength of the adoption motivators summarized in Table 4 for the dependent variables: Adopter, planner, and non-adopter.

The stepwise forward selection method was applied for the logistic regression analysis procedures. The level of inclusion and exclusion was set at the five-percent level. The stepwise method was chosen since it enables examination of the relative strengths of the variables in the model. This stepwise approach has another advantage. It is more intuitively appealing since it builds models in a sequential fashion and allows for examination of a collection of models (Hosmer & Lemeshow, 1989).

6.1.1   Binary logistic regression analysis for adopters

Table 5 lists the results of the logistic regression analysis for adopters. Two of the fifteen opinion data items had coefficients that were statistically significant.

Table 3. Logistic regression analysis for adopters

Maximum Likelihood Estimates			Odds Ratio Estimates		
Parameter	Estimate	Pr > ChiSq	Point Estimate	95% Wald Confidence Limits	
Intercept	-1.2413	0.0004			
O6, yes	0.5958	0.0546	3.292	0.977	11.097
E3, yes	1.1872	0.0008	10.744	2.700	42.747

Response profile		
Ordered value	Adopters	Total frequency
1	Yes	19
2	No	58

Hosmer and Lemeshow Goodness-of-Fit Test		
Chi-Square	DF	PR > ChiSq
4.3804	2	0.1119

Since the Hosmer and Lemeshow Goodness-of-Fit Test p-value (0.1119) is greater than 5 percent this supports the fit of the model.

6.1.2   Binary logistic regression analysis for planners

Table 6 lists the results of the logistic regression analysis for planners. Two of the fifteen opinion data items had coefficients that were statistically significant.

Table 4. Logistic regression analysis for planners

Maximum Likelihood Estimates			Odds Ratio Estimates		
Parameter	Estimate	Pr > ChiSq	Point Estimate	95% Wald Confidence Limits	
Intercept	-1.5351	0.0073			
E2, yes	1.1998	< .0001	11.019	3.492	34.764
E4, yes	-1.2237	0.0327	0.087	0.009	0.817

Response profile		
Ordered value	Planners	Total frequency
1	Yes	27
2	No	50

Hosmer and Lemeshow Goodness-of-Fit Test		
Chi-Square	DF	PR > ChiSq
0.1637	2	0.9214

Since the Hosmer and Lemeshow Goodness-of-Fit Test p-value (0.9214) is greater than 5 percent this supports the fit of the model.

6.1.3 Binary logistic regression analysis for non-adopters

Table 7 lists the results of the logistic regression analysis for Non-adopters. Three of the fifteen opinion data items had coefficients that were statistically significant.

Table 5. Logistic regression analysis for non-adopters

Maximum Likelihood Estimates			Odds Ratio Estimates		
Parameter	Estimate	Pr > ChiSq	Point Estimate	95% Wald Confidence Limits	
Intercept	- 2.2777	0.0011			
O6, yes	- 1.1409	0.0147	0.102	0.016	0.639
E2, yes	- 1.4475	0.0026	0.055	0.008	0.363
E3, yes	- 1.9272	0.0015	0.021	0.002	0.229

Response profile		
Ordered value	Non-adopters	Total frequency
1	Yes	27
2	No	45

Hosmer and Lemeshow Goodness-of-Fit Test

Chi-Square	DF	PR > ChiSq
1.5645	4	0.8152

Since the Hosmer and Lemeshow Goodness-of-Fit Test p-value (0.8152) is greater than 5 percent this supports the fit of the model.

Table 6. Overview of the odds ratio estimates for adopters, planners, and non-adopters

Proposition	Adopter	Planner	Non-adopter
O6: The notion that companies considering themselves to be well prepared for EDI are more likely to adopt EDI	3.292		0.102
E2: The prospect of increasing the company's market share is a motivator for EDI adoption		11.019	0.055
E3: The knowledge that several business partners already use EDI is a motivator for EDI adoption	10.744		0.021
E4: The fact that EDI has been recommended by others is a motivator for EDI adoption		0.087	

Propositions related to the technological context were not found to influence the motivation for adoption for any of the responders regardless of level of adoption. One reason could be that the opinion data items were not well-defined. As shown in the Cronbach's coefficient alpha test the construct of the technological context was not well-defined for any of the three adoption levels.

For adopters and non-adopters factors related to the organizational context and environmental context were found to explain motivation for EDI adoption or non-adoption. For planners opinion data related to the environmental context were found to explain the motivation for EDI adoption. In the following section a closer look at the significance of each of the explanatory opinion data items in relation to the adoption level is presented.

7.   DISCUSSION OF RESULTS

7.1   Factors motivating adoption

Two factors were found to motivate EDI adoption. The two propositions O6, "The notion that companies considering themselves to be well prepared for EDI are more likely to adopt EDI" and E3, "The knowledge that several business partners already use EDI is a motivator for EDI adoption" could not be rejected.

One could argue that proposition O6, "The notion that companies considering themselves to be well prepared for EDI are more likely to adopt EDI" from a managerial point of view comprises all the organizational context opinion items. When a company states, that it is well prepared for EDI adoption this implies that the remaining organizational context items in some way or other are covered.

Another interpretation of the outcome of the analysis of adopters could find support in the nature of the social system (Rogers, 1995). If the prevailing attitude in the social system is, that EDI adoption is the norm, then companies are likely to perceive themselves to be ready for adoption. The common problem, which according to Rogers, is one of the characteristics of the nature of the social system, is then related to efficiency. The mutual goal therefore is to improve efficiency through EDI adoption thereby creating interorganizational efficiency and network externalities. Next, the importance of organizational readiness could be a result of the influence from change agents' promotional efforts, which through campaigns have informed about the innovation. The importance of proposition O6 is according to this interpretation influenced by social processes and communicated information about the innovation. If this type of interpretation is used, then the knowledge that several business partners already use EDI (proposition E3) is supporting the notion of a social process attitude towards adoption even more strongly.

Amongst the environmental context opinion data items proposition E3 appears to be the most important statement. The awareness that business partners already use EDI induces the potential adopters to perceive adoption to be the norm. Another interpretation of the importance of proposition E3 for adopters can be supported by the exponential diffusion curve (Attewell, 1992). Adoption according to this view becomes more and more attractive the more people adopt the innovation. This is especially the case when interorganizational attributes are related to an innovation, where critical mass is important for benefits to accrue from the investment (Markus, 1987).

One aspect, which is important to consider, when interpreting the priorities indicated by the adopters, is that their responses reflect an ex post evaluation. The two propositions, O6 and E3, that were found to be statistical significant in the logistic regression analysis are less concrete and less specific and therefore of a more general nature than the rest of the propositions comprising the organizational context and the environmental context. Instead of specifically replying that the motivation for adoption was for example related to concrete attributes the motivation is expressed in more general terms.

7.2   Factors motivating companies planning to adopt EDI

Two factors were found to motivate companies considering adopting EDI. Propositions E2, "The prospect of increasing the company's market share is a motivator for EDI adoption" and E4, "The fact that EDI has been recommended by others is a motivator for EDI adoption" could not be rejected.

For planners the determining factors motivating EDI adoption are solely related to the environmental context. Here it should be noted that planners do not consider recommendations from others to be of any importance. This indicates that recommendations from other businesses and from business associations are of no importance when businesses decide to adopt EDI. This appears to be in contrast to the variables determining adoption defined by Rogers. As mentioned in relation to adopters the variables related to the nature of the social systems and change agents' promotion efforts were used as a suitable framework for understanding why these particular propositions were of relevance for adopters.

One interpretation is that rationality rather than social processes drive the motivation for EDI adoption amongst the responders that indicated that they plan EDI adoption. One reason could be that planners compared to adopters indicated contemporary adoption preferences contrary to the adopters who expressed an ex post evaluation of their motivation for adoption. The planners in contrast to the adopters indicate more concrete motivation priorities. This suggests that the planners independent of recommendations from change agents and norms in the social system consider adoption of EDI to improve the organizations' strategic performance, thereby leading to increased market shares.

7.3   Factors causing a non-adopting attitude towards EDI

Three factors were found to cause a non-adopting attitude towards EDI. Propositions O6, "The notion that companies considering themselves to be well prepared for EDI are more likely to adopt EDI" E2, "The prospect of increasing the company's market share is a motivator for EDI adoption", and E3, "The knowledge that several business partners already use EDI is a motivator for EDI adoption" could not be rejected.[2]

The environmental context seemed to be the dominant explanatory factor for responders remaining as non-adopters. One opinion data item related to the organizational context was however also found to be a significant explanatory factor for non-adopters. Proposition O6 was found to be of major importance for adopters. Non-adopters on the other hand stated that they did not consider organizational readiness to be of any importance with respect to EDI adoption. A similar pattern was found in relation to proposition E3. This opinion data item was of major importance for adopters, whereas it had no relevance for the non-adopters at all. However, there might be some logical explanation for this inconsistency of preferences amongst the two levels of adoption – what makes good sense for adopters and planners, does not appear to make sense for non-adopters. Common for all opinion data items for non-adopters was that they did not agree with any of these statements.

One explanation for the situation, that non-adopters do not find EDI attractive at all, might be found in the attributes of the non-adopting companies included in the analysis sample. The non-adopters were generally small and independent companies. Such companies are believed to have limited power to initiate an EDI partnership and they are most likely allotted the role of an EDI follower. It is generally found that followers do not derive the same benefits of EDI as initiators (Swatman & Swatman, 1992). Operational and strategic gains from EDI adoption for small companies might therefore be limited. This is also the case in relation to the five innovation attributes defined by Rogers (1995). The relative advantage of EDI for small adopters is limited in relation to the efforts required to set-up an EDI solution with a few business partners.

It was argued that one possible reason for adopters indicating that organizational readiness was a motivator for EDI adoption was to be found in the nature of the social system and the change agents' promotion efforts.

---

[2] In this respect it is important to remember that the questionnaire items for non-adopters were all phrased negatively.

Organizational readiness was according to the responders of no relevance for non-adopters. An interpretation of this outcome is that non-adopters did not consider themselves to be addressees of the EDI campaigns launched by change agents. Pedagogical intervention (Eckhoff, 1983; Henriksen, 2002) might therefore be of limited value for companies that postpone or reject adoption of EDI. Additionally, the norms of the social system, which they perceive themselves to belong to, may not attach any value to EDI.

The two opinion data items concerning the environmental context, which resulted from the logistic regression analysis for non-adopters, were related to a possible increase of the company's market-share due to EDI adoption and the awareness that several business partners were using EDI. Proposition E3 was considered as the mildest form of pressure leading to EDI adoption among the fifteen opinion data items. This external "community" pressure did not influence non-adopters. A rational interpretation could be that non-adopters could not foresee that they would reach a critical mass of business partners using EDI. An interpretation guided by social processes could be that non-adopters simply do not identify themselves with EDI adopters. Therefore, there is no basis for an imitation process. With respect to proposition E2 it could be argued that if the non-adopters had found that EDI adoption in any way would influence their market share in a positive way, then they probably already would have adopted EDI.

To sum up, it looks like the non-adopters think that they can do fine without this innovation. And therefore they do not agree with or show any sign of enthusiasm with respect to any of the defined motivators for EDI adoption.

8. CONCLUDING REMARKS

About 16% of the companies in the Danish steel and machinery industry had adopted EDI by the time of the survey. The national share was in comparison about 15% (Henriksen, 2002). Given the claimed strategic and operational advantages companies can derive from EDI adoption this low level of adoption is difficult to understand. In the reported survey fifteen propositions related to a mix of operational and strategic benefits of EDI were tested based on data from 247 Danish companies in the steel and machinery industry. The objective of the study was to uncover the patent priorities of the responders. Based on our analysis it was found that organizational and environmental attributes rather than technological attributes determined EDI adoption in this particular sector. Pressure and organizational readiness were found to be the primary motivating factors for EDI adoption and rejection of the innovation.

One of the lessons learned from the study is that there is a discrepancy in the way an innovation is presented by business associations and the way it is perceived in the business community. The business associations representing the steel and machinery industry communicated the innovation as a means for improving competitive advantage. However, businesses belonging to the categories of planners and non-adopters did not share this viewpoint. Planners were not open to recommendations from others but they did at the same see EDI as a competitive tool namely as a means for increasing their market share. What is remarkable in this survey is that what made adopters accept the innovation was exactly what influenced non-adopters not to adopt. They disagreed with the notion that organizational readiness or pressure would influence their decision to adopt. Similar to planners the opinion data item related to increase of market share was also one of the patent priorities of non-adopters. However, non-adopters did not find that this item would influence their decision to adopt.

It is surprising that the responders do not pay much attention to the technological attributes as promoters or inhibitors of adoption. One reason could be that the awareness campaigns on EDI after all have "de-mystified" the technological dimension of EDI. Another, and perhaps more plausible explanation is that the technological attributes related to an innovation determines the rate of adoption less than IS researchers normally expect.

REFERENCES

Abrahamson, E. (1996) Management Fashion. *The Academy of Management Review* 21, 254-285.

Andersen, K.V., Juul, N.C., Henriksen, H.Z., Bjorn-Andersen, N., & Bunker, D. (2000) *Business-to-Business E-commerce, Enterprises Facing a Turbulent World.* DJØF Publishers, Copenhagen.

Arunachalam, V. (1995) EDI: An Analysis of Adoption, Uses, Benefits and Barriers. *Journal of Systems Management* 60-64.

Attewell, P. (1992) Technology Diffusion and Organizational Learning: The Case of Business Computing. *Organization Science* 3, 1-19.

Bergeron, F. & Raymond, L. (1992) The Advantages of Electronic Data Interchange. *DATABASE* 23, 19-31.

Chatfield, A.T. & Bjorn-Andersen, N. (1997) The Impact of IOS-enabled Business Process Change on Business Outcomes: Transformation of the Value Chain of Japan Airlines. *Journal of Management Information Systems* 14, 13-40.

Chau, P.Y.K. (2001) Inhibitors to EDI Adoption in Small Businesses: An Empirical Investigation. *Journal of Electronic Commerce Research* 2, 1-19.

Chau, P.Y.K. & Tam, K.Y. (1997) Factors Affecting the Adoption of Open Systems: An Exploratory Study. *MIS Quarterly* March, 1-21.

Choudhury, V., Hartzel, K.S., & Konsynski, B.R. (1998) Uses and Consequences of Electronic Markets: An Empirical Investigation in the Aircraft Parts Industry. *MIS Quarterly* December, 471-507.

Clark, T. & Stoddard, D.B. (1996) Interorganizational Business Process Redesign: Merging Technological and Process Innovation. *Journal of Management Information Systems* 13, 9-28.

Cooper, R.B. & Zmud, R.W. (1990) Information Technology Implementation Research: A Technological Diffusion Approach. *Management Science* 36, 123-139.

Cox, B. & Ghoneim, S. (1996) Drivers and Barriers to Adopting EDI: A Sector Analysis of UK Industry. *European Journal of Information Systems* 5, 24-33.

Crum, M., Premkumar, G., & Ramamurthy, K. (1996) An Assessment of Motor Carrier Adoption, Use, and Satisfaction with EDI. *Transportation Journal* 35, 44.

Dearing, B. (1990) The Strategic Benefits of EDI. *The Journal of Business Strategy* January/February., 4-6.

Downs, G.W. & Mohr, L.B. (1976) Conceptual Issues in the Study of Innovation. *Administrative Science Quarterly* 21, 700-714.

Eckhoff, T. (1983) *Statens styringsmuligheter. Særlig i ressurs- og miljøspørsmål.* TANUM - NORLI, Oslo.

Grover, V. (1993) An Empirically Derived Model for the Adoption of Customer-based Interorganizational Systems. *Decision Sciences* 24, 603-640.

Harrison, D.A., Mykytyn, P.P., & Riemenschneider, C.K. (1997) Executive Decisions About Adoption of Information Technology in Small Business: Theory and Empirical Tests. *Information Systems Research* 8, 171-195.

Hart, P. & Saunders, C. (1998) Emerging Electronic Partnerships: Antecedents and Dimensions of EDI Use from the Supplier's Perspective. *Journal of Management Information Systems* 14 , 87-111.

Hatcher, L. & Stepanski, E. (1994) *A Step-by-Step Approach to Using the SAS System for Univariate and Multivariate Statistics.* SAS Institute Inc., Cary.

Henriksen, H.Z. (2002) *Performance, Pressures, and Politics: Motivators for Adoption of Interorganizational Information Systems.* Samfundslitteratur. URL: http://www.cbs.dk/staff/hzh/Dissertation.htm (accessed ultimo October 2003).

Hosmer, D.W. & Lemeshow, S. (1989) *Applied Logistic Regression.* John Wiley and Sons, New York.

Iacovou, C.L., Benbasat, I., & Dexter, A.S. (1995) Electronic Data Interchange and Small Organizations: Adoption and Impact of Technology. *MIS Quarterly* December., 465-485.

Jelassi, T. & Figon, O. (1994) Competing through EDI at Brun Passot: Achievements in France. *MIS Quarterly* 18, 337.

Kurnia, S. & Johnston, R.B. (2000) The Need of a Processual View of Inter-organizational Systems Adoption. *Journal of Strategic Information Systems* 9, 295-319.

Kwon, T.H. & Zmud, R.W. (1987) Unifying the Fragmented Models of Information Systems Implementation. In: *Critical Issues in Information Systems Research*, R.J. Boland and R.A. Hirschheim (ed), pp. 227-251. John Wiley & Sons Ltd..

Lai, V.S. & Guynes, J.L. (1997) An assessment of the influence of organizational characteristics on information technology adoption decision: A discriminative approach. *IEEE Transactions on Engineering Management* 44, 146-157.

Lee, H.G., Clark, T., & Tam, K.Y. (1999) Research Report: Can EDI Benefit Adopters? *Information Systems Research* 10, 186-195.

Markus, L.M. (1987) Toward a "Critical Mass" Theory of Interactive Media. Universal Access, Interdependence and Diffusion. *Communication Research* 14, 491-511.

Massetti, B. & Zmud, R.W. (1996) Measuring the Extent of EDI Usage in Complex Organizations: Strategies and Illustrative Examples. *MIS Quarterly* September, 331-345.

Masters, J.M., Allenby, G.M., LaLonde, B.J., & Maltz, A. (1992) On the Adoption of DRP. *Journal of Business Logistics* 13, 47.

Moore, G.C. & Benbasat, I. (1991) Development of an Instrument to Measure the Perceptions of Adopting an Information Technology Innovation. *Information Systems Research* 2, 192-222.

Newell, S., Swan, J., & Galliers, R. (2000) A Knowledge-focused Perspective on the Diffusion and Adoption of Complex Information Technologies: The BPR Example. *Information Systems Journal* 10, 239-259.

Nunnally, J.C. (1978) *Psycometric Theory*. McGraw-Hill, New York.

O'Callaghan, R. & Turner, J.A. (1995) Electronic Data Interchange - Concepts and Issues. In: *EDI in Europe: How it Works in Practice*, H. Kcmar, N. Bjorn-Andersen, and R. O'Callaghan (eds), pp. 1-19. John Wiley & Sons Ltd..

Premkumar, G. & Ramamurthy, K. (1995) The Role of Interorganizational and Organizational Factors on the Decision Mode for Adoption of Interorganizational Systems. *Decision Sciences* 26, 303-336.

Premkumar, G., Ramamurthy, K., & Nilakanta (1994) Implementation of Electronic Data Interchange: An Innovation Diffusion Perspective. *Journal of Management Information Systems* 11, 157-177.

Raymond, L. & Bergeron, F. (1996) EDI Success in Small and Medium-Sized Enterprises: A Field Study. *Journal of Organizational Computing and Electronic Commerce* 6, 161-172.

Rogers, E.M. (1995) *Diffusion of Innovations*. The Free Press.

Siegel, S. & Castellan, N.J. (1988) *Nonparametric Statistics for the Behavioral Sciences*. McGraw Hill, Boston.

Srinivasan, K., Kekre, S., & Mukhopadhyay, T. (1994) Impact of Electronic Data Interchange Technology on JIT Shipments. *Management Science* 40, 1291-1304.

Steinfield, C., Kraut, R., & Plummer, A. (1995) The Impact Of Interorganizational Networks On Buyer-Seller Relationships. *Journal of Computer-Mediated Communication* 1.

Swatman, P.M.C. & Swatman, P.A. (1992) EDI System Integration: A Definition and Literature Survey. *The Information Society* 8, 169-205.

Thong, J.Y.L. (1999) An Integrated Model for Information Systems Adoption in Small Businesses. *Journal of Management Information Systems* 15, 187-214.

Timmers, P. (2000) *Electronic Commerce: Strategies And Models For Business-To-Business Trading*. Wiley and Sons, New York.

Tornatzky, L.G. & Fleischer, M. (1990) *The Process of Technological Innovation*. Lexington Books.

Truman, G.E. (2000) Integration in Electronic Exchange Environments. *Journal of Management Information Systems* 17, 209-244.

Tuunainen, V.K. (1998) Opportunities of Effective Integration of EDI for Small Businesses in the Automotive Industry. *Information & Management* 34, 361-375.

Zaltman, G., Duncan, R., & Holbek, J. (1973) *Innovations and Organizations*. John Wiley & Sons, New York.

LIFTING THE BARRIERS TO INNOVATION
A Practical View from the Trenches

Jim Brown
Draeger Safety UK Ltd, Northumbria University, UK

Abstract: Draeger Safety UK based in Blyth Northumberland manufacturing breathing apparatus for the search and rescue market has been following the path to organizational learning for several years. So far, this path has entailed individual and group learning, leading to practical application of that learning and knowledge in innovative improvement projects and problem solving process. This paper examines the practical view from the trenches of the efforts of the organization at all levels to lift the barriers to innovation from several academic perspectives. These perspectives include Stafford Beer's Viable System Model, Peter Drucker's seven sources of innovation and Margaret Wheatly's argument that innovation means relying on the creativity of everyone within the organization. However, the purpose of Draeger Safety UK in cybernetic terms is survival through the generation of profit from its core activity of breathing apparatus manufacture. Therefore resources of both time and finance must remain balanced between the activities that produce today's profit and the innovation based forward drive to all round improvement that ensures tomorrows. Thus lifting the barriers to innovation is the conscious decision to utilise resources in equipping staff with the knowledge, tools and opportunities to enable innovation to take place. While at the same time the view from the trenches is the comparison of academic theory and practical reality.

Key Words: Knowledge, Organizational learning, Viable Systems, Innovation

1. INTRODUCTION

Some look on innovation as flashes of inspiration others just as hard work and planning. (Drucker, 2002) holds a position between the two leaning more to the hard work end of the spectrum. Arguing some innovation comes from flashes of genius, but most from a purposeful search for innovation opportunities. He holds the view that as with other organizational activity, innovation must be deliberately managed, targets set and results monitored. However he also maintains that innovation is more knowing than doing, being a special function of enterprise that must be committed to the systematic practice of innovation through knowledge.

Draeger Safety UK also holds this belief, having followed the path towards a learning organization for several years.

This paper examines the activities and results over these years to embrace principles of innovation through knowledge building, active participation and the development of trust at all levels within the workforce. The paper will utilise three themes to examine and place into academic context the practical activities carried out by the management and staff of Draeger Safety UK. The themes being Stafford Beer's Viable System Model, Peter Drucker's seven sources of innovation and Margaret Wheatly's argument that innovation means relying on the creativity of everyone within the organization. It is not the intention to give a definitive answer to organizational innovation; merely to examine the Draeger Safety UK approach. While this approach works for Draeger Safety UK within other organizations, it may not.

Innovation has many definitions including:

"Innovation is the successful exploitation of new ideas and is a vital ingredient for competitiveness, productivity and social gain within businesses and organizations."-**London Innovation definition**

"The introduction of something new"- **dictionary definition**

"An innovative business is one which lives and breathes 'outside the box'. It is not just good ideas, it is a combination of good ideas, motivated staff and an instinctive understanding of what your customer wants." **Richard Branson - DTI Innovation lecture, 1998**

"Once you've worked on a truly innovative project you realise how important transformation is to the success or failure of a project. Your way of thinking changes your priority changes, your company changes and your way of working changes forever. True innovation is not just about changing a product, a service or even a marketplace it's also about recognising and realizing the need to change yourself." **Ralph Ardill, Marketing & StrategicPlanning Director, Imagination - London Innovation Conference, 2003**

Therefore within this paper innovation will be looked upon as any new product, method, or process that brings improvement and competitive advantage.

Today we live in an ever-changing world; a world where the speed of change increases every year and the need for organizations to adapt to change is not only important but also imperative to continued prosperity. An era requiring rapid change where to stand still is to fall behind, left behind in the survival race overtaken by competitors reacting to the requirements of change. Organizations are finding themselves on an exponential curve requiring ever faster adaptation and change (Wheatley, 2001). To survive organizations must change and adapt with ever increasing rapidity. The simple result of this requirement is that the acquisition and use of knowledge by individuals, groups and organizations, as the basis for innovation is becoming, if not already has become, an essential organizational survival tool.

Social anthropologists (Morris, 1967, 1969) tell us that people naturally wish to learn, explore new challenges, investigate new environments and adapt to changing conditions. This is part of human nature; in other words, people are natural innovators. In the past organizations have deliberately stifled natural innovation in the name of standardization, efficiency, predictability, reduction of variability etc to such an extent they have become barriers that must be overcome in today's rapidly changing environment.

2. LIFTING THE BARRIERS

Throughout both the private and public sectors traditionally several barriers to innovation have been in place. The challenge to senior management is in lifting these barriers. Barriers such as time allocation, finance, and innovational demarcation in the past have prevented involvement of all members of an organization in innovational activity. While the three barriers mentioned are not a full list. They form major obstacles to innovational progress.

2.1 The Time Barrier

For individuals or groups to engage in learning and innovation, time allocation is required. If the organizational culture denies the need for allocation of time for such activities innovation is impossible.

Recently the move to the so-called lean mean organization has been in vogue. This has lead to in some cases to 'streamlining', 'downsizing' and other practices that in effect have reduced staff numbers and increased the workload for remaining staff. Some organizations have pursued this path to such an extent; reaction to change and innovation are virtually impossible.

Staff being more than fully occupied in meeting daily requirements. In fact this practice has brought reengineering into disrepute (Willets, 1996). Leaving staff highly suspicious of management motives when introducing efficiency and effectiveness programs, reducing trust and further stifling innovation.

'Lean is mean and doesn't even improve long term profits' (Mintzberg, 1996)

For all staff to become enthusiastically involved as (Willets, 1996) maintains is the way forward to rapid innovation. The establishment of a trusting cooperative environment is a prerequisite. However, for increased trust time saved and improvements must be seen as beneficial to the whole workforce and not just management and profit.

At Draeger Safety UK, time is available for individuals to be able to engage in training and educational activities. This then leads to better understanding and if innovation is more knowing than doing (Drucker, 2002), thereby increases the innovation opportunities available within the organization.

This has enabled an enforcing systems thinking (Senge, 1990) feedback loop (Figure 1) to develop within Draeger Safety UK. Thus enabling learning and innovation with all members of staff becoming involved and being allowed the time to do so.

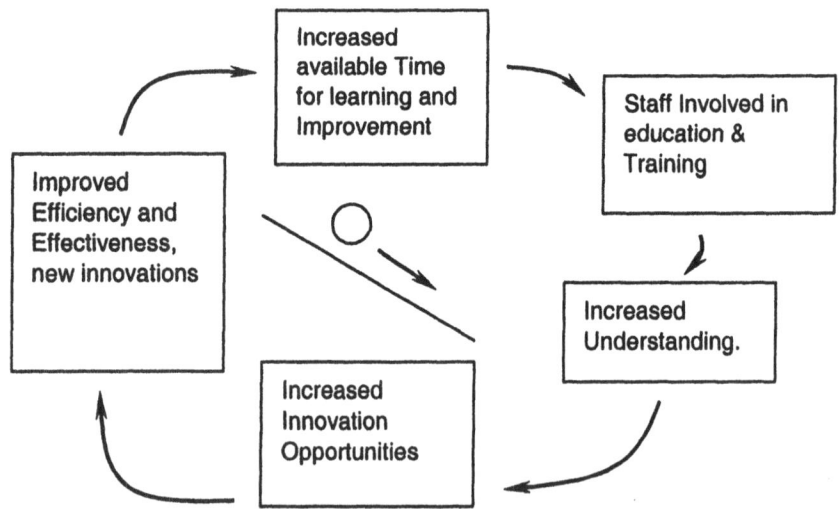

Figure 1. Positive Reinforcing Feedback Loop (Senge, 1990)

The Positive Reinforcing Feedback Loop shown above demonstrates the systems thinking thought processes behind the personal and organizational learning within Draeger Safety UK. Time is available for people to partake in training and education. Thus increasing understanding of general operational, managerial and improvement principles as well as specific knowledge of products equipment and operations etc, this leads to increased understanding providing increased innovational opportunities.

Draeger Safety UK and specifically the operations department capitalize on such opportunity via kaizen groups, problem solving teams Draeger Best teams, (Draeger Best is a form of the European Foundation for Quality Management Model and practice EFQM) and action teams. This has resulted in a more involved workforce, increased trust and the empowerment of individuals. In addition, the formation in January 2003 of a dedicated team entitled the "High Performance Work Team". Consisting of four individuals drawn from Quality, Design, Assembly Supervision and Industrial engineering, providing cross-functional abilities, has increased the innovational potential within the company

The team's purpose is one of dedicated attention to management identified areas of concern where improvement is either required or desirable.

Any positive response loop such as that described above also has a balancing loop that is often unseen, forgotten or ignored (Senge, 1990). For example in the loop shown in figure 1, available time and innovation can not grow indefinitely, sooner or later a state of balance or equilibrium will be established by management or will naturally establish itself. When the loop is implemented innovations may increase slowly until staff generally recognizes the benefit, then increase more rapidly as use is made of the learning innovation time loop. However as the easily attained innovations 'or low hanging fruit picked as some writers have termed it' are attained the pace of innovation will slow and find a stable level. This level can be controlled and maintained via time and resource allocation in a managed environment or be allowed to establish itself at some arbitrary level in an unmanaged one. In an uncontrolled environment, this level can also change for what seems no apparent reason. It is far more desirable from a systems control respect that management determines and controls this level to suit the needs of the organization rather than allow arbitrary events to have affects in ways sometimes not visualized or desired.

2.2 The Innovation Demarcation Barrier

Draeger Safety UK has also tackled the innovation demarcation barrier, the belief that "only designers, industrial engineers, managers, people with degrees etc are capable of innovative thinking" and therefore allowed the opportunity to become involved in innovation. Termed 'academic snobbery' by one senior manager. This belief stifles large possible sources of innovative creativity, can cause frustration and certainly lose organizations holding this view to innovation competitive advantage opportunities. The formation of the High Performance Work Team might form the impression that in fact Draeger Safety UK does believe in demarcation, where innovation and creative thinking is concerned. However, this team is in addition to, not instead of alternative innovational activities. In some cases with the 'High Performance Work Team' functioning as a resource or facilitation medium for other improvement teams. Management ensures that 'academic snobbery' in not considering suggestions and ideas from every level (especially operator's etc) is not allowed to happen.

Draeger Safety UK Operations manager (Vince Smith) has a favourite saying, *"with every pair of hands employed comes a free brain"*. This has become an operation department philosophy and using and improving those free brains through training and education forms a major part of the operations innovational drive. The realization that any individual can have innovational ideas that are worth pursuing is accepted and followed. People are encouraged to partake in educational and vocational courses, contribute to kaizen and action groups etc. Recognition given via for example an operator of the month award, carrying a small financial prize. Nominations for this award made not by management but operational staff. A committee made up of volunteer staffs, who decide on the recipient of the award, considers the nominations. Senior management also holds regular communication meetings (normally monthly) where staff can put forward ideas, suggestions and concerns. These meetings also allow senior management to signal the strategic direction, goals and priorities of the company. As a final example, a weekly brief contains financial information on sales and production output, Information of visitors to the company along with other general items of interest. Briefing sessions held by all supervisors and section managers again give staff an opportunity to give opinions, ideas or ask questions etc. These being directed to the appropriate member of staff or team for consideration and answer the following week. The discussed methods put into place by senior management. Ensure that staff has at least the opportunity to be informed as to company policy, strategic direction and has a platform for communication of ideas etc. that could lead to innovational improvements. These methods reduce to an extremely low level the possibility of 'academic snobbery' at Draeger Safety UK

2.3   The Finance Barrier

Innovation also requires finance, eventually any innovational activity brought to fruition and delivering benefit to the organization will require financial expenditure. This expenditure could be on equipment, material for trial production, external consultation etc. One of the functions of management is to make the decision as to which projects receive finance, and which do not. Draeger Safety UK has a philosophy of financing innovation whenever such ideas indicate possible improvement. However as with other organizations Draeger Safety UK must balance the finance required in pursuit of any project with the possible financial benefit that project will return. Management also has a responsibility to the originator of ideas etc rejected finance (or any other required resources) to communicate the reason for the idea not being pursued.

Quality also carries high priority; therefore, a project that maintains or improves quality receives finance, even if financial payback is not a result. An example of such a project was the development of an automated piece of equipment to insert high pressure sealing o-rings into a pressure reducer. Three colour-coded seals inserted in a specific order by hand had a possibility of error in insertion order. An action team formed from engineers and assembly operators being established to examine the situation and develop an answer that would eliminate the possible error condition.

The team first formulated the basic characteristics of the device. For example, it must be able to recognize colour, the order of the o-rings must be correct, the device should not insert the rings on error detection and all health and safety requirements must be established and met. This initial general specification then outlined the knowledge required, such as Programmable Logic Control (PLC) programming optical sensor use, control of pneumatic cylinders and knowledge of health and safety rules.

The requirement of PLC programming was unavailable within the organization and therefore required training and the purchase of software to enable generation of PLC programs via computer. While this required a financial expenditure, it brought new knowledge and the required capability to the company. In this manner, working through the requirements, discussing ideas and possibilities, a new process method and piece of equipment answered the initial quality sustainability question was developed by the team working together towards a common goal.

The final solution brought together pneumatics, programmable logic and optical equipment in a pokayoke device that recognizes colour and will not insert the seals, if they are in the wrong order or any are missing.

This device having no financial advantage in the form of improved cycle time, guarantees maintained quality of the operation and thus received the finance and resources of time and manpower required. This example is a demonstration of as (Weatley, 2001) suggests *"The human capacity to invent and create is universal"*. Draeger Safety UK recognizes that people have innate capacity to innovate. All they need is encouragement, backing and engagement in meaningful issues.

3.   THE SEVEN SOURCES OF INNOVATION

The question "is innovation inspiration or hard work", is posed by (Drucker, 2002)? With the conclusion that it is largely the latter. He identifies seven sources of innovation he comments " *In business innovation rarely springs from a flash of inspiration. It arises from a cold-eyed analysis of seven kinds of opportunities*". These opportunities being:

1. Unexpected Occurrences.
2. Incongruities.
3. Process Need.
4. Industry and Market Changes.
5. Demographic Changes.
6. Changes in Perception.
7. New Knowledge.

The seven sources will effect various departments of any organization somewhat. The effect being dependant upon and varying according to departmental function. For example, process needs could have high impact on production departments while for sales and marketing this opportunity will have low innovational potential. Thus for some organizational departments some sources will be irrelevant. However put together they account for the majority of innovation opportunities within the overall organization.

3.1   Unexpected Occurrence

As one example of unexpected occurrence, the recent heightened threat from terrorist attack has increased the awareness of search and rescue organization to the need for additional standard as well as specialized equipment. In the case of one piece of specialized equipment, a requirement existed for fully working prototype to be available three weeks from first inquiry. During discussions with the leader of the project team, it emerged that a consolidated effort was required from all sections of the company to achieve the three-week deadline using rapid prototyping techniques. Design used high-powered drawing and modeling software (pro-engineer) to use existing components in new configurations. Production engineering and assembly personnel in actually building the apparatus, purchasing working closely with suppliers to secure purchased components and sales clarifying requirements with the customer.

It was only with the cooperation of all departments and staff that allowed a working breathing set to be available within the time limit.

As Wheatley (2001) states, most people are very intelligent. They have figured out how to make things work when it seems impossible, they have invented ways to get around roadblocks, they have created their own networks to support and help them learn. These acquired personnel networks come into their own at times such as the development and manufacture of the example above. The usual methodical laid down methods and steps in the development of new products within Draeger Safety UK had to be temporarily circumvented, while maintaining control of design integrity, quality etc. in order to meet such a short deadline. Impromptu meetings being arranged agreements being reached and work carried out when required. This type of unsuspected occurrence requires rapid response and adaptability from the whole workforce, as disruption to normal activities is virtually inevitable under the circumstance. Management must rely on the ability of the staff putting trust in them to accomplish the task. In turn, the staff must use their knowledge and abilities to work in harmony putting trust in each other to achieve the innovational approach required.

3.2   Incongruities

Collins dictionary defines incongruous as *"inappropriate or out of place"*. However, Drucker (2002) in his description takes the definition a little further by encompassing such things as necessary operations or procedures performed as part of a production process disliked by the people performing the operation. One such within Draeger Safety UK concerns a piece of equipment used in the production of lung demand valves. The particular piece of equipment used to test the functioning of the valve while some years old is nevertheless extremely reliable. This type of equipment is also used by certification test house (such as Lloyds and TUV) in conformance test of equipment. Thus, saleable products undergo tests on equipment similar to that used to gain certification. While not difficult to use the equipment required several changes of supplied air pressure, air flow rates and respiration cycles per minute during the test. These changes being made by manual adjustment of valves, proving both tiring time consuming and a possible risk of strain to operators. The industrial engineering manager tasked an engineer to as far as possible automate these valve actions. This being completed some months ago with the use of electronic switching and programmable logic controls. To carry on this automation and incorporate computer control. A project team was formed recently and is exploring the possibility of working with local universities to design and develop a replacement breathing test facility that not only carries out the function of existing equipment but incorporates possible future requirements.

While still in the very early stages this project has the possibility of removing the incongruity completely, incorporating data base technology and simplifying data analysis.

3.3   Process Needs

The operations department has the responsibility of manufacture and therefore process need plays a major part in driving innovational change.

The example given above of the seal insertion device is a response to process need and requires no further discussion.

Another prime example of process need driving innovation is the development of a new generation of pneumatic test equipment. The realization of this particular need arose from a growing opinion that the existing test equipment developed some five years ago and state of the art at that time. Could today be improved via technologies not used in the original equipment. This opinion eventually being held by production management engineers and staff operating the equipment. Therefore, the decisions to invest time labour and finance in the development of new test equipment came almost by default. As normal within Draeger Safety UK, an action team established to undertake the development consisted of cross-functional abilities with members tasked to work together maximizing joint ability's to achieve the set objective. However it soon became apparent that while the general technical, abilities where present within the company collaboration with external organizations would be required for the specific detail of some technical aspects. Thus, the team sought assistance from suppliers to specify particular equipment best suited to the needs of the system. Working in collaboration with suppliers of such items as laminar flow equipment, computer digital and analogue signal boards with signal conditioning a detailed, specification of the test equipment resulted. The possibility of commissioning a software supplier to develop software to control the equipment after investigation lead to a decision to develop software internally. Total control of software intellectual property being the prime reason. Internal development also had the advantages of cost. The completed software being designed and programmed by engineers that knew and worked on the package as a whole not just the software. This allowed a system thinking holistic approach Senge (1990) instead of a reductionist approach taken by a company interested in software only. Thus the action team working with test operators suppliers and contractors specified developed and built the next generation of test equipment. An innovational approach bringing together hardware and software in ways new to the organization, providing direct reading of data by computer thereby minimizing risk of data entry error. The system also allowing subsequent data analysis of stored data thus providing information to direct future improvement.

3.4   Industry and Market Changes

Markets are customers driven and Draeger Safety UK responds to customer requirements through continuos product development.

Innovational drive in this area is by a combination of customer requests and internal development and enhancement. The latest developments to Draeger breathing equipment is the move to electronics from the traditional mechanical and pneumatic systems for air warning and pressure indication

These innovations arrived at through cooperation between the development department and suppliers of electronic equipment designing and developing new equipment to suit the applications and environments required.

3.5   Demographic changes

Market demographics have changed for not only Draeger Safety UK but also all organizations. The changes to the political scene in China opening markets that were none existent only a few years ago. In answer to these changes, breathing apparatus specifically designed by Draeger Safety UK to meet the requirements of the market is assembled in a recently established plant in China. Thereby using local market knowledge and contacts to expand sales at a higher rate than would be possible if bases within the UK or Europe were used to service the Chinese market.

3.6   Changes in Perception

Example already given can also come under the heading of changes in perception. The development of the seal pack from the perception that quality improvement would result. Development of new pneumatic test equipment from a growing realization that new technology existed that could improve this function. Breathing test equipment development from the change in perception towards automation and computer control instead of manual operation of valves and hand written records of test data can also come into the changing perception category.

3.7   New knowledge

"Application of new idea or principles based on totally new knowledge, these have long lead times but can be the history making innovations" (Drucker, 2002).

The examples of the seal pack and pneumatics test equipment are both of application of new knowledge. While not exactly in the history making division, for Draeger Safety UK such applications have supplied the innovational improvements allowing year on year achievement of improved overall results. Application of new knowledge is taking place on a daily basis within the organization ranging from knowledge gained from customer's opinions incorporated into products, to knowledge of new production techniques. Therefore, the knowledge gained by individuals from whatever quarter and used in continuing improvements is the innovational lifeblood of the company.

4.   INNOVATION IN TERMS OF THE VIABLE SYSTEMS MODEL

4.1   The Viable System and Innovation

· Organizations can be viewed as viable systems; such viable systems have the objective of survival, the requirement to carry on into the future (Beer, 2000). Therefore if innovation is essential to continued survival as argued by writers such as(Drucker, 2000), (Senge, 1990), (Wheatley, 2001), (Eppinger, 2001) it must form part of the viable system. The model has five sub systems one of which (system four) deals with the "outside and then" "Outside" being described as the problematic external environment and "then" referring to the future. Thus, system four deals with innovation when formulating new processes, markets, products etc. In short, innovation is the responsibility of system four within the viable system model. Thus as all viable systems must by definition have a system four innovation is possible within all viable systems.

4.2   The Recursion effect on innovation

Viable systems are also considered to be recursive, the recursive systems theory states *"In a recursive organizational structurally, any viable system contains and is contained in, a viable system"* (Beer 2000, p.118). Viable systems can be envisaged as an infinite set of Chinese boxes or Russian dolls each containing and being contained within another viable systems at higher and lower levels of recursion. These levels of recursion extend from the "macrocosmic to the microcosmic" (Beer, 2000, p.312). In an organizational context, recursive viable systems are considered as the various organizational levels from holding company through divisions to plants then departments on to departmental sections then to individual personnel. Each of these levels forms a recursive organizational viable system contained in the former and containing the latter. Therefore, if each recursion by definition must contain a system four then innovation is possible at any recursive level.

Figure 2 shows a partial view of the levels of recursion for Draeger safety in the Chinese box format. The recursive levels shown in Figure 2 are only one of many possibilities. For example at the Draeger Safety Level, Draeger Medical or Gas Detection exists with recursions applicable to those divisions.

When viewed as a viable system it becomes obvious that the focus of innovation will have a close relationship with the functions carried out by the system. Therefore, the level of recursion heavily effects the focus of innovational activity. It is also possible for strategy to pass down from level to level driving innovation in the desired direction at each level.

This trickle down effect is possible via a command channel running between each recursive level. It is via the command channel that one level influences and directs the general direction of the system occupying the level below. In the case of strategy the command channels between systems maintains the common general focus of all levels. The trickle down effect proved a driving force for improvement and cost reduction within the operations department of Draeger Safety, as outlined in section 4.3

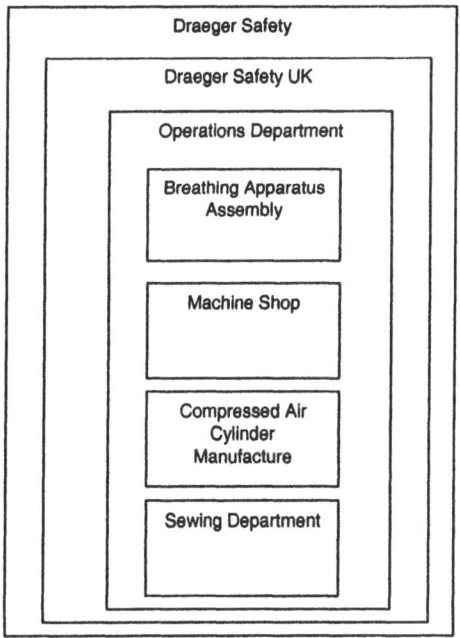

Figure 2. Levels of Recursion for Draeger Safety UK

4.3   Viable Systems and Innovation at Work.

In 1999, the board of Draeger (Parent Company of Draeger Safety UK) decided as part of corporate strategy to reduce the cost of poor quality by 50%. This became a directive and trickled down via the command channel from one level to the next. At each level while the goal was the same the innovation method used varied according to the function of the recursive level. Upon reaching the Operations department of Draeger safety, it was decided that an action group would be given the task of formulating methods and procedures to meet the directive. It is not the intention to detail the thinking and academic principles used by the group. A paper by Brown & Sice entitled "A Knowledge Management Approach to Reducing the Cost of Poor Quality-the Draeger Safety UK Ltd. Case" and presented at the UK Academy for Information Systems (UKAIS) conference 2003, detailed these methods and theories.

In short, it was decided to produce an IT system to collect data on items reworked and rejected. System output information being used to identify components and operations having high rework or reject costs. Once identified kaizen or action groups being formed to concentrate on improvements that would result in a reduction of poor quality cost. To enable production and operation of the IT system required the assistance and cooperation of assembly staff. Information on problems encountered and the time taken to rectify those problems being best obtained from the people actually dealing on a daily basis with rework. The principles of trust in each other that data given and collected was accurate and used only for constructive problem identification and resolution became an essential part of the successful operation of the Cost of Poor Quality (COPQ) program. Such trust and cooperation being essential in today's workplace if innovation is to play a part Wheatley (2001).

Using the IT system to identify high cost areas and teams to investigate and implement improvements resulted by the end of 2001 in a reduction of the cost of poor quality from 3.5% to 0.5% of manufacturing costs. The success of this approach can be attributed to the people at the operational level taking ownership of the IT system, examining the outputs and deciding on relevant actions. It was people that provided the solutions, people that implemented them and people that provided the innovation. The IT system providing a channel to gather data and convert that data to useful information in a form relevant to the people at the operational level of recursion. Thus in the case of COPQ an innovational method was devised, implemented and operated by the people at that level to answer a directive carried down the command channel from a higher level.

A final point exemplified by COPQ is that of data and information. As the people who would eventually operate the system were closely involved in development, the language and terminology used for data and information was the language of the production not the technical or IT department. Thus, this deliberate use of the production language while possibly not technically accurate had two beneficial effects. The first is the reduction of transaction ambiguities. Beer (2000) maintains in the 'third principle of organization' that whenever information transmitted on any channel crossing a boundary undergoes transaction. Such a boundary exists between data collection and information production in the form of the IT system. When the language of the IT system differs from the language of the users, interpretation can introduce ambiguities and translation of meaning. Resulting in misinterpretation of data collected and information then supplied. Tailoring the language of the system to the language of the user reduces this possibility. The second advantage is empowerment through involvement. Users immediately became owners after all, they had a role in the look and feel of the system, they decided how and what data was to be collected, it was the users that decided the format of output information. Therefore, the involvement and empowerment of the people that actually use the system played a substantial role in the success obtained.

5. CONCLUSION

That Draeger Safety UK believes that its people are the basis of innovation and future success is without doubt. The development of the operations training department, staff development via training and education coupled with involvement and empowerment all point in this direction. However, the all-important question is has this approach delivered innovation as defined in the introduction leading to organizational benefits? To answer this question results will be examined from two distinct views, the hard issues of figures and the soft issues of opinion.

Factual examples of hard issues show improvements in all directions. Within production, for example the lung demand assembly area some two years ago produced an average of between 200 and 210 units per day, while the present average is 300 to 330 units per day. A 50% increase in production with the same number of people. Sales and profit targets have increased and been achieved for the past several years. Profitability measures such as profit per employee, return on investment etc. are amongst the highest within the Draeger group. In addition, investment for the future is increasing with additional building to enable expansion under way. The staff of Draeger Safety has embraced the idea of vocational training and education, with 191 people achieving nationally recognized NVQ, management and other courses up to higher degree level in the last three years. With a total workforce of 356 this equates to 54% of the workforce engaged in educational improvement activities.

For the soft issues of opinion and perception, the following comments, and extract from written documents are indications of the beliefs and opinions prevalent within the staff of Draeger Safety UK and outside bodies associated with the company.

During a series of informal interviews on quality perception and inquiry, it emerged that the general opinion was in favour of the approach adopted. The people based approach had gained favour with the majority of staff, who felt that it had improved the quality of working life. Specific comments included

"I like working here, I feel respected as a person, things like that"

"This is a much better place than it was ten years ago"

"It is a far more relaxed atmosphere in the machine shop now and we are still getting the work done"

"I find people here are particularly happy, contented you know willing to talk to you"

"I like working with other people, I like being in control and achieving what I set out to do"

An NVQ examiner writing to a Draeger Safety UK training officers states:

"Your organization sees the ongoing training and development of staff as a key aspect of success in the marketplace"

While a member of staff, in an email to a course tutor writes:

"I have now realized that most folk who 'care' are passionate about their role in the work environment and desire to do well for their boss and the company"

The small selections of both hard and soft issues given above nevertheless are indicators of the success of the improvement through innovation via staff knowledge and involvement taken by Draeger Safety UK.

The answer to the question "has this approach delivered innovation as defined in the introduction leading to organizational benefits", is a resounding yes. However as innovation by definition is a never ending quest, then the aim of improvement through innovation is likewise never ending. As an organization Draeger Safety UK will not sit back and think we have done it we are there. For the simple reason, there is always another improvement awaiting discovery. Draeger Safety UK has chosen to put faith in its people via empowerment invest in staff development and make innovation and continuos improvement in all aspects of its business a cornerstone of it philosophy. For this organization, it has paid dividends financially and culturally; putting Draeger Safety UK well placed in today's rapidly changing market environment that requires innovational adaptation to survive and continue as an organizational viable system.

REFERENCES

Argyris C (1998), Empowerment the Emperors New Cloths, Harvard Business Review May – June 1998

Beer S. (1990): The Heart of Enterprise, JohnWiley & Sons

Bohm D, Factor D, Garrett P (1991), Dialouge – A Proposal,
URL http://www.muc.de/~heuvel/dialogue_proposal.html 21/10/03

Davenport T, & Prusak L (2000), Working Knowledge, How Organizations Manage What They Know, Harvard Business School Press

Drucker P (2002), The Discipline of Innovation, Harvard Business Review Aug 2002 Vol. 80

Eppinger S. D (20001), Innovation at the Speed of Information, Harvard Business Review Jan 2001 Vol. 79

Mintzberg H (1996) Ten Ideas Designed to Rile Everyone Who Cares About Management, Harvard Business Review July – August 1996

Morris D (1969),The Human Zoo, Jonathan Cape

Morris D (1967) The Naked Ape: A zoologists study of the human animal, Jonathan Cape

Senge P (1990), The Fifth Discipline: The Art & Practice of the Learning Organization, Bantam Doubleday

Handy C (1997), The Citizen Company, Source Extract from the Hungry Spirit 1997, pp179-204, in Creativity Innovation and Change The Open University

Teece D. J (2001) Startegies for Managing Knowledge Assets: the Role of Firm Structure and Industrial Contex, in Managing Industrial Knowledge ed Nonaka & Teece, Sage Publications

Teece D. J (2000), Managing Knowledge Assets in Diverse Industrial Context, in Knowledge Horizons: The Present and the Promise of Knowledge Management, Ed Despres C & Chauvel D.

Wheatley M J (2001), Innovation Means Relying on Everyone's Creativity. Leader to Leader No20 Spring 2001

Willets L. G (1996), The Best Ways to Survive Reengineering, Expert Tips on How to Reinvent Your Attitude
Url: http://www.reengineering.com/articles/sept96/reengsurvival.htm 16/10/03

DIFFUSION OF APPLICATION SERVICE PROVISION AMONG SMES
Who are the adopters?

Björn Johansson
Informatics, Jönköping International Business School, Jönköping University, Sweden

Abstract: This paper provides an exploratory empirical survey of the diffusion of Application Service Providers (ASPs). It does do so by presenting a web survey among small and medium-sized enterprises (SMEs). In addition to that it presents a study on one specific ASP and its clients. The research questions are: how is the ASP concept diffused among SMEs and who are the users among SMEs? The concept, ASP, includes software vendors and an ASP enterprise which acts as a third-party firm in a business model. For its clients the concept forms a strategy to rent applications and organise maintenance of their Information and Communication Technology (ICT). The web survey shows that some knowledge of the ASP concept is widely spread among SMEs. However, a closer examination shows that the knowledge of what the ASP concept actually means is limited. This is seen as one reason why the ASP concept has not taken off among SMEs. This reason is related to the question about trust; in order to trust an innovation there is a need to be knowledgeable about the innovation. The conclusion is that ASPs need to be more explicit about their offer. They also need to show how adoption of the ASP concept affects the adopting organisation.

Key words: Adoption, Diffusion, Application Service Provision, SMEs.

1. INTRODUCTION

The use of Application Service Providers (ASPs) as providers of Information and Communication Technology (ICT) has been expected to grow significantly. ASPs are often seen as a way for small and medium-sized enterprises (SMEs) to use ICT to increase their effectiveness and efficiency. The ASP concept can be seen as selective ICT outsourcing. The outsourcing market is vibrant and receives a great deal of attention. Outsourcing is not a new phenomenon. As early as the mid 1960s there were computer service bureaus which ran a variety of systems for external clients (McFarlan & Nolan, 1995). These systems were mainly financial and operational applications. And ever since Kodak decided to rent its ICT resources from an external partner in 1989, there has been a trend towards ICT outsourcing (Hirschheim & Lacity, 2000). Many large companies have decided to transfer their ICT assets, leases, and staff to a third party (Lacity & Hirschheim, 1993). The degree of ICT outsourcing varies considerably. Some companies outsource just a few ICT functions while others outsource their entire ICT operations (McLellan et al., 1998). In recent years we have seen in the outsourcing market a growing number of ASP start-ups and companies offering their products and services through the ASP concept.

At the same time ASPs struggle with the fact that their inflow of new customers is low. Most reports about the ASP concept are predictions about the growth of the market for ASPs. Kern et al. (2001), for instance, mention that there were over 1,000 companies that claimed to be ASPs during the second quarter of 2001. This can be compared with Lacity & Willcocks (2001) who say that only 200 firms fitted the ASP definition by mid-2000. Lacity & Willcocks (2001) predict that the ASP market will rise from US$ 150 million in 1999 to between US$ 11.3 billion and US$ 21 billion by 2003. Firms such as Gartner Group, IDC and Ovum forecast potential market sizes of up to US$132 billion by 2006 (Kern et al., 2001).

Although the ASP concept is often seen as an exhausted trend, it is commonly considered here to stay. But the frequently asked question is whether customers have a demand for ASPs. My impression is that there is a lack of customers. Gartner Group in 2000 (Kern et al., 2001) predicted that 60 per cent of the ASPs will be out of business. Jayatilaka et al. (2003) state that with the rise in the availability of scalable network technologies the ASP concept will become a more feasible source when it comes to the provision of ICT. They claim that the slow growth is due to the fact that business managers have misunderstood the ASP concept. It can be said to be more fragmented, more complex and riskier than they first thought. The actual adoption rate of the ASP concept is low; according to the ASP Industry Consortium only 8 per cent of firms worldwide use an ASP for the provision of ICT (Jayatilaka et al., 2003).

The core of the ASP concept is for the ASPs to offer applications to external customers. The applications can be IT-related, but such applications are not the only thing that ASPs can offer. An ASP enterprise can, for example, also be some kind of information broker. Puelz (2001) describes an ASP enterprise that benchmarks data from 16 financial institutions. However, the most common way to describe ASPs is as providers that offer software applications, which they manage and deliver to external clients (e.g., Cherry Tree, 2001; Kern et al., 2001; Currie & Seltsikas, 2000). The clients use the application in their own businesses, where the software applications are in areas such as website hosting, payroll/billing, e-mail, e-commerce and ERP applications.

Kern et al. (2002a) describe the difference between an ASP option and other ICT-sourcing options. They mention four general ICT-sourcing models: insourcing, buy-in, traditional outsourcing and the ASP option. The difference between an ASP option and the other ICT-sourcing models is that the resource ownership is on the supplier side in the ASP option. This option is also a one-to-many supplier-customer relationship. Kern et al. (2002b) select the term netsourcing as the overarching name, because the common element in the ASP option is the delivery of a product or service over a network. The primary product of an ASP enterprise is remotely managed business applications. This means that the ASP concept can be seen as an innovation and defined in the following way: an ASP is a third-party firm that supplies, manages and remotely hosts software applications through centrally located data centres on a pay-as-you-use basis. For the client the ASP business model is a strategy to rent software applications and organise its ICT maintenance.

Having defined ASPs and the ASP concept, the remaining question is: how is the ASP concept diffused among SMEs?

The paper presents and discusses findings from two studies, one study on diffusion of the ASP concept among SMEs, and one study investigating who the clients are and for what applications they use an ASP. The paper also specifies reasons given by ASPs why SMEs adopt or ignore the ASP concept. The findings are compared with reasons addressed in the literature. The reasons are used to discuss the diffusion of the ASP concept among SMEs.

The paper first discusses reasons given by SMEs to adopt or discard services from ASPs as reported in the literature. Section three presents a survey among SMEs in Sweden. Section four presents a study of a specific ASP and its clients. The final section summarises and discusses the results.

2. REASONS FOR ADOPTING OR NOT ADOPTING THE ASP CONCEPT IN SMES

Kern et al. (2001) point out three reasons why SMEs should adopt the ASP concept. First, even though a package software licence is cheaper than an in-house developed solution, it is still the case that many SMEs cannot afford the packaged solution costs. Second, an SME will be unable to attract and afford the necessary ICT staff. Third, the packaged applications require an established ICT infrastructure and connectivity to ensure optimal performance. For an SME it is difficult to retrieve the necessary human and financial resources to support and continually develop such ICT infrastructures. The ASP concept can also be seen as a way for SMEs to take advantage of the rapidly changing opportunities in ICT (e.g. Turban et al., 2001; Currie & Seltsikas, 2000; Currie & Seltsikas, 2001). An ASP enterprise can assist SMEs with ICT skills, especially in the development and software maintenance areas (Kern et al., 2001). Dewire (2001) argues that there are eight different reasons for an organisation to adopt the ASP concept. The reasons are: if there is a need for flexible ICT infrastructure, if the organisation cannot afford a huge ICT capital outlay, if it does not have the necessary capital resources, if it needs to scale its ICT infrastructure quickly, if it needs to switch to another environment in the near future, if it needs to deploy applications rapidly, if the organisation finds it difficult to attract and retain ICT staff, and finally if ICT is not a core competency.

The close connection between ICT outsourcing and the ASP concept makes it possible to increase the knowledge of reasons for adopting the ASP concept by comparing them with reasons for ICT outsourcing. One commonly quoted reason for ICT outsourcing is the provision of increased flexibility to cope with changes in technology and in the business environment. Paradoxically, the traditional ICT outsourcing agreement is based on long-term contracts that rather tend to inhibit than facilitate change (Shepherd, 1999). One of the ideas of the ASP concept is to make it possible to have a short-term agreement. According to Lee (2001), outsourcing is motivated by strategic, economic and technological benefits. Shepherd (1999) argues that for the majority of organisations the motives for ICT outsourcing could be summarised as a combination of financial restructuring, reducing or stabilising costs, overcoming cultural and organisational problems, concentrating on core competencies and accessing world-class expertise.

McLellan et al. (1998) give five reasons for ICT outsourcing: financial motivations, the internal IT department does not respond to organisational needs, strategic motivations, to improve long-term business performance and to facilitate strategic change. Whether an organisation should outsource its ICT or not is, according to Weill & Broadbent (1998), principally a question of what strategy the organisation has.

According to Loof (1995) external suppliers of ICT-sourcing predict large cost reduction, improvements in quality and higher responsiveness if customers hand over their ICT functions to them. He also states that reports on outsourcing are often overly optimistic. The result is that many organisations are in doubt whether there are any benefits for them in outsourcing. Udo (2000) says that there is a trend among organisations today to classify ICT functions into two categories according to the kind of services they deliver, commodity services and strategic services. He (2000) maintains that commodity services can be outsourced without having doubts, but strategic services should never be outsourced. However, Udo refers to Lacity & Hirschheim (1993) who mention that this categorisation can lead to serious problems for an organisation. The reason is that commodity services at present time can be of high strategic importance for the organisation in the future.

According to Udo (2000) providers purpose the following benefits:
- A predictable ICT budget is gained by tying it to actual requirements.
- Lower costs for ICT, which means cost savings compared with both current and future expenditures on ICT equipment.
- Increased access to technological resources and technologically skilled personnel.
- The organisation can focus on core products and services.
- The organisation's fixed costs for ICT can be changed into variable costs.
- The risks in the development of ICT applications are spread.

But, as stated above, Udo (2000) also claims that outsourcing has more disadvantages than advantages. The following potential disadvantages are reported:
- A lack of chemistry between the partners.
- Dependency on another party for the organisation's critical information.
- Loss of ICT capability.
- Loss of control of ICT assets.
- Threat of opportunism from the supplier.
- Loss of flexibility.
- Loss of competitive advantage in information management.
- Loss of ICT expertise and as such loss of memory in the organisation.
- A decline in morale and performance of the remaining employees.
- No guarantee for long-term cost savings.

Baldwin et al. (2001) argue that a selective sourcing approach with the opportunity to use several different suppliers is an increasingly popular strategy to minimise risks, maximise benefits and reduce costs. Jurison (1995) states that there are primarily economic reasons, i.e. economic consideration in different forms is the primary driver for an ICT-outsourcing decision. He mentions risks as the primary reason for not choosing an ICT-outsourcing option, where the irreversibility of the decision is seen as the dominant risk.

3.   A SURVEY OF THE DIFFUSION OF ASP AMONG SMES

The question in this paper is how the ASP concept is diffused among SMEs. In order to answer that question, results from two different studies are used. This section presents one of them. It was conducted by Isaksson & Linderoth and is presented in their master thesis (Isaksson & Linderoth, 2003). The questionnaire was submitted as a web survey to 400 SMEs in Sweden, and data were collected in a database. SME was defined to be an organisation with 10 to 250 employees and an annual turnover not exceeding EUR 40 million (EU Office in Blekinge, 1999). The base for this population was searched out from a list of joint-stock companies. There were 27 838 companies that fitted the description of an SME. From the total number of SMEs the survey was submitted to every 70th SME on that list. The total number of responding SMEs was 119. This means that response rate was 30 per cent. However, there were in a follow-up 17 organisations that did not fit the definition of an SME. These respondents' answers were excluded from the study.

The questionnaire consisted of fifteen questions, grouped into three categories. These categories were characteristics of the respondent, grade of diffusion of the ASP concept, and attitudes to the ASP concept. In this paper questions connected with the grade of diffusion of the ASP concept among SMEs are discussed. In the questionnaire there were four questions dealing with this. These questions are presented and the result from them is delivered in sections 3.1 to 3.4.

3.1   First question: When did you hear about the ASP concept?

The first question asked was: when did you first hear about the ASP concept? This question asks two things; first, if the SMEs have heard about the concept as such and second, when they were made aware about it. The respondents could choose from the options shown in Table 1, which also shows the distribution of the answers.

Table 1. Answers to the first question: When did you first hear about the ASP concept?

When did you first hear about the ASP concept?	Distribution of respondents' answers	*Per cent of SMEs participating in the questionaire*
Before 1998	14	*14 per cent*
1998 – 2000	43	*42 per cent*
2001 – 2003	32	*32 per cent*
Have not heard about the ASP concept	12	*12 per cent*

The findings in the table are interesting when compared with, for instance, Jayatilaka et al. (2003) who claim that the ASP concept is a new concept, and most reports on ASP state that the concept was created at the end of the 1990s. According to Kern et al. (2001) this kind of service provision evolved in the late 1990s. IDC, an American analyst company, claims that the idea originated from them in 1998, while the company TeleComputing declares that they coined the expression (Elerud et al., 2001). Despite this 14 per cent say that they heard about the concept before 1998. Irrespective of who coined the term, ever since the late 1990s there has been an increasing interest both among practitioners and researchers in the phenomenon. However, Jayatilaka et al. (2003) say that the concept as such dates back to the 1960s, but then labelled as application hosting. When summing up the numbers in Table 1, 88 per cent said they have heard about the ASP concept. From answers to questions later in the study by some of these respondents it appears that they might have heard about the concept but not really understood what ASP in this context is about. One statement was for instance that ASP is an easy programming language, which indicates that the respondent thought it was about active server pages code. The conclusion that can be made from these numbers is that a majority of SMEs have heard about the ASP concept. Only 12 per cent of the respondent said that they had never heard about it.

3.2 Second question: When did you start to use ASP?

The second question asked the ones that had heard about the ASP concept: When did you start to use ASP? The answers to this question are shown in Table 2. Also this question was a multiple-choice question, and the possible options to choose among are shown in the table.

Table 2. Answers to the question: When did you start to use ASP?

When did you start to use ASP?	Distribution of respondents' answers	*Per cent of SMEs participating in the questionaire*
Before 2000	5	*6 per cent*
During 2000	6	*7 per cent*
During 2001	3	*3 per cent*
After 2001	8	*9 percent*
Do not use ASP	67	*75 per cent*

The numbers show that among the SMEs that had heard about ASP most do not use ASP. The SMEs that use ASP are distributed fairly equally over time. However, there is a drop during 2001, which is amazing since during this time heavy marketing was done about the ASP concept. An explanation of the drop in 2001 can be that ASPs had a very negative performance on the stock market during that time (Cherry Tree & Co, 2001). This had an impact on clients' purchasing patterns. The numbers reveal that 79 per cent do not use the ASP concept and 21 per cent uses the ASP concept. The question is to what degree they use ASP. This will be discussed in the next section.

3.3   Third question: To what degree is the ASP concept used?

The third question asked was: To what degree is your organisation's ICT provided through the ASP concept? This question was answered by the organisations who in the answer to the second question stated that they used the ASP concept. The statements and the answers are shown in Table 3.

Table 3. Answers to the question: To what degree is your organisation using the ASP concept?

To what degree is your organisation's ICT provided through the ASP concept?	Distribution of respondents' answers	Per cent of SMEs participating in the questionnaire
Less than 20 per cent	14	65 per cent
20 – 40 per cent	6	27 per cent
40 – 60 per cent	0	
60 – 80 per cent	1	4 per cent
80 – 100 per cent	1	4 per cent

The answers show that 65 per cent of the SMEs use the ASP concept for less than 20 per cent of their ICT. This means that in this case the ASP concept can be compared to selective outsourcing. According to Willcocks (1994) selective outsourcing is when an organisation outsources less than 80 per cent of their ICT budget. Selective outsourcing consequently means a mix of outsourcing and internally managed ICT. The findings also show that only four per cent use the ASP concept for the provision of their ICT to a degree of 80 per cent. This implies that ASP in this case can not be seen as total outsorcing. Total outsourcing is defined by Willcocks (1994) as when at least 80 per cent of an organisation's ICT budget is spent on external partners. The results build on a fairly small number of SMEs. However, they give some indication of how the ASP concept is used. Another indication is obtained by looking at what applications the SMEs buy as an ASP service. That is the question presented in the next section.

3.4   Fourth question: What applications are provided through the ASP concept?

The fourth and final question presented from the survey among SMEs is a question that deals with what kind of applications are provided through the ASP concept. The SMEs that answered this question were those who stated they were using ASP. There were ten different systems or categories of systems that the respondents had to choose among. There was also a possibility to describe other kinds of systems they rented from an ASP. The most used systems among the ASP customers are financial systems and tools for publishing on the web. These two categories of systems were used by 37 per cent of the SMEs that used an ASP for the provision of ICT. An interesting finding is that office suites were provided as an ASP service to 16 per cent of the ASP customers. This is interesting because when discussing with ASPs on what applications they deliver, an office suite is seen as the basic offer that all customers need and also have.

To deepen the findings from the web survey they are compared with findings from the second study, which is presented below. The second study was conducted among customers to a specific ASP and part of an exploratory study of the ASP concept.

4. A STUDY OF AN ASP AND ITS CUSTOMERS

The second study was directed to customers of a specific ASP, acting as a horizontal ASP. According to Currie & Seltsikas (2000) an ASP is categorised as a horizontal ASP if it offers collaboration tools and other applications to a broad base of customers'. The focus is mainly on business processes. The ASP studied fits this description insofar as it does not focus on a specific industry. Instead it tries to support its customers with all ICT applications they need. The first part of the study of the horizontal ASP, which is presented in Section 4.1, can be described as an exploratory study. It was conducted as open-ended, semi-structured interviews with the service provider. Two interviews with sales managers at the service provider were done (November 2001 and February 2003). The interviews were tape-recorded and transcribed. There were two overall questions in the interview with the ASP: 1) what do they deliver? and 2) why should SMEs adopt the ASP concept?

4.1 The ASP company

When the first interview was conducted the ASP was a consulting firm located in Denmark, Norway and Sweden. The company was the result of mergers of three different companies; the three companies were an Internet Service Provider (ISP), an IT-consultant, and an ASP-firm. The history of the organisation can be seen as an expression of the still ongoing consolidation among service provider organisations. The organisation conducted its business under one name for almost a year before it went bankrupt. Almost directly it was restarted as a new firm. According to the interviewee in November 2001the company was a leading ASP actor in Scandinavia with a steady inflow of new customers. The organisation's data centre in combination with ISP service and IT-consultancy experiences makes the company well prepared to become a competitive player in the ASP market. Currently (November 2003) the organisation has 45 employees located at five places. According to the interviewee the background of the ASP offers a possibility for the company to provide their customers with flexible solutions.

4.1.1   Applications and services provided by the ASP

There are two different types of services provided—a consultancy and a hosting part. The type examined here is the hosting part. It is described as an offer of a flexible solution to customers. This flexible solution consists of a base block and a block of customer-specified solutions. The base block consists of Microsoft's Outlook, Explorer, Office, Project, and WinZip. The base block is needed for the service provider to increase its volumes and to make a profit. The second part is customer-specific applications that customers either already have or want to have. These solutions can, for example, comprise payroll/billing, e-commerce and ERP applications. The ASP portfolio of the ASP studied includes between 80 and 90 different software applications.

There are two ways of implementing the applications, dedicated servers or one-to-many servers. Currently the firm focuses on finding applications aimed at being suitable for the one-to-many solution. According to the interviewees the ASP concept has had some problems. One of the problems was that too much was promised in the first stage. It was stated that the ASP concept should be able to take care of all ICT including all services and support for their clients. This has not been possible to fulfil due to the fact that it is not possible to handle all applications as an externally hosted application. The ASP has tackled this situation by taking a complete view of their customers and now it develops and manages some applications at their customers' places as well. The aim is to move these customer applications to the data centre of the ASP. The aim of the ASP is to be a provider of all necessary ICT and ICT services for their customers.

4.1.2   Why become a customer?

The ASP company's customer segment is small and medium-sized enterprises (SMEs), where the ordinary customer has an e-mail system and an office packet as its basic ICT. In addition, a customer often rents a customer relationship (CRM), an accounting and/or an enterprise resource planning (ERP) system. The main reasons for adopting the ASP concept are, as stated by the interviewees in the ASP:

- A customer does not want to have extensive ICT-competence.
- Getting 24/7 accessibility to applications—according to the interviewees, SMEs have problems with providing this amount of accessibility.
- Access to a helpdesk function.
- To get control of their ICT-costs, customers want to know the ICT cost per user every month.
- To get full control of ICT investments.
- To increase their ICT security—help in tackling spam and virus problems as well as internal security issues, for example, backups.
- To get rid of problems with software upgrades.

Two main reasons for not adopting the ASP concept emerged from the interviews:

- Losing control of data by having data at the ASP's data centre.
- Cost - many presumptive customers say that, based on a calculation from the ASP, ICT and associated services will cost too much.

Both these reasons for not adopting the ASP concept are handled by the ASP. The fear of losing control is addressed by showing how the concept works. The ASP uses a specific model showing how the customer is "implemented" in the data centre and the customer can see how security is handled and how the company's system is protected against intruders. The cost reason is addressed by doing a total cost of ownership (TCO) analysis. If the customer permits the ASP to do this TCO analysis it will get a good picture of its current ICT costs and these costs can be compared with the ASP fee.

4.2 The survey of the ASP customers

The second part in the study was a survey of the ASP's customers. The questionnaire used consisted of twenty-one questions. This report focuses on three questions: two related to who the customer is and one related to what applications are rented from the ASP. The questionnaire was sent to all customers identified as ASP customers. In this study, an ASP customer is defined as an organisation using at least one rented software application in its organisation. Using the definition, there were 17 ASP customers identified. Fourteen of these customers responded to the questionnaire. In developing the questionnaire results from the interviews as well as from two other studies were used. Currie & Seltsikas (2000) developed a framework for evaluating ASP offers. The framework consists of four different categories and related performance criteria. Susarla et al (2003) did an empirical analysis of ASP satisfaction. The study focused on four different areas: 1) satisfaction in terms of organisational attitudes, 2) expectations about the service, 3) perceived disconfirmation and 4) perceived provider performance. The questions used in this study focus on who the customers are and what applications are provided from the ASP.

4.2.1   Who are the customers of the ASP?

The questionnaire started with questions about who the customers are and some characteristics of the customers. The first question asked was: Which option describes your organisation best? The answer is shown in Figure 1.

A majority of the customers describe themselves as service organisations. However, to deepen this description, the second question asked was: Which option describes your customers best? The answer is shown in Figure 2. A majority of the ASP customers identified their customers as other organisations. 77 per cent said that the option "mainly other companies" described their customers best.

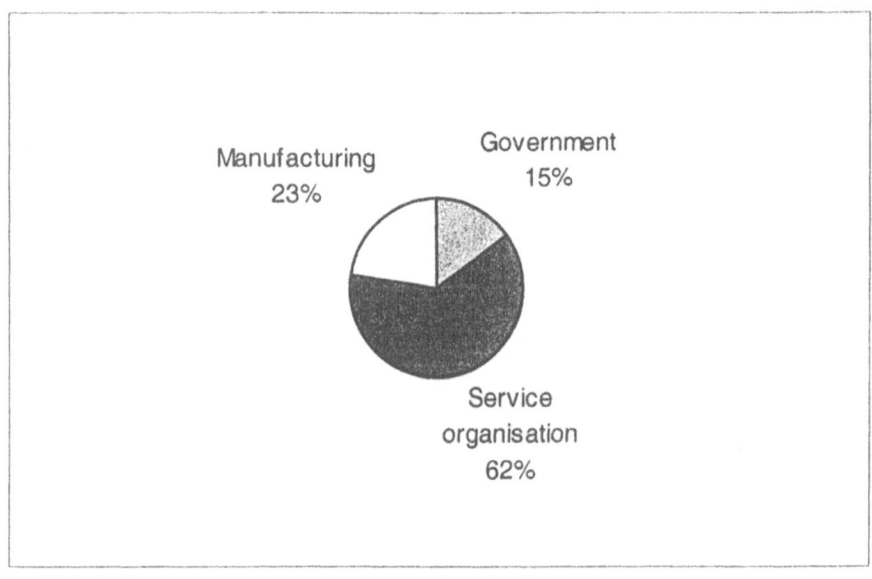

Figure 1. Answer to the question: Which option describes your organisation best?

The answers to these two questions show what kind of customers the horizontal ASP has. Analysing the two questions together leads to interesting conclusions. One conclusion is that a large number of the customer states that they are service organisations that have other companies as their customers. This implies that the ASP customers are used to cooperating with other companies. Another possible conclusion is that the ASP concept is more used among organisations that provide their customers with information. This is in contradiction to a reason that is often put forward for adopting the ASP concept – to focus on core competence. In the study only 23 per cent of the organisations identified themselves as a manufacturing industry, and for those organisations ICT could certainly be said not to be their core competence.

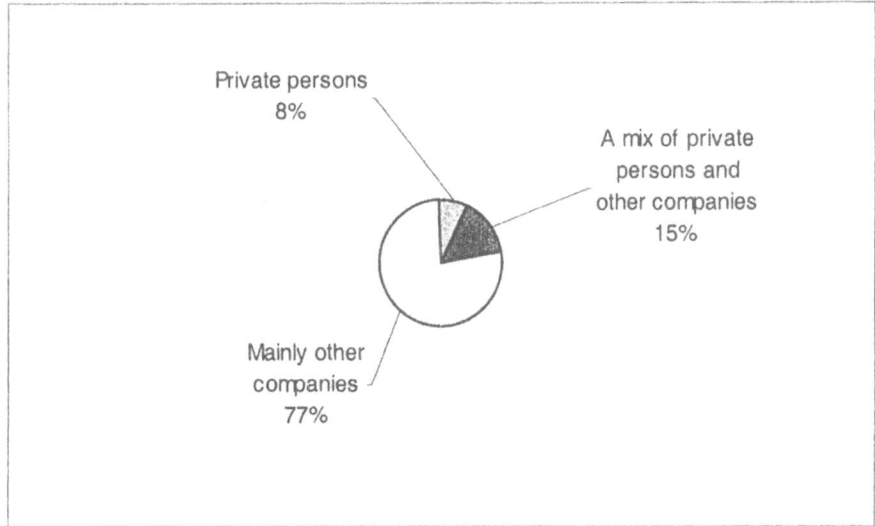

Figure 2. Answer to the question: Which option describes your organisation's customer best.

A statement often made concerning who the ASP concept fits, is that the ASP concept demands a network connection and if the organisation does not already have this the ASP concept is not suitable. That ASP demands a network connection is so self-evident that it does not have to be questioned. However, the statement from the interviewees at the service provider organisation declares that it only fits organisations that are located at different places. This statement was tested by asking the customers of the horizontal ASP if they are located in more than one place.

The result shows that 62 per cent of the organisations are located in more than one place. However, the rest of the customers are located at only one place which contradicts the statement made by the service provider. In order to explain this finding it is fruitful to look at who the users are and what function they have in the organisation.

4.2.2   Who uses the applications provided by the ASP?

The survey established which categories of employees use applications provided by the service provider. The result is shown in Table 4.

Table 4. Categories of employees using applications provided by the ASP.

Category of employees	Number of organisations with users in the respective category (total number of organisations)	Per cent of organisations that use rented applications in the respective category	Per cent of organisations that do not use rented applications in the respective category
Administrator	12 (13)	92 per cent	8 per cent
Service staff	6 (8)	75 per cent	25 per cent
Factory hand	2 (3)	67 per cent	33 per cent
Other employees	6 (13)	46 per cent	54 per cent

One finding is that 46 per cent of the organisations have "other employees" that use applications from the service provider. The majority of employees in this group are salesmen and consultants. Five organisations out of six describe "other employees" as salesmen and consultants. These employees need to have access to the organisation's system when they are outside the organisation's premises. This can maybe explain the finding that 38 per cent of the customers are only located at one geographical place. The service providers claim that there are two basic characteristics of the SMEs, at least one of which has to be fulfilled if the ASP concept should be interesting for SMEs. One characteristic is that the SMEs should be located at several places. The other characteristic is that they have employees that need to have access to applications from various locations. The first interpretation of why organisations located at one single place have adopted the ASP concept is that they probably have sales personnel that need to have access to the organisation's applications from various places. However, this statement is not supported when the data are examined in detail. Instead customer organisations that have sales personnel are also located at several places. Four out of the eight customer organisations located at several places state that they have sales staff. Only one customer organisation is located at one place and has sales staff. The possibility to have access from anywhere is something that service providers emphasise as one benefit of the ASP concept. If an organisation decides on using the ASP concept the organisation also obtains increased accessibility. The only thing needed according to the service providers is a connection to the Internet.

4.2.3 What applications are rented from the ASP?

To get an understanding of what applications are rented from the ASP the study investigated what applications the customers rent, what applications they do not rent and what applications they do not use. The study included twelve different categories of software applications. The result is shown in Table 5.

Table 5. Applications in the ASP customer organisations.

Software application	Rented	Not rented	Not used
Office suites (e.g. MS Office)	11	2	
e-mail systems	11	2	
ERP systems	9		4
Website	9	1	
Finance systems	9	3	1
Accounting systems	8	2	3
Payroll systems	6	4	2
MPS systems	4		8
CRM systems	3	1	8
CAD/CAM systems	2	2	8
e-commerce	1	1	8
Video conferencing systems	1		9
Other systems	3		

Table 5 shows how many organisations use and rent, use but do not rent respectively do not use categorised software applications. The sum of each row in the table should be thirteen. However, some of the respondents did not give answers in all categories. In all likelihood respondents who did not mention some applications do not use those applications in the organisation. The result shows that the most commonly used and rented applications are office suites (MS Office) and e-mail systems. These are software applications that are very common in organisations today. These two applications are also seen by ASPs as the basic option and it is more or less compulsory to rent them.

Two of the organisations have chosen not to rent an office suite, despite the fact that they use one. One of the organisations only rents an ERP system including accounting and payroll. The other organisation has its website located at the service provider and rents a payroll and an accounting system from the service provider. It can be concluded from this that they do not really utilise all the advantages the ASP concept can offer, for instance the advantages of easily upgraded versions. One explanation for why the two organisations do not rent the office suite can be that they already have paid for the licence and do not see any reason for transferring these systems to an ASP. This explanation is supported in the interviews when the ASPs claim that one of the difficulties in the ASP concept is how to handle the licence fee.

More advanced applications such as customer relationship management (CRM) and e-commerce are not so widely used among the customers. One reason for this is perhaps the fact that these applications have not yet been discovered by SMEs or maybe they were discovered but are not needed in SMEs. Another explanation is that SMEs first rent basic applications and when they have more experience from the provider and the concept as such they expand their use of ASP. This latest explanation is supported by the ASPs, who claim that often a customer starts by renting, for instance, Microsoft Office and later on, when they are familiar with the concept and the service provider is familiar with the customer, more customer-adjusted applications are rented. If this is the case there should be a correlation between when the customer adopted the ASP concept and what applications they use. When searching the data for this correlation it was found that among organisations that use software applications that are not rented non-rented applications were equally distributed among the customers.

5.   CONCLUSIONS

The paper describes what the ASP concept is used for and who the ASP customers are. It can be concluded that the ASP concept and the services delivered from an ASP do not focus on a specific customer segment. The basic idea of the ASP concept and the business model of the investigated ASP to aim at being a complete provider of software applications, has not been found to be fulfilled. Instead the usage is more in line with selective outsourcing as described by Baldwin et al. (2001). The web-based survey of SMEs found that financial systems and tools for publishing on the web were the most rented applications. This was not supported in the findings from the study of ASP customers of the horizontal ASP. The most common applications rented from this service provider are office suites (MS Office) and e-mail systems.

There are several reported reasons why an SME should adopt the ASP concept. At the same time it can be stated that SMEs know what ASP is about. However, it can also be stated that the knowledge and deep understanding of what ASP exactly means is inadequate. This could be seen as one reason why the diffusion and adoption of the ASP concept is reported as limited.

If the reported benefits are true, why do SMEs not adopt the ASP concept? One explanation is that the reported benefits can be seen as strategic. By strategic is meant that the benefits are seen as a way of improving the organisation over a longer period.

The reported risks on the other hand can be seen as more direct and affect the organisation in a more direct manner. Does this have anything to do with the diffusion of the ASP concept among SMEs? It can be concluded that it has if the SMEs are seen as more operational in their decision-making. It can be stated that managers in SMEs need to solve their organisation's problems at once and do not have enough time or energy to always make strategic decisions. This implies that the risks have a stronger impact on the decision to adopt or not. The SMEs need to solve the problems at hand and if an adoption of the ASP concept does not solve the existing problems, they will probably not even think of adopting something that could imply new problems even if it would solve the problems in a longer term perspective.

The service provider states that the ASP concept is suitable only if the SMEs are located at several places and have a network communication already. The reason they give for this statement is that otherwise it will be too costly to implement and use ASP for SMEs. In this study it was not found to be the case. Instead 38 per cent of the customers have decided on using the ASP for the provision of their ICT, despite the fact that they are only located at one place. The reason for this could be that these customers have employees that need to have access to their applications at a distance. The need of having access to applications on a distance is another statement the provider gives for what organisation the ASP concept fits. However, this was not supported in the study. It was found that the customer located at one place did not have employees that needed distance access.

When the first interview was conducted in November 2001 the provider stated that only 10 per cent of their customers used the Internet for network communication. Since then this have changed and at the moment (October 2003) the Internet is used by 35 per cent of their customers. One reason why the use of the Internet has increased is probably increased trust in the Internet. Lack of trust is probably one reason why the ASP concept has not taken off. The conclusion that the ASP concept has not taken off can be drawn despite the fact that 88 per cent of the SMEs know about the concept. The limited use of the ASP concept in SMEs has probably several different reasons. One of the most commonly reported reasons is, as stated above, trust, both when it comes to trust in the technology as such and trust in the partner to cooperate with.

According to the service provider one reason for the low uptake of ASP can be that it promised too much in the beginning. This statement can be compared to Hagel's (2002) statement that the ASP concept was introduced too soon. He claims that one reason for the low uptake was that the offer from ASPs was not developed for the ASP business model. Many applications were not adapted to network use. The result was that ASPs started to deliver ASP services as a service hosted at the customer's place, which means that the original ASP concept became blurred.

Another reason why the diffusion of ASP has been limited can be an effect of the dotcom collapse and the impact this has had on investments in ICT. If this is the reason one can conjecture that it will change and that we will see an increase in the use of the ASP concept in the future, but this will only happen if the ASPs show profit and if the applications they offer are needed by SMEs. However, it also requires trust in the new ways of providing ICT at a distance over a network.

The main finding in this paper is the fact that SMEs state that they know what ASP is. However, the study shows that the knowledge of the ASP concept among SMEs is limited. When asked, ASP customers gave various interpretations of the ASP concept. This is also indicated in a study by Isaksson & Linderoth (2003) who claim that 88 per cent of SMEs say they know what ASP is, but at the same time their study shows that there is a lack of knowledge among SMEs about what the ASP concept actually means. However, answering the question whether they were familiar with the ASP concept, all customers said they were, which is not surprising since the question was put to ASP customers. More interesting are the different explanations of what the concept means to them. To some customers the ASP concept means that they just rent storage. To other customers the ASP concept means that the service provider reengineers the organisation's business processes. The most common statement is that the ASP concept means that an external provider takes care of the management of software applications and makes sure they work. It can be concluded from this that the ASP concept from the customers' view is seen as something operational and not something strategic. This conclusion is based on the fact that customers do not describe the ASP concept as a way of improving their business over a longer period. It can also be concluded that the adoption or non-adoption of the ASP concept depends on whether the potential customer has trust in the concept. And in order to have trust there is a need to be knowledgeable. The implication of this is that ASPs need to be clearer about what their offer consists of. They also need to show how the adoption of the ASP concept affects the adopting organisation.

REFERENCES

Baldwin, LP., Irani, Z., & Love, PED. (2001). Outsourcing information systems: drawing lessons from a banking case study. *European Journal of Information Systems* Vol. 10 pp. 15 – 24.

Cherry Tree & Co. (2001). Trends in Outsourcing: Strong, Sustainable and Growing, Spotlight Report, July.

Currie, W.L. & Seltsikas, P. (2000). Evaluating the application service provider (ASP) business model, Executive Publication Series CSIS2000/004, Centre for Strategic Information Systems, Department of Information Systems & Computing, Brunel University, Uxbridge, UK.

Currie, W.L. & Seltsikas, P. (2001). Exploring the supply-side of IT outsourcing: evaluating the emerging role of application service providers, *European Journal of Information Systems*, Vol 10, pp 123-134.

Dewire, D. T. (2000). Application service providers, *Information Systems Management*, Vol 17, No 4, pp 14-19.

Elerud, F., Gustafsson, T., & Jusufagic, M. (2001). ASP – Framtidens IT-drift? Ett strategiskt perspektiv. Master thesis, Företagsekonomiska Institutionen, Lunds Universitet.

EU Office in Blekinge. (1999). EU-bidrag: korta råd och tips om projektbidrag från EU. http://eu-kontoret.softcenter.se/hursok.htm 2003-03-14.

Hagel, J. (2002). *Out of the Box*. Harvard Business School Press, Boston, MA.

Hirschheim, R. & Lacity M. (2000). The myths and realities of information technology insourcing, *Communications of the ACM*, Vol 43, No 2, pp 99-107.

Isaksson, R., & Linderoth, P. (2001). Application Service Provision for SMEs: Vision or reality? Master thesis in Informatics, Jönköping International Business School, Jönköping University.

Jayatilaka, B., Schwarz, A., & Hirschheim, R. (2003). Determinants of ASP choice: an integrated perspective. *European Journal of Information Systems*, Vol. 12 pp 210 – 224.

Jurison, J. (1995). The role of risk and return in information technology outsourcing decisions. *Journal of Information Technology*, 10, pp 239-247

Kern, T., Lacity, M., Willcocks. L. (2002b). Netsourcing: Renting Business Applications and Services Over a Network. Financial Times Prentice Hall, Upper Saddle River, New Jersey.

Kern, T., Lacity, M., Willcocks. L., Zuiderwijk, R. & Teunissen, W. (2001). ASP Market Space Report 2001. Mastering the Customers Expectations, GMG report.

Kern, T., Willcocks, L. & Lacity, M. (2002a). Application Service Provision: Risk Assessment and Mitigation. *MIS Quarterly Executive* 1 (2) pp 113-126.

Lacity, M. & Hirschheim, R. (1993). Information Systems Outsourcing: Myths, Metaphors and Reality, John Wiley & Sons, Chicester.

Lacity, M. & Willcocks, L. (2001). Global Information Technology Outsourcing: In Search of Business Advantage, John Wiley & Sons, Chichester.

Lee, J-N. (2001). The impact of knowledge sharing, organizational capability and partnership quality on IS outsourcing success, *Information & Management*, Vol 38, pp 323-335.

Looff, L.A. de (1995). Information systems outsourcing decision making: a framework, organizational theories and case studies. *Journal of Information Technology*, 10, pp 281-297.

McFarlan, F.W. & Nolan, R.L. (1995). How to manage an IT outsourcing alliance, *Sloan Management Review*, Winter, pp 9-23.

McLellan, K., Marcolin, B.L. & Baemish, P.W. (1998). Financial and strategic motivations behind IS outsourcing, in *Strategic Sourcing of Information Systems*, L.P. Willcocks & M.C. Lacity (Eds), John Wiley & Sons, Chicester, pp 207-248.

Puelz, R. (2001). Entrepreneurship and an ASP in financial services, *Journal of Business & Entrepreneurship*, Vol 13, Special Issue, October, pp 33-55.

Shepherd, A. (1999). Outsourcing IT in a Changing World. *European Management Journal*, vol.17, no. 1. pp 64-80.

Susarla, A., Barua, A. & Whinston, A.B. (2003). "Understanding the service component of application service provision: an empirical analysis of satisfaction with ASP services", *MIS Quarterly*, Vol 27, No1, pp 91-123.

Turban, E. McLean, E. & Wetherbe, J. (2001). *Information technology for management: Making Connections for Strategic Advantage*, second edition. John Wiley & Sons Ltd, Chicester.

Udo, G.G. (2000). Using analytic hierarchy process to analyze the informationtechnology outsourcing decision. *Industrial Management & Data Systems*. MCB University press 100/9 pp. 421 – 429. http://www.emerald-library.com

Weill, P. & Broadbent, M. (1998). Leveraging the New Infrastructure. Harvard Business School Press, Boston

Willcocks, L. (1994). Collaborating to Compete: Towards Strategic Partnerships In IT Outsourcing. Oxford Institute of Information Management, RDP94/11. Templeton College, Oxford.

PART V

NEW ENVIRONMENTS, NEW INNOVATION PRACTICES

DIGITAL GAMING: ORGANIZING FOR SUSTAINABLE INNOVATION

J. P. Allen[1] and Jeffrey Kim[2]
[1]Masagung Graduate School of Management, University of San Francisco, USA; [2]Information School, University of Washington, USA

1. INTRODUCTION

Digital gaming has caught the attention of the management world because of its amazing growth rate—the largest massively multiplayer online games, running 24 hours a day, now have millions of subscribers, and host hundreds of thousands of players simultaneously—and perhaps because of its 'sexy' subject matter. But digital gaming also offers one of the most extreme examples of the need for continuous and rapid IT innovation. Games must constantly innovate to preserve the long-term interest of paying, voluntary players that seek a vibrant community and a compelling experience.

In this paper, we discuss digital gaming as an example of 'experiential' IT (Rosenbloom, 2003) that likely will have applications beyond today's gaming industry. We outline a research agenda around the question of how digital games are able to change in order to stay alive and vibrant over the long term—how do the relevant social groups organize themselves for sustainable innovation? We propose a theoretical approach based on sociotechnical change theory. A sociotechnical change approach attempts to unpack the amazing amount of social and technological diversity associated with digital gaming, in a way that will move the research literature beyond a discussion of 'technological features' on one hand, and the 'impacts' of games on the other.

2.   DIGITAL GAMING AND IT INNOVATION

The innovation demands of digital gaming are particularly intense. This section of the paper defines more precisely what those innovation demands are, the nature of the business problem faced by gaming organizations, and how we define a research agenda to investigate these challenges.

2.1   Innovation demands of gaming

Digital gaming has been defined as 'experiential' IT (Rosenbloom, 2003). Our typical view of IT is to design a technology for specific information processing functions, then have it adopted and used by end-users. "In experiential environments, users apply their senses directly to address information related to an event (such as a business transaction or even a college football game)." (Rosenbloom, 2003; p. 29) Digital games, as an experiential IT, succeed by creating environments for exploration and interaction that immerse players in an artificial world. Classic literary techniques such as scenario, character, and storyline are combined with IT to make digital gaming memorable, engaging, and even compelling (Bushnell, 1996).

A new and difficult challenge of experiential IT is the demand for constant change, over a relatively long period of time, in order to preserve an engaging and compelling experience. "The driving force in these immersive, realistic environments is the user's experience, not merely a specification." (Rosenbloom, 2003, p. 31) The relationship between IT producers and customers in experiential IT takes on a whole new flavor. Producers and users constantly interact to produce the day-to-day innovation that will keep a community of digital players vibrant and alive. Experiential IT demands sustainable innovation—not just one-time during specification and design, or even in occasional new releases.

2.2   The business challenge of sustainable innovation

The business challenged faced by organizations trying to produce digital gaming is to keep customers playing (and paying) as long as possible. State-of-the-art experiential IT does not come cheap. A successful multiplayer, online gaming environment is the equivalent of a "moonshot," according to Game Developer magazine (February, 2003). "It's expensive, technically difficult, and can take years to complete, and yet everyone wants to give it a try."

After a difficult and expensive beginning, then the need for sustainable innovation arises over the longer term: a steady stream of new content, new storylines, new challenges, and new capabilities, along with the need to adjust technology, regulations, and even the interactions between players. The players, as they try to create vibrant communities, have a tremendous influence on the future directions of the digital game. The need for continuous innovation in experiential IT blurs the line between product and service.

The scale and speed of interaction among players is another distinctive characteristic of digital gaming. The largest M.M.O.R.P.G. ("massively multiplayer online role-playing game" such as Lineage in South Korea, or Everquest in the US) can have millions of subscribers and hold as much as 100,000 players in the same virtual game space at the same time. Players interact within the game in sophisticated and complex ways. "Game designers...have found that their models are realistic enough to engender many of the same phenomena that exist in the real world, including market pricing, civic organization, friendship, environmental shortages, hyperinflation, theft, murder, and inheritance." (Lovejoy, 2002) In a vibrant game such as Everquest, players average over 5 hours per day playing time, and spend on average more than 800 hours developing their virtual characters. (Lovejoy, 2002)

Players can also interact outside of the game. Interactive web games, for example, attract players who will volunteer to solve puzzles in the game collectively. In the case of one interactive web game (The Beast), players created their own website (www.cloudmakers.org) to exchange suggestions and possible solutions to the game puzzles. As players put together the clues left in the game space and solved the puzzles, new puzzles were added by the main creator of the game story (a science fiction author). As the players banded together, puzzles were solved more quickly, increasing even further the rate at which new game innovations had to be introduced. After trying to keep up with a community of 7,000 players over three months, the main storyteller confessed that it was one of the most intense and demanding writing experiences of his life (personal communication, September 2003).

If these digital games are any indication, experiential IT will require sustained innovation to cope with the active role that 'users' play in the evolution of the product/service itself. Already, the 'customer service' needs of online, multiplayer gaming are proving somewhat unique. Sony, the developer of Everquest, hires over 100 expert gamers to constantly intervene, appearing in the game as characters to help organize players and resolve disputes, or to "go invisible" and "investigate those suspected of violating the rules or wreaking virtual havoc" (Walker, 2003). Dozens of programmers "continually add fresh weaponry, skills and adventures to each game. Unlike most short-lived PC games, massively multiplayer online games constantly evolve in an attempt to keep players engaged." Thanks to this constant innovation, the average Everquest player stays at least 10 months.

2.3   The research agenda

Our research agenda is to better understand how to produce experiential IT. In particular, how do producers organize their work activities to create the sustainable innovation that experiential IT seems to require? For digital gaming specifically, we want to better understand how producers organize the work of changing the game over time to create a vibrant, compelling experience that keeps players coming back for more.

We suspect that the unique features of experiential IT innovation will reveal themselves along the following work dimensions:

- *Developer-Player*: 'Users' are intimately involved in the direction of innovation, and interact among themselves.
- *Developer-Developer*: The range of talent required to develop digital gaming is diverse. It includes programmers and network specialists, but also artists, writers, and producers. How will members of these different communities interact to produce IT-based innovation?
- *Developer-Sponsor*: Business strategists, investors, and outside clients will place their own demands on digital gaming. How will the need for cost controls, for example, impose itself on the creative innovation required for experiential IT?

3.   SOCIOTECHNICAL CHANGE THEORY

Our approach to studying sustainable innovation is based on sociotechnical change theory. In this section of the paper, we present our theoretical approach, why it fits with the problem of IT innovation in gaming, and discuss how we believe our approach will move the research literature forward.

3.1   A sociotechnical view of gaming

Sociotechnical change theory, as defined by Bijker (1995), refers to a school of technological change theory that has emerged from the sociology of technology community. Sociotechnical change theory includes a number of different research traditions (including, for example, social construction of technology, and actor-network theory) that share common features. Sociotechnical change theory views technology as an accomplishment, whose successful stabilization into a particular useful form needs to be explained as a combination of social and technical elements, and not simply as a victory of 'superior' technology.

We argue that any attempt to describe how modern digital games are created, stabilized, and evolve would benefit from this sociotechnical approach (Bijker et al, 1987; Bijker and Law, 1992). Sociotechnical change theory highlights the diverse interactions of the groups involved in game evolution, including artists, programmers, technicians, business strategists, expert players, casual players, activists and hackers. This sociotechnical approach focuses on gaming itself as a process involving complex associations between different social/technical elements, creating a mix of digital technology, product, and service that itself provides a venue for social interaction. The questions asked by a sociotechnical change study include: "Why did designers think this way rather than that? What assumptions did the engineers, or the business people, or the politicians, make about the kinds of roles that people—or indeed machines—might play in the brave new worlds they sought to design and assemble? What constraints did they think about—or indeed run into—as they built or deployed their technologies? What were the uses—or abuses—to which the technologies were put by their users once they were deployed? How, in other words, did users themselves reshape their technologies? And how did the users and their technologies shape and influence future social, economic, and technical decisions?" (Bijker et al, 1992)

Research informed by sociotechnical change theory must study the interactions between these diverse groups that lead to specific technological decisions. It also focuses on the assumptions held by different participants about the proposed roles for people and technology in a newly proposed system. The set of concerns in digital game design span traditional technical problems, such as allocating network resources (Smed et al, 2002), and traditional social problems, such as preventing technically sophisticated players from cheating. In reality, these problems are difficult to separate into distinct social and technical issues.

3.2   Moving the literature forward

The current literature on digital gaming and innovation is mainly in two areas. The first literature focuses on game development, and which specific technology features lead to successful games. The second focuses on the psychological or cultural impact of games. Both literatures rely on technology-centric, even technological determinist assumptions with well-known limitations for understanding complex sociotechnical situations (e.g., Kling, 1996).

Our approach, based on sociotechnical change theory, tries to move beyond discussions of technology features and technology impacts. It views digital gaming as a rapidly changing production system in which the IT product and service are a complex ensemble of social and technical elements. We believe that success with experiential IT will improve with a more systematic knowledge of its sociotechnical production process.

4.   POSSIBLE CONTRIBUTIONS

As we write this paper, our study of digital gaming is just starting its fieldwork stage. This section of the paper discusses the kinds of results we anticipate, and the potential wider significance of our work.

4.1   Structure of the study

We begin our study with an initial in-depth case study of digital gaming. The focus will be on changes made to the game, particularly after its initial release. We will investigate the interactions that led to each change, and how specific social/technical decisions were shaped by the problematizations employed by different social groups (Callon, 1987).

Techniques for mapping and coding sociotechnical networks are still in their infancy. We hope to use some variant of the script idea from actor-network theory (Bijker and Law, 1992) to reduce participant interactions to a manageable level of complexity. The scripts specify the roles that different network elements, both human and non-human, are supposed to play in a proposed system. Another possibility is to use the technological frame concept (Bijker, 1995) to capture the problems, solutions, and problem solving techniques employed by different technological communities. The inclusion of individuals in different technological frames helps explain why technological innovation proceeds in some directions rather than others.

4.2   Wider significance of digital gaming innovation

While digital gaming has an intrinsic interest for students of IT, organizations, and society (Poole 2000), we believe that elements of digital gaming will find their way to other kinds of important technologies, products, and services (Rosenbloom 2003). Features of experiential IT usually associated with gaming are being introduced into other kinds of IT products and services in areas such as advertising, education, and organizational collaboration. Bushnell (1996; p. 34) asks "is there a new piece of game software that when played well will create a business plan for the player? Is there a game that trains a sales force on a newly released product? Are there games that can become the answer to declining scores and capabilities of the nation's school children?" Knowledge of more user-driven production processes, already seen in gaming applications (e.g., Kushner, 2003), could be essential for deploying experiential IT in other domains.

The demands of organizing for sustainable innovation challenge some of our traditional assumptions about the nature of IT innovation. The IT literature has stressed the importance of clear specifications, listening to users, and even observing users. We argue that this view does not begin to capture the diversity of roles that game players assume in the production of experiential IT. A better understanding of innovation processes for experiential IT will hopefully lead us to an improved understanding of IT innovation as a whole.

REFERENCES

Bijker, W.E. *Of Bicycles, Bakelite, and Bulbs: Towards a Theory of Sociotechnical Change.* MIT Press, Cambridge, Massachusetts, 1995.

Bijker, W.E., Hughes, T.P., and Pinch, T. (eds.) *The Social Construction of Technological Systems: New Directions in the Sociology and History of Technology.* MIT Press, Cambridge, Massachusetts, 1987.

Bijker, W.E., and Law, J. (eds.) *Shaping Technology/Building Society: Studies in Sociotechnical Change.* MIT Press, Cambridge, Massachusetts, 1992.

Bushnell, N. "Relationships between fun and the computer business," *Communications of the ACM* (39:8) 1996, pp. 31-38.

Callon, M. "Society in the Making: The Study of Technology as a Tool for Sociological Analysis," in Bijker et al., 1987.

Kling, R. *Computerization and Controversy: Value Conflicts and Social Choices.* Academic Press, San Diego, 1996.

Kushner, D. *Masters of Doom: How Two Guys Created an Empire and Transformed Pop Culture.* Random House, New York, 2003.

Lovejoy, K. "Playing for real," *Regional Review - Federal Reserve Bank of Boston* (12:4) 2002, p. 1.

Poole, S. *Trigger Happy: Videogames and the Entertainment Revolution* Arcade Publishing, New York, 2000.

Rosenbloom, A. "A game experience in every application: Introduction," *Communications of the ACM* (46:7) 2003, pp 28-31.

Smed, J., Kaukoranta, T., and Hakonen, H. "Aspects of networking in multiplayer computer games," *The Electronic Library* (20:2) 2002, pp. 87-97.

Walker, L. "Game worlds in creation," *The Washington Post* June 26, 2003, p. E1.

INTRODUCING MOBILITY: THE MPOLICE PROJECT

Michael Ney[1], Bernhard Schätz[2], Joachim Höck[3], and Christian Salzmann[4]

[1]Fakultät für Wirtschaftswissenschaften, TU München, Germany; [2]Fakultät für Informatik, TU München, Germany; [3]Polizei Oberbayern, München, Germany; [4]BMW Car IT GmbH, München, Germany

Abstract: Mobile IT is standardly claimed to be a key innovation for modern organizations. But how can this technology be adopted to improve the business processes of an organization? How to assess the benefits of a possible resulting organizational change? How to ensure the adequacy of the technologically driven solution and its adoption? We show how a systematic innovation process can be applied to an organization of the public sector for introducing mobile IT. In the first step, we identify gaps in the business process that can be overcome by mobile IT. In second step, enhanced economic evaluation is used to assess the adequacy of the technical solution. Based on this evaluation, in the final step, suitable technology-driven changes of the organizational business processes are identified.

Key words: Mobility, IT, public sector, extended effectiveness analysis, Business Process Analysis, scenario analysis, assessment, process improvement.

1.   INTRODUCTION

With the growing acceptance of its' "always and everywhere" paradigm mobile computing is one of the most promising approaches in today's IT market. In all sectors of economy and private use, mobile devices like cellular phones, PDAs or notebooks replace traditional media like letters, notepads, etc. Many enterprises use already mobile computing to improve their business processes: Parcel Services use mobile technology to integrate their field staff.

To get from bare potentials of new technologies to the realization and use of innovations Rogers points out, that we have to realize the effectiveness of new technologies: As shown in other domains, the impact on work plans and organization as well on future rate of adoption depends on "(...) a combination of an innovation's profitability plus its observability (...)" [Rog95], i.e. economic efficiency. In this paper we focus on the police: To achieve its central function – to serve and protect – the police is depending on the mobility of its units. Therefore mobile communication is an essential functionality – using mobile radio for the support of officers on site. Since the police is a non profit organization (with non profit targets i.e. safety for society) the possible impact is even harder to assess.

While mobile technology has evolved rapidly in the past years, the research of the potential of possible application in coherence with economic efficiency of investments has not kept up. When the Police of Upper Bavaria was introducing mobile IT, its organization was facing the challenges of analyzing and evaluating the benefits of mobile services:

– How does mobile computing possibly influence the work of the police, on the level of the workplace as much as on an organizational level?
– How does it pay off, in terms of the different stakeholders (the officer, the management, and even the society)?

Therefore the task force mPolice was established as a joint project between the Police of Upper Bavaria, Siemens Business Services, and the TU München: The main targets of mPolice is a holistic analysis addressing IT as well as economical issues: On the one hand, to establish the influence of mobile IT on the business processes of an organization from the Computer Science point of view to get a picture of its possible influence; and, on the other hand, to get an understanding of the possible benefits (or draw-backs) for all stakeholders involved from the economical point of view.

In the following we present the results of an enhanced economic evaluation approach, which supports the multilayered task of evaluating mobile services tailored for the specific needs of the police. A set of business processes of the police was analyzed for applicability of mobile computing. After the technical analysis a system was proposed improving these processes. To measure the benefits of an approach, we used the integrated approach of the enhanced economic evaluation. The remainder of this article is structured as follows: In the following section we describe the domain of application, the police. We explain some of the processes as they are practiced. We then explain the methodology to detect the gaps within theses business processes, followed by the methodology to measure the benefits of a possible solution for the public sector. In the final section we present the technical solution we proposed and found it on the expected social and economic benefits. We close with a conclusion that sketches the next steps that follow the results.

2.   CONTEXT: MOBILE IT AND THE PUBLIC SECTOR

Upper Bavaria with almost four million inhabitants is the largest operational area of the police within Bavaria. The tasks handled by the police reach from shop lifting to murder, plane crashes, and train accidents. Surely the use of mobile end devices could simplify the work of the police considerably. However, the differences between optimizing in the public sector and private sector have to be considered. The goal of an enterprise is normally to make the most profit of the invested capital. So a most important goal must be to lower the costs to raise the efficiency. Police work however isn't profit orientated. Its goal consists of services such as the recording of traffic accidents, to avert, to prevent, and guarantee general safety in the population. The most important goal is to improve the quality of the services of the police, for example to catch an escaped person quicker or to patrol a critical area with more coverage. Obviously, it is hard to measure the benefits of prevented crimes with the same optimization measures as you do in case of a private enterprise.

2.1   Application Domain: The police of Upper Bavaria

Each year the statistics of crime and the clarification quota are published by the Bavarian Police Department. The number of crimes in Bavaria, in comparison with the preceding year, rose by 8.100 to 401.380 or by 2.1 %, and in Upper Bavaria by 1.641 crimes to the value of 72.880. I.e. despite of the rise of crimes the number of clarified cases has risen. This is surprising because the number of public employees of the Bavarian Police Department has hardly changed in comparison with the preceding year. Apart from the high motivation of Bavarian officers, the increasing accouterments of technology have a great effect. More than 15 years ago, all operations were managed manually, using typewrites. The Bavarian Police Department now has an electronic integration method based on the principle of singular recording. The officer records all data at his disposal only once into the databank and can use it to create complaints or notices of accidents according to the requirements. Also the management of operations is largely automated. Thus the handling time for a complaint has reduced from 74 minutes to 24 minutes. However, this case handling is only possible from one Net-PC in the headquarters. Thereby capacities for prevention and enforced investigation were gained. Two thirds of approximately 33.000 employees of the Bavarian Police Department do shift work, which, in 60 % cases, takes place outside of headquarters. Here there are two different forms of field service. In most cases motorized patrols in both civil and uniformed vehicles. Momentarily the Bavarian Police Department has 9.000 vehicles in use. The Police Department in Upper Bavaria has approximately 500 uniformed and 650 neutral vehicles at its disposal. In patrol duty the police officers only use a 4m-radio unit, which is part of the vehicle to communicate with headquarters or operation base and with other patrol vehicles. For close-up range to headquarters there is also a 2m-radio unit. In few exceptions the officers additionally carry a cellular phone. At present in whole Bavaria there are additional 400 mobile tracing clients, which access the Central Computer of the Bavarian LKA (regional bureau of crime investigation) over the GPRS-Network of Dt. Telekom, to enquire tracing data. Presently 1.000.000 persons and affairs per month are checked over these clients. Furthermore data, which is filed on the central computer, e.g. formats of driver's licenses, forging characteristics of ID papers or car documentation, can be enquired.

However these tracing clients who are built into the vehicles cannot be used for case handling of operations. As the majority of police operations take place outside of the vehicle, the officers only dispose of the 2m-radio unit, which only has a limited range. Thus the question for the use of mobile end devices for the support of the officers on duty automatically arises. As such, in which sectors and operations is this support possible? As an example we would like to mention the recording of accident notices and the deployment of a hundred Bavarian police officers for the search of a forest area for a searched person. Should an accident occur in duty range of a police officer, more patrols, according to the severity of the accident, are ordered to the place of accident? During the trip the officers receive first information by radio about number of injured or casualties, whether emergency physician has been called, whether the fire brigade is necessary to salvage injured. After the arrival of the crew the rescue of human life resp. the treatment of the injured is first priority. Only after that the actual police work is commenced, e.g. hearing of witnesses, recording of the vehicle data and the checking particulars of the involved, measurement of the site of accident. This data has to be, like in the old days, noted on paper by the officers because there aren't any electronic aids at their disposal. Also the check of the received particulars and the data of the identity papers and car documentation can only be done by radio. Unfortunately this is interfered with other prior radio transmissions and comprises the danger of transmission errors.

2.2   Structured Approach towards Innovation

As described in the previous subsections, experiences concerning the introduction of mobile IT gained in other organizations do not readily carry over to public organization like the police due to essential differences concerning their organizational and operational structure and goal; nevertheless, adapting mobile IT innovations seemed to be a promising possibility in both domains. However, it remained largely unclear, how innovation might effect the organization, how much benefit could be expected from those innovations, and finally, which possible innovations should be adapted on to ensure maximum benefit. Therefore, a structured, transparent approach to assessing the possible impacts of innovation was needed. To avoid erratic guesses about the impact of the technology, a structured approach is needed that does not leave out important possibilities of improvement by mobile IT. To this end, a structured Business Process Analysis approach especially tailored towards mobile IT as describe in Section 3 is used. Furthermore, to ensure the adaptation by the user at the workplace, to help the management to decide on possible alternatives and necessary organizational changes, and to quantify the payback on the society, a structured and reproducible approach is needed to assess the benefits of the innovation for all stakeholders in a transparent and communicatable manner. To this end, multi-layered economical efficiency analysis as describe in section 4 is applied.

3. DETECTING THE GAPS: ANALYZING PROCESSES

To assess the disadvantages and advantages that IT innovation can have for the organization under investigation, we first have to get a clear understanding where new IT technologies will affect the organization most. In the following subsections we explain how those leverage points can be detected in a methodical fashion, by introducing related approaches in Subsection 3.1, show how those approaches can be adapted to mobile IT infrastructure in Subsection 3.2, and finally give a few short examples in Subsection 3.3.

3.1 Related Approaches from Computer Science

From the technical point of view, we use Business Process Analysis to analyze where IT innovation can influence the workflow of the organization. Our approach to identifying process improvement is similar to the complete Business Reengineering approach as suggested, e.g., by [HC94]. Like in the approach used there, we use a structured analysis of the existing processes concerning using IT as driving force of change. However, in contrast to complete reengineering, we are rather interested in a stepwise improvement of the business process. As explained in Section 2, this is mainly due to the fact that organization of the public sector are not driven by the main factors customer, competition, and change as much as in the private sector; consequently, radical organizational change is much more likely not to be adapted than in the private sector. Therefore it is much more important to identify the current process as they are "actually performed" in the organization rather than they are "supposed to be performed". Additionally, when describing the actual processes based on the activity profile we also document why these activities are performed in that fashion (especially including the "positive and negative aspects" as perceived by at the workplace). Thus, when analyzing the processes we will start out from a description of the "workplaces" and combine these to form a picture of the actually performed business processes on the "process level" as well as on the "organizational level".

This evaluation of an organization from different points of views (work place level, process level) becomes even more important when assessing the effects of the improvements inspired by an IT-driven analysis of the business process. Consequently, our incremental approach suggested here is similar to Total Quality Management (TQM) [Mad98]; as shown in Section 4 we will also use a structured analysis similar to Quality Function Deployment (QFD) to assess the benefits; however, in our approach we focus on the IT impact as the driving force to direct the change of the business processes. Furthermore, we use a four-level approach to quantify the impacts on the work-place level, the process level, the organization level, and even the society level, as discussed in detail in Section 4. During assessment, especially the information obtained on the workplace level is reused to ensure the adoption of the process by the end-user.

3.2   Technology-Driven Process Analysis

As mentioned above, we are a priori not interested in defining completely new business processes or changing the organizational structure. We are much more interested in "systematically" detecting gaps in the processes which can be "closed using mobile IT infrastructure". Since mobile terminals introduce three different aspects of operation not available with classical IT infrastructure ("mobile information processing", "mobile communication", "constant reachability"), we are especially interested in three kinds of gaps that can be overcome using this technology:

1. Gaps caused by "access centralization": Information is only available at certain locations or steps in the process; the infrastructure supports on-line access only by means of indirection (e.g., by telephone operators).

2. Gaps caused by "data replication": Information is reproduced at different steps in the process; the infrastructure does not allow reusing the produced information.

3. Gaps caused by "passive processes": Information is only available at request ("information pull"); the infra structure does not actively distribute information to possible clients ("information push")

To obtain a description of the business processes actually performed, as mentioned above the activities as performed on the individual workplaces (e.g., officer in the field, officer in the control room) were described. Gaps in the business processes show up on the level or the processes and on the level of the organization, depending on their type:

– "Process level": Since in a non-mobile environment access is generally centralized, processes carried out in different workplaces have a tendency to show gaps caused by access centralization. Furthermore, since work places tend in a non-mobile environment generally offer different infrastructure, information is reproduced to cover with these differences.

– "Organization level": Since processes often are not integrated on an organizational level, gaps by information replication are introduced between those processes. Even if processes use shared information, information is not actively forwarded, leading to gaps by passive processes.

To analyze the business process where mobile IT infrastructure can be applied, a four-step approach was used: In the first step, the general tasks of the police and the overall business task including the workplace, the processes, and the organization were examined to describe the overall structure of the organization and the main information flows. In the next step, a more in-depth analysis of the actually performed activities was carried out. The purpose of this step is to identify activities accessing or producing information and to analyze whether this information is available on-line (e.g., by means of interface installed in the car, by radio communication, or by telephone) or off-line (e.g., maps, forms); furthermore, the participation of the system in distributing the information was assessed ("information pull" vs. "information push"). In the second step, those activities in the identified business processes requiring access to information were analyzed. For those activities using on-line access to information, the scenarios describing the interaction including the involved parties were defined. For those activities involving off-line information, additionally, the required or produced information was defined. As a part of the scenario, the sources and the sinks of this information were identified. Additionally - for the "information push/pull" analysis - the initiating events for the generation and the use of the information were identified. In the final step, business processes on the process and on the organizational level were constructed from the description of those activities by linking the workplace processes based on the described information flow. By evaluating those links in the defined processes, possible gaps where detected and documented.

3.3   Application Examples

To describe the relevant processes in sufficient detail for assessing the IT impact, an approach analogue to [JEJ94] was used, however extended to fit the UML description techniques [FS99] as used in the Rational Unified Process [Kru00]. To identify and structure the overall tasks and the general business processes, use cases structured in form of function trees with actors and interactions were used. Once, the main processes were identified, activity diagrams were used to identify the main activities of a process. For those activities accessing or producing information, detailed sequence diagrams describing the interaction and exchanged information were defined. We illustrate the process analysis using the business process "Recoding an accident" as example. This process includes the activities "Dispatching Car", "Checking IDs" and "In-the-Field Protocol". The first requires on-line information using information pull; the second requires access to on-line information; and the third produces off-line information. The analysis of those activities is explained in more detail in the following.

Figure 1 shows the process of checking IDs of a person involved in an accident. The description of this process is obtained by the combination of the description of the work places "Officer" and "Control Room"; the linking data access is "on-line" ID information obtained by demand via radio. According to the isolated view of this workplace in the current, non-optimized version, after obtaining the ID of the involved person, the officer waits until a radio connection to an operator in the control room is available. After stating his request and identifying himself, he transmits ID information of the involved person to the operator. The operator then checks the transmitted data with the registry via the intranet and relays the obtained information back to the officers. The officer acknowledges the receipt of the data and returns the ID of the person involved. Analyzing the activities of the control room in this workplace-focused view reveals that it only acts as a relay to the registry concerning the information flow of the "Checking Ids" activity.

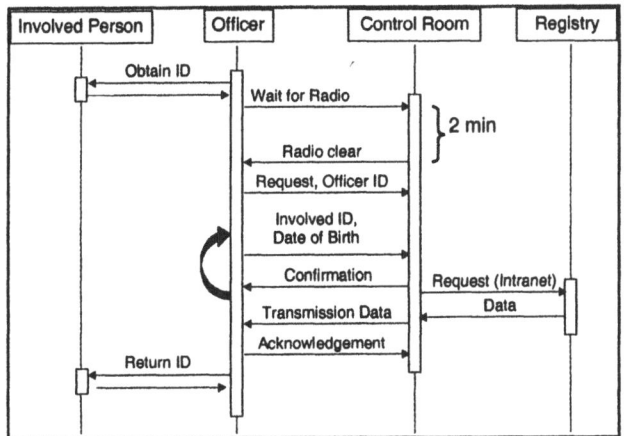

Figure 1. Data Flow for "Controlling ID Papers" (Pre Optimization)

Combing these activities on the process layer by means of their common accessed information (ID data) shows that here a gap caused by centralized access is introduced in the process. As a result, by verbal communication to the operator errors can be introduced and therefore a protocol to avoid those is introduced: information transmitted to the other party is sent back in form of a confirmation; if confirmation and sent data differ, the data is retransmitted (as indicted by the backward loop in Figure 1). Obviously, this gap results in additional delays, either when waiting for an available operator or when retransmitting misunderstood information. Additional to the gap due to data centralization mentioned above we can identify gaps of data replication by linking the activity "In-Field Protocol" to the business process. Once the place of the accident got secured the police officer usually starts to prepare a protocol. During the activity, information is produced in form of hand-written protocols (including the data of the ID check done before. This information is therefore re-entered into the information system upon return to the precinct. Finally, when combining processes on the organizational level, additionally, gaps by passive processes can be identified. During the "Dispatching Car" activity, the control room actively has to obtain information about which available car is in the vicinity of the accident, i.e. query the cars about their position and their status. Only after this information is available, a corresponding car can be assigned to the task.

4. QUANTIFYING IMPROVEMENTS

Today many IT projects analyze innovative applications solely orientated on monetary values and isolated from the organizational context. This traditional investment calculation process makes sense for many operations side companies. But the police is not operating profit-oriented, but offering public services. Furthermore the traditional calculation makes sense as a pre-stage for outsourcing, but it's not helpful for introducing innovative work plans and organizational changes. To get an integrated evaluation of mobile solutions, it's not sufficient to analyze the cost effects only on the workplace. An implementation of mobile service in an organization usually has a widely ramified and multi layered impact to many divisions inside and outside an organization.

In the following chapter we introduce the Enhanced Economic Evaluation approach that supports the following options [RHW96, PRW03]: In addition to the consideration of "qualitative" (non monetary) and "quantitative" criteria this approach uses an integrated view of a decision's impact to an organization. The application of the approach is discussed in detail in Section 5.

4.1   Related Approaches from Business Administration

To systemize and to point out impacts, our approach uses a multilayer reflection to integrate different point of views into one analysis. This idea - originally used in IT and Telecom domain - characterizes a way to get an integrated view on organizational issues [Ant95; Bau96; Bod84; Nie88; PR87; RHW96; RMES00; Zan88] and differs the following layers:
- Layer 1: Workplace: Here we analyze impacts on a specific workplace or workstation that arise from the introduction of mobile services.
- Layer 2: Processes: Here we focus on the effects of the work flow in the organization (e.g. preliminary process steps).
- Layer 3: Organization: On this level we gather primary the value effects of mobile services to the long term efficiency of an organization.
- Layer 4: Society: In this section we analyze the impacts of mobile services on the labor market and the assurance of the location of the industry and economics.

These layers must not be analyzed separately, but tightly linked. They represent different views to the economic actions (i.e. the introduction of mobile services). For this reason we use them as an analytical separation, to point out the correlations and network effects of mobile services: They show i.e. the links between human capital and organization, the process of value creation, the relation between the organization and its environment. Using this multilayer approach we can apply to different views within the police: On layer 1 the policeman, on layer 2 the supervisor and on layer 3 the ministry of interior. The last layer can operationalize the society's impressions of the police i.e. the perceptions of the population.

4.2   Integration of Stakeholders

The utilization of distributed information is the main problem of implementing new technologies in organizations. A large percentage of information – i.e. information regarding the usage of radio communication of the field service – is already available within the organization but distributed in different locations [Ant95, Dek03, PRW03]. Owner of these information are employees in a variety of functions and hierarchies. However, to integrate processes, this information (and above all the availability of this information) becomes increasingly important [RMSE00]. In enhanced economic evaluation the merge of the distributed information proceeds as a moderated group process with the participation of employees from all relevant layers. In order to minimize the risk to adulterate the results it's necessary for the discussion to be moderated by a person outside the organization. Aggravating circumstances are the conventional information flows in the hierarchical structure of the police that handicaps efficient extracting and forwarding of information. The decision quality that can be archived with this approach depends on the consensus of decisions made across the different layers. Therefore the part of the moderator is very important: The quality of the consensus falls if one perception stands up to the rest.

4.3   Criteria Catalogs

A basic characteristic of the enhanced economic evaluation is a comprehensive analysis and consideration of the specific situation of an organization. According to recent publications in business administration we deny the option of an objective and universally valid evaluation. Every decision in an organization is related only to a specific situation: The potentials of processes, organizational structure and human resources demand an evaluation approach that considers the "subjective" needs and specific targets of an organization. The enhanced economic value approach uses a criteria catalog to support the evaluation process in an efficient way and to present details of the specific situation clearly showing the decisions, which led to the obtained results. This catalog consists of different basic sets of criteria and can be modified to match the specific needs: i.e. the predefined catalog consists of the 6 basic criteria sets. Due to the safety-critical tasks of the police, the representatives of the police decided to integrate the criteria "safety" into the decision catalog.

4.4   Value Benefit Analysis

For aggregating all criteria containing several layers it turned out to be useful to apply the value benefit analysis: the different criteria get assigned a weight according to their importance to the participants of the workshops. Furthermore, the criteria are linked to a scale that helps to attach the parameter value of criteria to an achievement of objectives (that is aligned to a certain amount of points). To get a integrated evaluation – meeting the specific needs and including the distinctions of an organization – the points of criteria are multiplied with their weight and summed up to get the result of the evaluation.

5.   RESULTS

From March to August 2002 the Police headquarter Oberbayern started a pilot study in cooperation of the Siemens Business Services and Technische Universität München, to analyze the applicability of mobile services to enhance efficiency of police's processes.

5.1   Project Steps

To analyze the current processes of the police, to identify improvements, and to assess the impact, the project included four steps:

1st "Monitoring Shift Work" to assess the current situation at the Police: In the preliminary stages of mPolice we first had to get an overview about the current process situation. Therefore we attended several work shifts in differrent police departments. As a result of several interviews with police officers we got knowledge about the current process situation and an overview about potential gaps within these processes that might be closed with mobile services

2nd "Expert Interviews" to survey possible process improvements: After the monitoring of the shift work at the police we focused on the potential gaps within the police processes. As a basic principle it is useful to select participants with different backgrounds and a variety of functions and hierarchies (e.g., k-9, criminal investigation, traffic). Within 15 expert sessions we demonstrated possible mobile solutions (including latest mobile technology) to get a first idea of possible applications. Many officers started to integrate their vision of innovative mobile services. After that we matched the different potentials of mobile services to the potential process gaps. In these expert interviews we primarily focused on concerns police officers have using this new technology. In the introductions we focused on advantages that are given by insights to organizational coherences.

3rd "Focus Group Workshops" to combine the upper results and aggregate them to a criteria catalog: Within the previous process steps we identified dedicated police officers interested in the domain of mobile services. We invited these officers to our focus group workshops. The starting point of these workshops was the development of global targets based on the strategic goal of the police. We used the results of the pre-analysis based the preparatory work mentioned above as starting point. The results of step 1 and 2 included a variety of information, like detailed process analysis, software and hardware requirements used by the participants to fix the following global target categories for the evaluation: Even though the "cost" aspect is also important for a non-profit organization, it is obvious that the police doesn't decide only based on monetary values. Thus, the officers in charge attached importance further aspects influenced by mobile technology: The first is to enhance the "safety" of field service by raising the "quality" of the cooperation of field service and the operator. The second is the increase the "flexibility" of Field service in information retrievals. And third, the reductions of throughput time by optimizing the information flow from the field service to the departments. By minimizing redundant work and the through put time several interviews with officers of the police presented the acceptance of the employees (i.e. "human situation") and of the "society" as general condition. By aggregating these results we got a first level catalog including the following dimensions: Human Situation, Costs, Time, Quality, Flexibility and Safety.

We then integrated the three different dimensions: On the first dimension we had to consider the results from the Business Process Analysis (as mentioned in chapter 3) that analyzed the potentials of mobile services by the existing gaps in the process flow. The Business Administration (chapter 4.1) showed us that mobile services had impacts not only on the specific workplace but the whole organization. Finally we had to integrate the specific (hard and soft facts based) strategic goals of the police (as described above). Aggregating these three dimensions we came up to a cube that represents with each dimension one single point of view.

Prepared with this model the participants could discuss different targets and to select relevant suggestions without being afraid of forgetting something. So this discussion established a basis for an integrated view to organizational coherences. The Mobility Cube including the origin, impact and dimension of mobile potentials represents the most important criteria and illustrates the base for the construction of the criteria catalog. It turned out, that participants rated some aspects with a large variation:; i.e., depending on the function of the participant, the rating of global criteria like safety was either attached with more or less importance. Finally the participants agreed to the following result regarding the weight of the different dimensions: The most important factors according to the workshops are an increase in safety and quality of cooperation: According to the weight of safety and quality these criteria represent already 46 % of the whole evaluation (23% each). Including the criteria of enhancements of flexibility (15%) and improvements of the human situation (15%) these factors represent more than 75% of the result. The feedback of the society as well as impacts in Cost and Time Savings are comparatively unimportant for the evaluation with a weight summed up to less than 25%. Summarizing the results: The participants of several workshops rated the so called soft facts more important than the traditional monetary based hard facts.

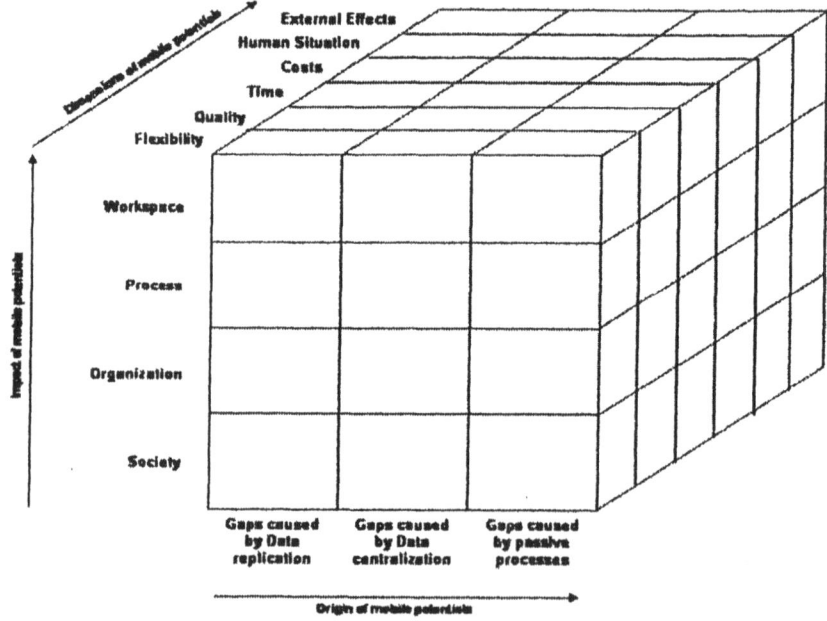

Figure 2. Mobility Cube

4th "Design and Implementation" of a exemplary analysis of mobile solution's economic efficiency for.the police: The cube was not uniformly distributed, i.e. the different potentials of mobile technology affect the organization in different ways:

– On the "work place layer" basically potentials of gaps caused by data centralization have an influence to the field service: In the pilot study we observed significant timesaving originated by the use of a non-limited communication channel (in contrast to the delay with radio communication). Furthermore we noticed the realization of potentials caused by data replication: By reducing redundant work (i.e. the information retrieved by the officer in field are reused for the protocol) we assessed an increase in the flexibility and a decrease in the through put time of some processes.

– On the "process layer" we observed advantages caused by the overcoming the gaps of data centralization in conjunction with gaps of data replication. By integrating the field service and department based on the level of information, the information exchange between the central and decentral units gets accelerated: e.g. the information about the possible victims of a traffic accident are available to the operator as soon as the field service retrieved the IDs from the participants.

– On the "organization layer" finally we found mainly potentials by overcoming the gaps of passive processes: By supplying the field service with current information (i.e. push services) the safety of the police officers gets boosted: i.e. by sending accurate information about a menace the infield service can handle situations on a better level of information.

As mentioned above by closing gaps through data replication we get a reduction in through put time that leads to an increase of customer satisfaction. Additionally by overcoming the lacks of data centralization the police can minimize the manual administrative task of information supply and focus on their core tasks.

5.2 Identifying Solutions

As described in section 3.3, by the usage of Business Process Analysis we obtain a detailed description of the interactions performed during in-field business processes. By focusing on the activities *generating* and *using information* (e.g., accessing a data base, writing a report) and later on identifying classes of coherent information (e.g., accident report), we are able to extract the *data flow* behind those processes. This method leads to an *information centric view* instead of a *process centric view*. Using the information centric view, the interactions from the business processes were used to develop a life history of the different classes of information including its generation, transformation/update, use, and destruction. Using the life histories, gaps in the business process and possible optimizations were identified:

– Gaps caused by *access centralization*: For process steps accessing centralized information (e.g., terminal-based access to data bases), mobile terminals are used to substitute proxies (e.g., the control room operator). If information is distributed with different access points (e.g., different data bases for different crimes), a new access point serving as a common front-end (e.g., common query interface) is added.

– Gaps caused by *data replication*: For process steps producing off-line information (e.g. accident reports) that are later added manually to the system, a suitable mobile terminal is identified; furthermore, if no online connection is suitable for this interaction, automatic integration of the generated information from the terminal to the system via defined access points (e.g., docking stations) are defined.

– Gaps caused by *passive processes*: For process steps requiring an active interaction to obtain up-to-date information, the processes are adapted to transfer the responsibility of initiating the update to the producer of the information and to automate the forwarding of updated information to the possible users.

Using these strategies, optimized versions of the business process can be constructed. Using the example of the previous sections, the following optimizations can be applied to close the identified gaps:

1. Checking ID: During "Checking ID", the optimized version using a mobile terminal (e.g., a PDA with GPRS connection) avoids gap introduced by data centralization and therefore leads to a leaner process. Here, after obtain the ID information and performing a local logon to the PDA, the officer enters the information into the PDA and transmits it to the registry via the GPRS connection. Since the information is transmitted as entered by the officer, no cross-checking is required. While the sequence diagrams look similar on first sight (we left out the two-way radio in the first diagram), the optimized process eliminates the delays and iterations introduced by the relay function of the control room. The reduced complexity of the interaction results in a shorter execution time, less possibilities to insert errors, and therefore to improved reliability of performance and less distraction for the officer.

2. In-Field-Protocol: During the "In-Field Protocol" activity, information can be added into a mask and entered online into the system for further editing by the officer once back at the precinct.

3. Dispatching Car: The gap identified in the "Dispatching Car" activity can be closed by adding information push, using a GPS module and remote access to the availability database through the car terminal. By actively supplying this information to the system, the system can identify cars available is in the vicinity of the accident.

Note that these optimizations depend on the available technical infra structure applied. E.g., in case of passive processes, the possible substitution depends on how much information can be generated/forwarded automatically: a car equipped with a GPS module can transfer its location information; by adding a mobile terminal, availability changes can be actively announced by the car personnel. Assuming different possible mobile IT infra structures, accordingly different possible optimized processes can be defined. However, the optimized business processes describe a system that is optimized with respect to the "functional data flow" of the organization. For the adaptability of this mission critical solution, "non-functional aspects" (like direct and indirect costs of ownership, change in the organizational structure), of the solution often are of even bigger impact then the functional aspects. As mentioned above, using the extended efficiency analysis, we obtain a metrics to quantify the impact of a solution on the individual as well as on the organizational level by including different aspects like cost, time and quality, as well as flexibility and personal situation, or public appearance and security.

To quantify these different factors, the enhanced economic evaluation is used [RHW96]. To apply this technique, different improvements are described in form of configurations. The definition of a configuration includes:
1. Selecting processes to be substituted and substituting processes (since in different configurations different organizational changes are assessed)
2. Identifying the necessary IT infrastructure (possibly fixing possible variants like bandwidth, reaction time, size of display)
3. Developing usage profiles (number of users, number of request, amount of data, etc.)

Using these configurations, a possible optimized system can be described in sufficient detail to have it evaluated using the enhanced economic evaluation. Accordingly, business process optimized with respect to information flow may show drawback concerning other evaluation aspects, leading to a reduced overall result. For example, the optimized access to the ID database introduced in section 3 has several different consequences:
- Positive factors:
  - "Flexibility": Access to data base is independent of availability of operator in control room
  - "Quality": Possible errors in the transmission are minimized
  - "Time": Time to obtain information is reduced
  - "Public appearance": Minimized interaction time with involved parties and higher quality of response lead to better acceptance
- Negative factors:
  - "Security": Distraction (loss of eye contact when switching from radio to terminal), loss of surveillance (operator in control room as third party)

Thus, defining different configurations and applying the metrics of the enhanced economic evaluation to each of them can be used to identify the overall optimized solution to the problem. Accordingly, the method can be applied to pre-estimate the influences of technological impacts and increase the possibility of adoption by decision-makers as well as users, leading to improved system acceptance.

5.3   Quantified Expectations

To illustrate the results, we use a limited mobility cube, reducing the dimensions to costs and quality. [RNW02]. In the following we exemplify an evaluation on a fictive mobile system. Therefore we analyze two alternatives of mobile services: The first alternative contains the latest and current mobile hardware using software developed for the mass market. This system represents – due to the only minor differences in design – an economically priced system with a large range of functionalities that is limited to enhancements. The second alternative consists of a mobile system specially developed for the police (based on a Business Process Analysis): e.g. the migration of the existing functionalities of the Car PC to mobile devices. The hardware offers lower performance but is provably reliable. As a consequence to the special development for the police this system is high priced and comes with limited functionalities (but due to the open software architecture open for enhancements).

Figure 3 shows the results for the costs criterion. The first alternative performs better than the second one due to the costs for the acquisition and the maintenance for the system. Due to the extra design and development of this system it is obvious, that alternative 2 costs more than the first one only composed of mass market hardware.

Criteria	Weight	Alternative 1		Alternative 2	
		Score	Sum	Score	Sum
Costs					
planning reliability	58.33%	0	0.00	0	0.00
Costs for Training	13.89%	-1	-0.01	2	0.02
Costs for Acquisition	12.22%	3	0.03	-5	-0.05
Costs for Maintenance	15.56%	3	0.04	-2	-0.03
Sum	100.00%		0.06		-0.06

Figure 3. Exemplary evaluation on a fictive mobile system: Costs

Concerning quality we can observe a different result (see figure 4): As a result of the used mass market conform standard software we observe that the introduction of mobile services doesn't increase the information level for the 1st alternative. In contrast the use of the 2nd alternative – including the migration of known functionalities from the Car PC – increases the information level. Additionally, the willingness to adopt the second alternative is much higher, because it is based on an established system. Thus the preparatory training is limited to handle the mobile hardware in contrast to the first alternative that takes an extended training including the use of the new software. But even after an adaptation phase the system alternative 1 differs by the limitation of functionalities that can't be enhanced.

As mentioned in chapter 5.1 the criteria quality plays a part of 23 % and costs 8 %. Aggregating the results of the exemplary evaluation of the fictive mobile system, we get a surprising result: Summing up the results in quality the score of the first alternative is − 0.20 and the score of the second one is + 0.26. Even though the hard facts seem to be promising for the 1st alternative, the so called soft facts determine the result: The second alternative − even though it's more expensive and equipped with less functions reaches a better evaluation than the first alternative.

Criteria	Weight	Alternative 1		Alternative 2	
		Score	Sum	Score	Sum
Quality					
Qualification of Employees	9.67%	0.00	0.00	0.00	0.00
Level of Information	11.33%	-4.00	-0.10	4.00	0.10
Availability of Extern Know How	11.33%	1.00	0.02	2.00	0.05
Enhancement with additional functionalities	14.67%	-3.00	-0.10	3.00	0.10
Error ratio on the man/machine interface	8.00%	-3.00	-0.05	2.00	0.03
Conditions of Employment					
Usability	16.11%	0.00	0.00	0.00	0.00
Robustness of hardware	16.11%	0.00	0.00	1.00	0.03
Reliability of the software	12.78%	-2.00	-0.06	2.00	0.06
Summe	100.00%		-0.29		0.37

Figure 4. Exemplary evaluation on a fictive mobile system: Quality

6. RELATED WORK

The possibilities of mobile information technology with respect to improvements of the workplace, its role in improving overall business processes, and its forms of impact have been studied before:

−   [BB96] shows the two decisive factors of mobile technology for supporting the workflow: mobility to support access to share resources ("immediate access"), as well as mobility to be available at any time ("constant reachability"). However, in contrast to our approach, they focus on a common (virtual) workspace with immediate interaction of the participants and the collaborative aspect rather than considering individual business processes.

−   [LH98] points out that the central aspect of mobility in most work places is not so much the immediate interaction of all participants but much more the integration of the individual interactions performed at each workplace to support a non-interrupted workflow of the participants. In contrast to our approach, however, they do not use an explicit model of the workflow to discover gaps in the business processes performed.

- [KS02] shows the consequences of the always access/always on paradigm by introducing three aspects of mobility: The possibility to receive the necessary information on the spot of activity ("spatial mobility"), the possibility to get immediate response ("temporal mobility") and to support context awareness ("contextual mobility"). However, while they focus on the classification, we focus on using these aspects to identify gaps in business processes and propose solutions to overcome those gaps.

In general, while those approaches discuss fundamental aspects of integrating mobile technology in the workflow, we concentrate on the application of those aspects to create innovative business processes: The main focus of our approach lies on "systematically" analyzing business processes to identify possible innovations that can be realized using mobile IT and that are likely to be accepted by the users. By combining Business Process Analysis (BPA) and Enhanced Economic Evaluation (EEE) we ensure that a balanced solution considering optimization with respect to workflow as well as with respect to socio-economical factors is identified.

7.   CONCLUSION

In this paper we presented our experiences in analyzing the options of mobile computing in a concrete application domain. We presented a way to determine the benefits of such an application, tailored specially for the needs of the public sector, including the points of view from computer science and business administration. The Enhanced Economic Evaluation (EEE) assists the introduction of innovative mobile services in cooperation with the employees as an incremental innovation, particularly with respect to the non-quantitative goals of non-profit. By adapting the criteria weights of the EEE, the approach presented here can also be adapted to the needs of organizations of the private sector.

As mentioned, the deployment of the enhanced economic evaluation approach aims on different perspectives: The computer science with its Business Process Analysis helps to analyze media breaks within a process or between processes (gaps caused by passive processes, access centralization, and data replication) and to suggest solutions using mobile infrastructure effectiveness. But this increase of effectiveness doesn't implicate an economical surplus of an innovation: Besides optimization of the process (i.e. reduction of media breaks and improvements to the different analyzed gaps) organizations have to make sure that they get to know the impact of an innovation to the established organization (on processes, organization,) and society, and their related aspects (e.g., security, human situation).

In the domain of organizational mobile services there's a change in innovation efforts towards the organization - employee interface to establish successful mobile business models. Therefore the computer science as well as the business administration perspective has to develop suitable approaches and techniques to tie an integrated evaluation system to the implementation of innovations. The instruments of the Business Process Analysis combined with Enhanced Economic Evaluation seem to be promising approach for this purpose.

REFERENCES

[Ant95] Antweiler, J. Wirtschaftlichkeitsanalyse von Informations- und Kommunikationssystemen, Datakontext, 1995.

[Bau96] Bauer, R. Die Methodik der Erweiterten Wirtschaftlichkeitsbetrachtung, in: Reichwald, R Henning, K. FAMO, Augustinus, 1995.

[BB96] Bellotti, V. and Bly, S. Walking away from the desktop computer: Distributed collaboration and mobility in a product design team. Proceedings of the CSCW '96, Cambridge, MA, 1996, pp. 209-218.

[Bod84] Kommunikationstechnik und Wirtschaftlichkeit, München 1984.

[Dek03] Deking, I. Management des Intellectual Capital, Gabler 2003.

[HC93] Hammer, M., Chamy, I. Reengineering the Cooperation: A Manifesto for Business Revolution. Harper Collins, 1993.

[JEJ94] Jacobson, I., Ericsson, M., Jacobson, A. The Object Advantage – Business Process reengineering with Object technology. Addison Wesley, 1994.

[FS99] Fowler, M. Scott, K. UML Distilled. Addison Wesley, 1999.

[Kru00] Kruchten, P. The Rational Unified Process. Addison Wesley Series, 1999.

[KS02] Kakihara, M. and Sorensen, C. Mobiliy: An extended perspective. Proceedings of the 35th Hawaii International Conference on System Sciences, 2002

[Mad98] Madu, C. (ed.) Handbook of Total Quality Management Kluwer Academic Publishers, 1998.

[LH98] Luff, P. and Heath, C. Mobility in Collaboration. Proceedings of the CSCW '98, Seattle, WA, 1998, pp. 305-314.

[Nie88] Konzepte der Wirtschaftlichkeitsberechnung bei integrierten Informationssystemen, in Horvath, P Wirtschaftlichkeit neuer Produktions- und Informationstechnologien, Schäfer-Poeschel, 1988.

[PR87] Picot, A. Reichwald, R. Bürokommunikation, 3rd Edition, Hallbergmoos, 1987.

[PRW03] Picot, A. Reichwald, R. Wigand, U. Die grenzenlose Unternehmung, 2003.

[Rog95] Rogers, E. The Diffusion of Innovation, Berkeley, 1996.

[RHW96] Reichwald, R. Höfer, C. Weichselbaumer, J. Erfolg von Reorganisationsprozessen, Schäfer-Poeschel, 1996.

[RMSE00] Reichwald, R. Möslein, K. Sachenbacher, H. Englberger, H Telekooperation, 2nd Edition, Springer, 2000.

[RNW02] Reichwald, R Ney, M Wagner, M. Kundenintegrierte Entwicklung von mobilen Diensten, in: Reichwald, R. Mobile Kommunikation, Gabler 2002, p. 347 – 359.

[Zan88] Zangl, H. CIM Konzepte und Wirtschaftlichkeit, in OM, 36 Jg. (1988).

SUPPORTING THE RE-EMERGENCE OF HUMAN AGENCY IN THE WORKPLACE
A New Role for Information System Technologies

Tony Salvador and Kenneth T. Anderson
Intel Research, Intel Corporation

Abstract: Based on a multi-nation ethnographic study of retail environments, we consider the role of information technologies, specifically ubiquitous computing, in the context of worker agency in the workplace. Agency (not software agents) is defined as the ability of the retail worker to act appropriately but relatively unencumbered by the dictates and wishes of the establishment; in a retail setting, it can mean making decisions based on an understanding of the business. However, rather than supporting workers making decisions, information technologies have been used primarily to rationalize operations (streamline, eliminate variability and gain efficiency) with concomitant reductions of worker agency. We found that an Actor-Network Theory analysis/synthesis of our data indicated opportunities to reestablish agency. By adapting ubiquitous computing technologies in the context of enhancing agency, we identify and present new and viable interaction paradigms, simultaneously attending to operations imperatives. We argue that new, emerging technologies can and should be designed explicitly to support the emergence of worker agency.

Key words: agency, actor-network theory, ethnography, information systems, ubiquitous computing

1. INTRODUCTION

Technological innovations, from large glass panes, steel and elevators, point-of-sale (POS) systems and the internet, have enabled a great majority of retail innovations in the last 150 years that have mostly added to – rather than replaced - the panoply of retail categories available to many of us. One of the most recent (last 50 years) technological innovations has been the introduction of information technologies especially in the context of streamlining or, rationalizing, retail operations (Meffert, 2000). As technology continues to progress and change, one can only expect retailers to continue along the historical trajectory of incorporating new technologies to further rationalize operations (cf., Williams & Larsen, 1999).

An intense focus on operations has resulted in obvious advantages for retailers with - perhaps unintended – side effects. Our research suggests that as operations become increasingly efficient and as variability in the system is increasingly controlled, these endeavors tend also to impinge on *the retail experience* for both the worker and the customer, inadvertently rationalizing that experience in addition to their operations. Rigid and predictable transactions – from the perspective of both the customer and the worker – become *de rigueur*. Human computer *interaction* becomes increasingly limited to human computer *reaction*. In short, there's less expression of *agency*, which can be roughly defined as the ability to impose one's own will in a given network of interacting actors (Latour, 1999).

There has been considerable work examining consumer agency, per se (cf., Sherry et al., 2001; Kozinets et al., unpublished manuscript), There has also been some earlier work on the role of workplace agency such as local innovations by workers sometimes kept secret from management (cf., Orr, 1996; Suchman, 1995). However, there's been very little work on the possibility of technological innovations specifically designed to enhance worker agency while maintaining the rationalization imperative of information technologies necessary for competitive retail environments.

In particular, we find an inverse correlation between the amount of information technology and the expression of worker agency in retail environments. That is, workers, as a part of the retail system, are themselves increasingly rationalized. To remain employed, it's necessary for them to become just another 'cog' (perhaps, 'bit' is more appropriate) in the (information) system. Rationalizing the human worker is, on the face of it, a natural extension of the prevailing role of information technology in retail environments. But it shouldn't be like this; humans should interact with, not simply react to information systems. What're needed are new interaction paradigms, new ways to actively consider the role of information systems in relation to human actors.

The emergence of increasingly viable ubiquitous computing technologies suggests possibilities. Weiser's (1991) paper is perhaps most well known for introducing the concept of ubiquitous computing. Since then, these technologies, have come to include three main capabilities: a) the ability to 'label' people, places and things with everything from passive (no power of their own) radio-frequency identification 'tags', b) the ability to distribute remote and independent 'sensors' to detect environmental stimuli, such as noise, temperature, emissions, etc. and c) the ability to combine these two capabilities into small fully functional computing devices that can form networks of communicating sensors and tags (cf., Sakamura, 2002). The general point from the technological perspective is that computing capabilities formerly confined to specific locations, can now be distributed through out the environment, making it possible, for example, to continuously track products from manufacture (indeed, before manufacture) through to the customer purchase with the envisioned further rationalization that inevitably would manifest itself.

In this paper, we consider these technologies in the context of the human worker. That is, while the development of ubiquitous computing technologies continues relatively unabated, its consideration merely as a further tool of rationalization, e.g., supply chain management, obscures the liberating and differentiating potential of the technology that suggests new and novel forms of interaction in the very same environments shared by the very same increasingly rationalized supply chains. By focusing on the human part of the human-computer interaction, we find opportunities for technology to contribute to the reintroduction of human agency in the workplace.

We also limit our discussion to the role of these particular applications of technology in the context of the individual worker as opposed to the role of the technology in the context of the most general business operations, or of the corporation's adoption of these particular capabilities, per se. We feel that further evaluation of this work from a purely business perspective is required before we can begin to comprehend a framework for organization-wide diffusion of these capabilities (Scheepers & Rose, 2000; Lyytinen & Damsgaard, 2001).

We present evidence from long-term, multi-nation ethnographic work in a variety of retail establishments and settings. We find a general, perhaps counterintuitive, pattern: workers in establishments that rely more heavily on information systems seem to express less agency then those that haven't incorporated such systems. We consider our evidence within the context an actor-network theoretic approach advocated by Callon (1986) and Latour (1999) and use this theoretical approach to define agency in terms of polimorphic and mimeomorphic actions that technologies can encourage or prohibit (Collins & Kusch, 1998). By way of illustration we also offer specific technological implementations and discuss their advantages and disadvantages in the retail environments of the future.

2.   METHODOLOGY

This paper draws on results of two distinct ethnographic studies of retail environments in addition to ongoing, multi-site, concurrent ethnographic exercises. The first ethnographic study of retail environments (Salvador, Bell, Anderson, 1999) focused on consumers from which we generated four abstractions of shopping experiences along two dimensions. We employed a variety of techniques including observation, participant observation, interviews and auto-ethnography. The second study, with fieldwork conducted in the winter of 2003, was a primarily observational study of a variety of retail environments along with interviews. In addition, we held innovation and design exercises *in situ* and used the results of these sessions themselves as data. In this latter exercise, we drew out patterns of innovation that seemed to fit most closely with certain retail environments and less with others. With these, we conducted additional interviews with retail staff. Finally, as a team of six social scientists almost continuously engaged in ethnographic work around the world, it is common practice for us constantly to record images and notes of a wide variety of retail environments.

3.   HABERMAS, LATOUR, AGENCY & RETAIL

We start this discussion with the example of a small motorcycle rental shop for tourists in a small town on a Greek island and continue weaving in several more examples of increasing size into our discussion ending with "Red" (not the real name), a very large discount retail chain.

This happened: Two friends were to leave the next morning from Skala Eressos to Sigri, the first leg of a week's trekking holiday around the Greek island of Lesvos. After dinner, walking along the main street, they saw a tourist shop renting motorcycles and thought it wouldn't be at all a bad idea to leave the packs, walk to Sigri and then rent motorcycles to come back and retrieve the packs. It'd be fun anyway. One approached the clerk: "Excuse me, but you do rent these, don't you?" "Yes, certainly." "Do you also rent them in Sigri?" "Why don't you tell me exactly what you are thinking." He told him. "We can do that. We can bring the scooter to Sigri and pick it up later. It will cost you $20 euros extra", which opened the negotiation for price.

We've defined agency as the ability of the retail worker to act appropriately but relatively unencumbered by the dictates and wishes of the establishment. Our Greek clerk did just that. Such a delivery service was not a formal part of the shop's operations, but given the time of year

(Spring), and low demand for that time of the week (Wednesday), the clerk knew his capabilities and was able to make an offer to the tourist and initiate a negotiation.

There's real power in being able to custom design a service for a particular client on the spot. And there are loads of similar examples: special orders in the fish markets in Nantes France, intense negotiations for prices and packages at Tech Mart in Seoul, South Korea and free beverages for special guests in the ferias of Santiago de Chile.

In all of these examples, the clerks had no access to any information technologies in support of their ad hoc arrangements. We note, therefore, that agency flourishes in the absence of information technologies. Do we find more agency or less with larger retailers replete with information systems? What we actually find is that higher prevalence of information technologies coincides with a lower prevalence of expressed agency. It's not that the technology limits the expression of agency by design, but rather, the design of the technology is to rationalize operations, supporting its management designers; that agency is limited as a byproduct of that design is unfortunate.

The issue has theoretical importance. In the past, Habermas (1970), as discussed by Whitley (1999), argued that humans alone possess and can express agency, and information systems (machines) do and cannot. Moreover, information systems (machines) are to be made merely to serve human needs. This is precisely the way information systems, for the most part, are designed today- especially in retail. But they don't necessarily have to be. It's a matter of adopting a deliberately different outlook.

Actor-Network Theory (ANT) offers such an alternative view. ANT asserts that in any system there are only "actors" and that agency is expressed as an emergent property of the interaction among the actors expressing their will one on the other (Latour, 1999). Phrased slightly differently, agency is not expressed at the "nodes", i.e., (only) by the actors, but rather as transformations that result from the interaction of all the actors in the network, including the information systems. Agency is, therefore, an emergent property of the total system. It's less important for the design of interaction whether the theory is "true" or not, but whether it is useful. ANT is a very useful theory for thinking about the designs of information technologies.

Since actors act, and since we've defined agency as actions of a particular type, we must also define what it means "to act". Identifying what action means will directly inform the sorts of technological innovations that can enhance or hinder agency. For this, we turn to some recent work by Collins & Kusch (1998) in which they define two types of actions: polimorphic and mimeomorphic. Basically, polimorphic actions (*poli-* from the root for *polis*, of the people, and not only from the root *poly-*, as in many) are those that can only be understood – or done – by other people who are functionally members of the same society. Mimeomorphic actions are those that if replicated by someone who didn't understand the action would still look the same as if someone who did understand was performing the actions. Put in their terms: polimorphic actions are defined such that only the enculturated can see sameness whereas enculturated and non-enculturated can see sameness in mimeomorphic actions. Action and behavior map onto social and natural kinds. Polimorphic actions can only be described in terms of social kinds. Behaviors – non-intentional acts, such as blinking your eye – can be described in terms of natural kinds.

In retail, the social kinds are inscribed by the actor-network of the retail establishment and by the mores of commercial/market transactions in general. What a customer might see as reasonable and what the retail worker may see as reasonable are trumped by the power of the corporate retailer. The corporate retailer actor has the power to impose its will on the actors in the store and the information systems are not only a means of translating the corporate retailer's will onto the workers and customers, but exert their own will in addition. Subversive activities of customers (e.g., shoplifting) or clerks (e.g., workarounds) are their means of translating their will onto the network.

We can, thus, modify our definition of agency as "the ability of the retail worker to act appropriately but relatively unencumbered by the dictates and wishes of the establishment *by engaging in a greater proportion of polimorphic actions as compared with mimeomorphic actions"*. Our question, then is to consider the types of technological innovations that explicitly support polimorphic actions – that is, those actions based on an understanding of the relevant culture/society, i.e., the retail environment.

Consider the example of Juliet, an interior designer with "Blue" (not the real name), a large chain of high-end furniture stores. Blue has a fairly unique business in that they control the design, manufacture, distribution, sales and financing of their furniture. One implication, for example, is that many of the 1500 different fabrics can be upholstered onto any piece of furniture, where many retailers limit the selection from three to perhaps a dozen.

Most, if not all, of the clerks in Blue stores are interior designers, called design consultants, who work entirely on commission. Blue provides them with, in essence, an office, equipment and a default client list. When we arrived, Juliet was with a client together at a computer. To assist the design consultant and the customers, the corporation provides a computer system that, in the Juliet's words, "enhances the customer's imagination" by being able to show images of all furniture in all fabrics, rendered in real time and visible on the computer monitor. Juliet, on the whole, was content with the system and found it very helpful when talking with her clients.

We also learned that she often meets customers in their homes to consult on the interior design and consider possible Blue products. As a designer, it often helps to see the physical space and she often brings printed copies of items she's assembled (fabrics rendered on a particular furnishing) with her when she goes into clients' homes. While Juliet is pleased with the computer system, she expressed interest in being able to download and use the "imagination" software on a laptop at the customer's residence to, "further help the customer imagine the room with new furniture in it". Recognizing the company was unlikely to provide a laptop, she offered to use her own personally owned computer, but at the time of our engagement, the company was unwilling to make such provisions and it was unclear that it could for policy, technical or legal reasons.

Juliet actually has quite a large degree of agency. However, rather than enhance her agency, this otherwise liberating software conforms to the rational demands of the company's operations and forces *her* to modify and restrict *her* behavior, relegating her to an increasingly mechanistic role – in this case simply by restricting her movement to being in front of a particular computer with the customer at a particular time. Her frustration, expressed through her affect more than through her words made it clear that the system was not designed by her or even with her assistance.

Finally, we consider the worker agency at Red discount department stores, certainly the retailer mentioned in this paper most well endowed with information technologies as well as the largest overall, with more than 1100 stores in more than 45 US States. In this case, we observed in several stores, interviewed two former store managers and went on an "ethnographic tour" with one to both a Red store and a primary competitor.

The information system pervades the entire store, linking the store to headquarters, to other sites, such as distribution warehouses, and going so far as to guide the day to day and, at times, minute to minute activities of most store employees. The primary role of the information system is emphatically to manage operations- inventory, shelf stock, sales, distribution, etc. The systems offer corporate management a large degree of control over a far-flung enterprise. Red stores, for example, designed according to particular models drawn from among several types with all specific instances of each type being nearly identical one to the other.

The store design includes the physical structure as well as such details as the exact location of shelving units and even product placement, including shelf number, the depth of product on the shelf (how many deep), the number of product along the aisle (facing), etc. The information system accounts for example, for a particular model of soccer ball that belongs on a particular shelf, that are stocked three deep and four across and that there should be exactly 12 on the shelves – unless there's a change, in which case the information system will inform the store manager and his/her crew.

Shelf stocks are refreshed on the floor three times as day at prescribed times and workers are required to restock at that time – a goal being to keep as much of the store operations like restocking "off stage" as much as possible. The information system accounts for that, embodying and extending the power of corporate management to control the operations of each store and to eliminate as much variability as possible, homogenizing each store for a fabulous brand experience. The ideal is that the workers will conform to the demands of the system, a proxy for corporate management, of whom their store manager is, him/herself just a proxy as well.

Of course, the ideal is not always achieved. And there's ample evidence in the literature to suggest that workers create their own, albeit private, innovations (Suchman, 1995). However, the store managers indicated that a large part of their job is to maintain the desired order in the store not only as a means of establishing order, but also as a means of establishing the company's brand identity and distinguishing itself relative to other similar retailers.   One large discount store on its own website states that differentiating their brands is "...central to [their] ability to continue to generate sales and profit growth at each of [their] divisions", (Target, 2003). The order demanded of the information system comprises more than simply efficiency, it's a part of the core identity.

There is, therefore, a certain amount of risk when violating the system. For example, one store manager told us that while he has some leeway to go off-grid, as it were, like the ability to put shovels in a main aisle during and after a snowstorm, even though according to the rules, product should not ever be placed in the main aisle. While snow shovels in plain view during a snowstorm seems obvious, placing other items in temporary displays does not conform to the requirements expressed by the information system and therefore is undesirable in general – even if it too makes sense. According to the store manager, "You can do it, but if you do and get caught, it better have been profitable. And even then, it's questionable."

The Red information system is designed specifically to rationalize the company's operations. But even Blue's system, though designed to aid the imagination, confines the worker to the desktop. Overall, there is either relatively little expression of worker agency or agency is constrained by the demands of information systems or both. Walmart, at this time generally regarded to be at least among the largest retailers in the world – if not the largest – even goes so far as to use it's market strength to encourage other corporations to conform to their information systems, reducing even the expression of agency by corporate management from other companies. Is it possible to have both efficient and rational operations with a greater degree of worker agency? It's this question we address in the last section.

4.   EXAMPLE CONCEPTS

Our goal is to identify technological innovations that accommodate the business requirements of the retailer while also providing an appropriate foundation for encouraging the emergence of agency. In particular, we advocate developing technologies that encourage an increasing proportion of polimorphic actions for retail workers. We've already seen several examples of polimorphic actions our ethnographic portraits: the Greek motorcycle entrepreneur making ad hoc arrangements for the trekkers, putting shovels in the aisle during a snowstorm at Red, Juliet heading to her client's homes and offering her own laptop for "corporate" use. Here we rely on the emergence of ubiquitous computing technologies to begin developing alternative interaction paradigms for retail environments that encourage a rise in the proportion of polimorphic actions by clerks. We must caution that nearly anyone reading this paper will have his/her own experiences with retail environments. In addition, retail environments are highly variable. What might work in one, might fail miserably in another. One must be careful not to evaluate the ideas relative to one's personal feelings, but in light of the current context of supporting agency.

4.1   Yield Management

While yield management (variable pricing according to demand) has been prevalent in the airline industry for some time, it's only recently beginning to be tried in retail establishments. For example, Stelios Haji-Ioannou, who, among other endeavors, began a string of companies beginning with the prefix "easy", easyJet, easyEverything, etc., has recently announced easyCinema, which will offer theatre seating at variable rates depending on the movie, the time you buy it, the demand, and other variables; easyEverything varied internet access pricing by time of day.

One possible application of ubiquitous computing technologies would be to encourage real time yield management akin to the Santiago's feria or the Greek motorcycle shop. Mimeomorphically, by using ubiquitous computing technologies (e.g., sensors & tags) and sophisticated algorithms, one can offer dynamic pricing depending on where people are in the store or the time of day, or perhaps better, the time of day and the number of people in the store. The sensors can report to the information system about what's been picked up, for example, the number of times, etc. The clerk, who sees people gathered in a spot could see a potential opportunity and engage in a dialog with the information system, suggesting that an immediate price drop might be appropriate given the current conditions. The system could provide probabilistic expected returns, and the clerk can consider these, but also consider the human elements in the store – what's been said, the behaviors of the customers, who they are or look like they are, etc., and make a decision. Of course, the decisions are tracked and the clerk's performance is evaluated.

It's a very different sort of interaction. One imagines the dialog happening discretely, perhaps with a handheld device, or ever verbally though one of the increasingly prevalent headsets seen on clerks. It also permits the clerk to initiate an action in conjunction with algorithmic interpretation, combining positive attributes of both information systems and human awareness, or, I daresay, intuition. It also allows the clerk to initiate a positive personal interaction with some customers.

Of course, in some contexts, this arrangement might be inappropriate. However, a hallmark finding of both our work and the intentions of the retail industry, as we discussed especially with regard to Red, is idiosyncraticity and diversity to create identity, brand and distinctiveness. It's not difficult to imagine a wide variety of sensors and tags deployed for a wide variety of purposes and supported by an even wider variety of algorithms

For example in one variation, suppose there's a group of people in the store and the clerk can tell – you know - there's a certain buzz, people "want to buy" but are looking for some prompting, something to get them over the decision. Clerks could, for example, offer an instant sale – maybe one of a variety that are vetted by the information system: "If we sell 10 pairs of jeans in the next 20 minutes, you all get 20% off." Or suppose sensors detected that a fair number of people have looked at a certain item in the shop, but then returned it to the rack or shelf. Clerks, having been alerted by the information system, could make (subtle) inquiries and could, if appropriate, initiate an instant sale.

Of course, information systems as actors in the network could do something like this sort of thing on their own, but these systems would miss "the buzz", they would rely on attributes available to computing systems which are not at all the same sorts of attributes available to human actors. They have – and will continue to have - immense difficulty in making subtle and suitable inquiries and detecting meanings in subtle human behaviors – and likely as not, there'd be an unacceptable number of false positives and perhaps more insidiously, false negatives.

In addition, having clerks actually get involved and make the decisions, albeit with information systems support, not only invests them in their job, but also encourages a greater relationship between the consumers and the retailer (Fournier, 1998) and between the retailer and the clerk. In these examples, the clerks had to agree to the initiating the "instant sale"; one imagines that their performance would be kept and one could link their bonus pay to their performance, sharing the profits with them. Weak performance might limit their participation in the program or increase their training. Responsibilities come with agency.

4.2   Clerk Relations

Orr (1996) demonstrated that copy machine mechanics swapped stories of repairs at coffee breaks. He went on to provide the ability for copy mechanics to be in continuous contact with one another so that should one request help on a particular issue another might be able to assist. One can imagine similar capabilities for clerks in retail establishments with communications carried through the IT systems in the store and between stores.

There's a lot happening in a large retail establishment. It's difficult for clerks to be everywhere and to help everyone and still take care of the minute-to-minute store operations. It's also difficult for clerks to learn from each other and aid each other on a moment's notice. Clerks should be able to talk with one another in the store requesting and providing assistance as necessary. Some retailers already do this, providing for a private communication network rather than broadcasting requests over a public address system as well as maintaining increased surveillance.

One can enhance such systems for a wider variety of purposes. In a particular shop, clerks could offer assistance to others when customers have questions about products. Or the clerks could be connected to other shops in the same corporation and request assistance on the spot. Or, the information system can track the sorts of expertise clerks have and make the connections for a clerk that requests assistance – sort of like a "match making" service, but matching clerks to customers, drawing immediately on the knowledge spread through the corporation. Clerks might, over time, gain a reputation.

If we imagine that shelves are provided with electronic displays, clerks could also get increasingly involved in making recommendations. At Powell's Used and New Bookstore, in Portland, Oregon, employees write short reviews of books they read and put a small placard along the edge of the shelf near the book. Many purchases are heavily influenced by stories of individual experience – of the customer, of the employees and second hand as reported by other customers, employees and their friends (Keller & Berry, 2003). Capturing these stories at the point of sale, editing and attributing them to particular clerks can provide an ongoing narrative associated with a particular product. Customers should also be able to leave their experiences, again, further involving the customer and clerk actors in the network. In one instantiation, this information could be provided on hand-held or stationary displays/tablets. Customers can find clerks' or other expert customers' comments at these stations – maybe by passing the product near a stationary display or, if the display is also a reader, passing it near the products, which have the tags.

5.   SUMMARY

All of these sorts of ideas challenge the extant models of in-store shopping. Shopping is indeed a learned activity (Cohen, 2003) and we are not suggesting that such changes happen overnight, nor that these particular ideas should become manifest. Rather, we are proposing these concepts as particular examples of technologies that are not necessarily geared toward improving the efficiency of the retailer's operations, but toward the notion of a different direction for technological innovation and importantly, a different type of human-computer interaction with information systems.

In these two examples, we demonstrate the possibility of information technologies, specifically ubiquitous computing, being differently designed to support the emergence of agency in an actor-network, in this case, in retail establishments. We specifically attempt to show that when consciously considered, it's possible to imagine technologies supporting polimorphic actions in concert with an information system as agentic partners in the actor-network.

Would retailers endorse these sorts of innovations? Would customers participate? One can argue if there's a viable business model, they very well might. Of course, retail spaces are different, one from another. An innovation in one establishment would be a distraction in another. But differentiation is a key component of the retailers' endeavors. And information systems applied for the continued rationalization of their operations can only serve to rationalize more and more of the retail experience. Ubiquitous computing technologies, far from being limited to improvements in supply chain operations, can and should be actively and consciously considered as new technologies that provide for new and innovative interaction paradigms.

REFERENCES

Callon, M., *The sociology of an actor-network*, in M. Callon, J. Law & A. Rip, Mapping the dynamics of science and technology, Macmillan, London, 1986.

Cohen, L., *A consumers' republic*. Alfred A. Knopf, New York, 2003.

Collins, H. & Kusch. M., *The shape of actions: What humans and machines can do*. MIT Press, London, 1998.

Fournier, S., Consumers and their brands: Developing relationship theory in consumer research. *Journal of Consumer Research*, 24, (1998),343-373.

Habermas, J., *Toward a rational society: Student protest, science and politics* (Jeremy J Shapiro, Tranlation), Beacon Press, Boston, 1970.

Hernandez, M & Iyengar, S.S., What drives whom? A cultural perspective on human agency, *Social Cognition*, 19(3), (2001), pp. 269-94 http://www.columbia.edu/ ~ss957/ whatdrives. html

Keller, E. Berry, J., *The Influentials: One American in Ten Tells the Other Nine How to Vote, Where to Eat, and What to Buy*, Simon & Schuster, New York, 2003.

Kozinets, R.V., Sherry, J.F., Storm, D., Duhachek, A., Nuttavithisit, K., Deberry-Spence, B. The screen play of media spectacle: Dimensions of ludic consumption at *ESPN Zone Chicago*, unpublished manuscript.

Latour, B. On recalling ANT, in Law, J. & Hassard, J. *Actor Network Theory and After*, Blackwell Publishing, Oxford, 1999.

Lyytinen, K. & Damsgaard, J. What's wrong with the diffusion of innovation theory, in *Proceedings of the IFIP TC8 WG8.1 Fourth Working Conference on Diffusing Software Products and Process Innovations*, Kluwer, Netherlands, 2001.

Meffert, H. Consumers Tomorrow, *ECR Academic Report* (2000), 68-71. http://www.ecrjournal.org/partnership/publications/report2000/10_Consumers.pdf

Orr, J. E., *Talking About Machines: an ethnography of a modern job*, IRL press, Cornell University Press, Ithaca, 1996.

Sakamura, K. Making computers invisible, IEEE Micro 22(6), (2002), 7-11.

Salvador, T., Bell, G. & Anderson, K., Design Ethnography. *Design Management Journal*, 10(4), (1999).

Scheepers, R. & Rose, J. Understanding ubiquitous IT in organizations: the case of intranet introduction, in Svenson, L., Snis, U., Soresnen, C., Fagerlind, H., Lindroth, T, Magnusson, M & Ostlund, C. (Eds), *Proceedings of the 23rd Information systems Research In Scandinavia (IRIS)*, Lingatan, Sweden, 2000, 597-609.

Sherry, J.F., Kozinets, R.V., Storm, D., Duhachek, A., Nuttavithisit, K., Deberry-Spence, B. Being in the Zone: Staging retail theatre at ESPN Zone Chicago, *Journal of Contemporary Ethnography*, 30(4), (2001), 465-510.

Suchman, L. Making work visible. *Communications of the ACM*, 38 (9), (1995), 56-64.

Target Website http://www.targetcorp.com/targetcorp_group/companies/companies.jhtml

Weiser, M. The computer for the 21st century. *Scientific American*, 265(3), (1991), 94-101.

Whitley, E.A., Habermas and the Non-Humans: Towards a Critical Theory for the New Collective. *Critical Management Studies conference*, Manchester, (1999). http://www.mngt.waikato.ac.nz/ejrot/cmsconference/documents/Information%20Tech/Habermas %20and%20the%20non-humans.pdf

LEARNING MANAGEMENT SYSTEMS: A NEW OPPORTUNITY

Audrey Dunne and Tom Butler
Business Information Systems, University College Cork, Ireland

Abstract: In an intensely competitive, rapidly evolving, and increasingly knowledge-based IT sector, the ability to learn becomes critical to the success of IT organizations. In today's knowledge economy, a firm's intellectual capital represents the only sustainable source of competitive advantage. This intellectual capital manifests itself, predominantly, in the form of both the individual and the collective competencies of employees within an organization. The knowledge management approach which seeks to facilitate the sharing and integration of knowledge has had limited success, primarily because of its focus on 'knowledge as a resource' rather than on 'learning as a people process'. A strategic 'people-oriented' approach to the management of learning is now emerging in many organizations and this has, in turn, led to the appearance of a new breed of Information Systems (IS) known as 'Learning Management Systems' (LMS). Based on a case study of the implementation of an LMS by a major multinational IT enterprise, this paper proposes an empirically tested framework for 'learning in organizations' and highlights the roles that LMS can play in the continued commercial success of IT organizations.

Key words: Learning Management, Theory Building, IS Research, Knowledge, Learning in Organizations, Organizational Learning.

1. INTRODUCTION

The critical role that organizational learning plays in IT organizations has long been noted; indeed Ray Stata, the founder and former CEO of Analog Devices Inc., argued that it was the key to the management of innovation (Stata, 1989). Accordingly, IT organizations such as Siemens and Analog Devices Inc. have leveraged the power of Knowledge Management Systems

(KMS) to this end; however, the limitations of such systems has been noted (Butler, 2003).

A new breed of Information Systems (IS) known as Learning Management Systems (LMS) are evolving to enable learning in organizations (Brennan, Funke and Andersen, 2001; Hall, 2001; Nichani, 2001; Greenberg, 2002). The need for such systems is reflected by the fact that IT organizations such as Cisco Systems, VERITAS Software, Alcatel and Xilinx are employing learning management to foster and manage learning within their organizations[1]. In essence, LMS replace isolated and fragmented learning programs with a systematic means of assessing and raising competency and performance levels throughout the organization, by offering a strategic solution for planning, delivering and managing all learning events including both online and classroom-based learning (Greenberg, 2002). The present focus on 'Learning Management' has been led primarily by industry practitioners and industry vendors and there is little empirical research in this area. Hence, the principal objective of this study is to deepen the IS field's understanding of the contribution of these new systems to learning within organizations.

The remainder of this paper is structured as follows: The next section considers the motivation for the study; following this, the research approach is outlined. The penultimate section details the overall findings of the study and presents an empirically tested framework for learning in organizations that incorporates the new phenomenon of LMS. The paper then discusses some of the opportunities and challenges presented to us by LMS and offers concluding suggestions for further research in the area.

2.   MOTIVATION FOR THE STUDY

2.1   The Importance of Learning in Organizations

Many definitions of 'organizational learning' have been articulated. Perhaps, the most succinct of these is that of Fiol and Lyles (1985, p.803) who state that *"Organizational Learning means the process of improving actions through better knowledge and understanding."* The importance of facilitating and managing learning within organizations is well accepted. Zuboff (1988), for example, argues that learning, integration and communication are critical to leveraging employee knowledge; accordingly,

---

[1] http://www.saba.com/english/customers/index.htm

she maintains that managers must switch from being drivers of people to being drivers of learning. Harvey and Denton (1999) identify several antecedents which help to explain the rise to prominence of organizational learning, viz.

- The shift in the relative importance of factors of production away from capital towards labour, particularly in the case of knowledge workers.
- The ever more rapid pace of change in the business environment.
- Wide acceptance of knowledge as a prime source of competitive advantage.
- The greater demands being placed on all businesses by customers.
- Increasing dissatisfaction among managers and employees with the traditional 'command control' management paradigm.
- The intensely competitive nature of global business.

2.2 Deficiencies in the Knowledge Management Approach

During the 1990s, there was a major shift in focus from organizational learning to knowledge management, in both applied and theoretical contexts (Easterby-Smith, Crossan and Nicolini, 2000; Scarbrough and Swan, 2001; Alvesson and Kärreman, 2001). Knowledge Management Systems (KMS) seek to facilitate the sharing and integration of knowledge (Alavi and Leidner, 1999; Chait, 1999; Garavelli, Gorgoglione and Scozzi, 2002). However, these systems have had limited success (Shultz and Boland, 2000) with reported failure rates of over 80% (Storey and Barnett 2000). This is because many of them are still, for the most part, used to support data and information processing, rather than knowledge management (Borghoff and Pareschi, 1999; Sutton, 2001; Hendricks, 2001; Garavelli *et al.*, 2002; Butler, 2003) and also because many implementations neglect the social, cultural and motivational issues that are critical to their success (McDermott, 1999; Schultze and Boland, 2000; Huber, 2001). Indeed knowledge management may be more of a new 'fashion' or 'fad' that has been embraced by the IS field (Swan, Scarborough and Preston, 1999; Butler, 2000; Galliers and Newell, 2001) and its popularity may be heightened by glossing over the complex and intangible aspects of human behavior (Scarborough and Swan, 2001).

2.3 New Potential Offered by Learning Management Systems

It is perhaps time to admit that neither the 'learning organization' concept, which is people oriented and focuses on learning as a process, nor the knowledge management concept, which focuses on knowledge as a resource, can stand alone. These concepts compliment each other, in that the learning process is of no value without an outcome, while knowledge is too intangible, dynamic and contextual to allow it to be managed as a tangible resource (Rowley, 2001). She emphasizes that successful knowledge management needs to couple a concern for systems with an awareness of how organizations learn. Researchers believe that what is needed is to better manage the flow of information through and around the "bottlenecks" of personal attention and learning capacity (Wagner, 2000; Brennan *et al.*, 2001) and to design systems where technology is in service to and supports diverse learners and diverse learning contexts (McCombs, 2000). In response to this need, another breed of systems known as Learning Management Systems (LMS) have evolved and many firms are now using this technology to take a new approach to learning within organizations. This new 'learning management' approach has been led primarily by a number of practitioners and IT vendors and there is a dearth of empirical research in this area. Therefore, an important challenge for the IS field is to better understand LMS and to examine the roles and relationships of these new systems within organizations.

3.   RESEARCH APPROACH

This study's primary objective is to examine how LMS may be utilized in an organizational context to facilitate and manage organizational learning. A conceptual model and related research framework is employed to guide the conduct of the investigation. A 'single' case study approach was selected for three key reasons. Firstly, the case study research method is particularly suited to IS research (Benbasat, Goldstein and Mead, 1987; Myers, 1997), since the objective is the study of IS in organizations and *"interest has shifted to organizational rather than technical issues"* (Benbasat *et al.*, 1987). Case research, with its emphasis on understanding empirical data in natural settings (Eisenhardt, 1989) is an appropriate method for studying IS issues and practices. Furthermore, Benbasat *et al.* (1987) maintain that IS researchers should learn and theorize primarily from studying systems in practice, as much IS research trails behind practitioners' knowledge. Indeed, this is the case with respect to the availability of empirical research on LMS.

Secondly, the objective of the research is exploratory in nature and the single case study is considered to be a potentially rich and valuable source of data and is well suited to exploring relationships between variables in their given context, as required by exploratory research (Pettigrew, 1985; Benbasat *et al.*, 1987; Yin, 1994; Stake, 1994).

Thirdly, the main argument against single cases has been answered by Lee (1989). He points out that single cases differ from multiple cases only in their degree of generalizability and in this sense, the 'lessons' learned from our case have been formulated as postulates, with specific view to their validity being confirmed, or otherwise, in future research.

3.1   Conceptual Model and Research Framework

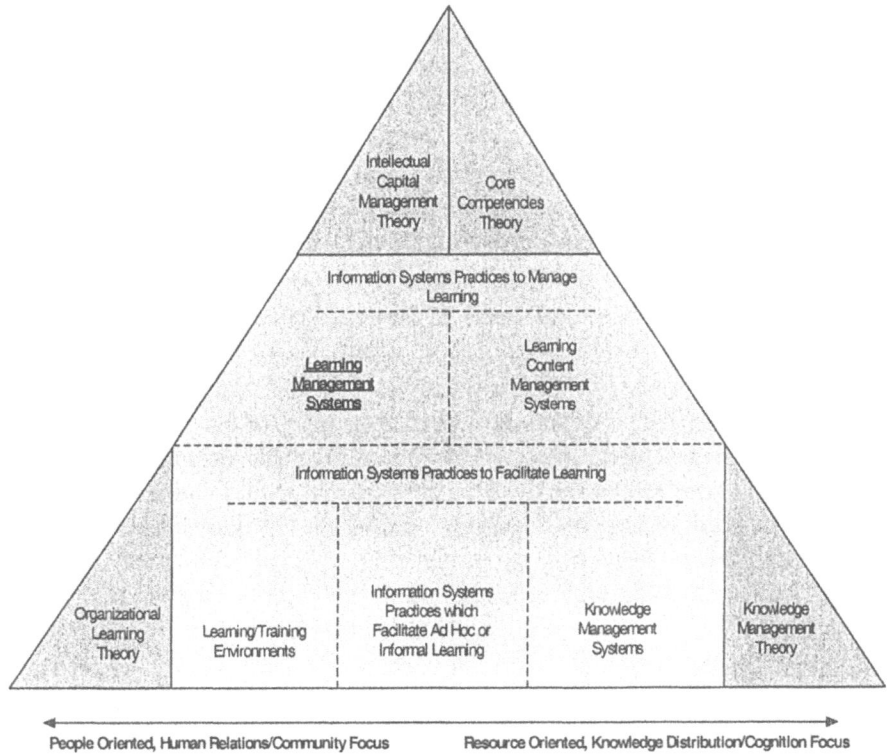

Figure 1. Learning in Organizations: Conceptual Model

The conceptual model and related research framework for the study were drawn from the literature to help guide the conduct of the study. The conceptual model depicted in Figure 1 establishes the boundaries of this research study. The relevant theoretical influences are shown on the periphery of the model and their positioning on the x-axis indicates the degree to which they are oriented towards a people or community focus at one end of the continuum, or a resource or knowledge distribution focus on the other end of the continuum. These theories are: (1) organizational learning theory; (2) competence theory; (3) intellectual capital theory; and

(4) knowledge management/knowledge transfer theory. Organizational learning theory is primarily people oriented; while knowledge management theory is primarily resource oriented; however intellectual capital theory and competency theory are both people and resource focused. The major categories of IS which support learning in organizations are all represented at the core of the model and are separated only by dotted lines, indicating that there are a lot of interconnections and indeed, intersections between them. These categories are: (1) Learning Management Systems; (2) Learning Content Management Systems; (3) Learning/Training environments; (4) Knowledge Management Systems; and (5) Information Systems that facilitate ad hoc and informal learning. It is proposed that learning and training environments and LMS tend to be people oriented, while KMS and LCMS concentrate more on knowledge resources and learning content. Organizational systems which facilitate ad hoc or informal learning vary considerably in their attention to people issues versus resource issues. Finally, the model distinguishes between IS that is used by management to manage learning within organizations and IS that facilitate learning and that are experienced at the cold-face of everyday life in organizations.

The research framework for this study was generated from a review of extant theory and research and depicts the topology of the key theories and categories of IS that underpin learning in organizations (see Figure 2). The framework illustrates that the theories have helped mould and shape the way that IS are used to support learning. Links drawn between the theories indicates that they have had an influencing role on each other. The IS category of LMS is highlighted within the framework to emphasize that this is the area of focus for the study. Links drawn from one systems category to another signifies potential interrelationships between them. The main premises which underpin this framework are:

- LMS can play a critical role in organizations by facilitating the 'people centered' management of learning.
- IS that facilitate and promote either formal or informal learning (or a combination of both) in an organization, possess interrelationships with each other; furthermore, the roles played by these systems is greater than the sum of their parts.
- Key theories of learning within an organizational context have influenced the way in which IS are used to facilitate and promote learning in organizations.

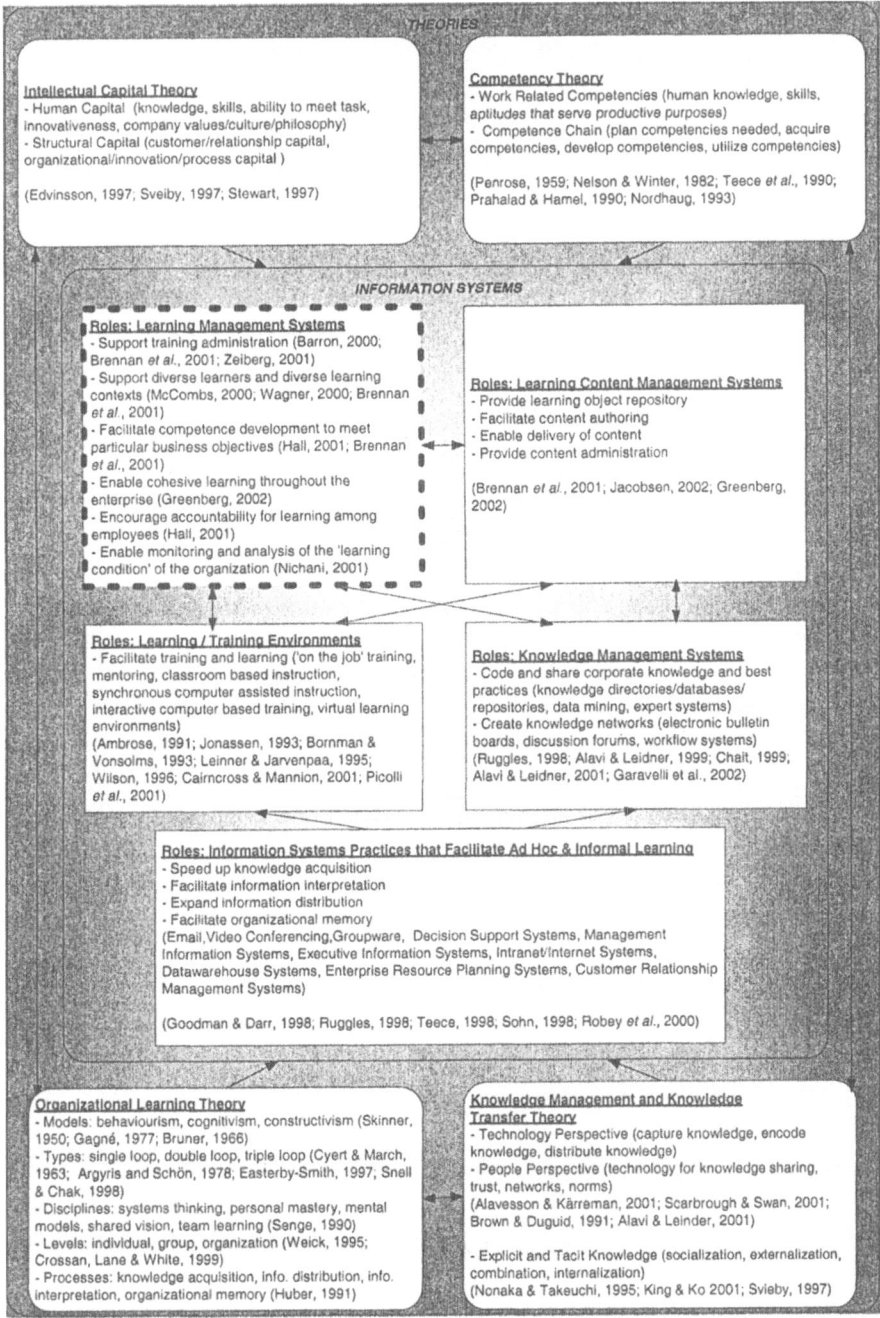

Figure 2. Learning in Organizations: Research Framework

Within the context of this framework, three research questions were formulated to help attain this study's research objective:

Research Question 1: What are the roles of the Learning Management System in managing learning within the organization?

Research Question 2: What is the relationship between the Learning Management System and the other Information Systems that support learning in the organization?

Research Question 3: What theories have influenced how the organization uses its Learning Management System to manage learning within the organization?

3.2   Implemention of the Research Method

The organization chosen for study was a major US multinational supplier of data storage solutions to the IT market that had recently implemented an LMS as part of its learning management strategy. For reasons of confidentiality, this organization cannot be identified and will be referred to as CEM Corporation. The LMS in use at CEM was complex and multifaceted system; hence, it was necessary to conduct several exploratory interviews with the subject matter expert. Five such site visits occurred over a six month period and each meeting lasted between one and one and a half hours. This type of elite interviewing (Marshall and Rossman, 1989) is sometimes necessary to investigate little understood phenomena. In one of these sessions, a detailed demonstration of how the system operates was provided by the Training Manager. A second demonstration of the system was subsequently obtained from a Training Specialist within CEM Corporation. This provided the researchers an understanding of the system's capabilities and insights into how the system is used on a day to day basis. The Human Resources Manager was also interviewed at this stage. Subsequently, the researcher carried out eight semi-structured interviews. Appendix A provides an outline of the interview guide as well as the profiles of the interviewees.

4.   LMS: TOWARDS A BETTER UNDERSTANDING

In February 2001, CEM Corporation deployed a Learning Management System known as Saba Learning Enterprise™ to employees across the entire enterprise, as well as to CEM customers and business partners. The business drivers for deploying this enterprise learning solution were:
– Decrease time-to-competency.

- Develop and manage skill sets for all employees.
- Leverage global, repeatable and predictable curriculum.
- Integrate competency assessments to development plans.
- Accelerate the transfer of knowledge to employees, partners, and customers.
- Provide a single learning interface for all internal and external users.

Currently, this corporate-based system is used by training managers primarily within the manufacturing and customer services organizations of CEM Corporation to deliver and track both technical and business training programs. Training administrators within these divisions use the system to administer and manage training courses, for example – they publish and manage learning content; they manage a catalogue of courses; and they create reports on learning activities. The system is also used by other training managers to formulate additional types of training and learning across the entire corporation, including individual personal development programs and management training.

Many employees within CEM Corporation are using the LMS to manage their own learning processes, for example – they use the system to enrol in classroom courses, to search for learning material, to engage in online learning activities, and to look at what development options are suitable for their role within the organization. Furthermore, the LMS facilitates a competency assessment process which enables employees to identify gaps or shortfalls in their competency levels with respect to a competency model defined for their own specific job role. Both business managers and technical managers within CEM Corporation use the system to manage the learning processes of their employees, for example – they examine the status of learning activities for their employees; they assign learning initiatives to be carried out by their employees; and they generate reports on these learning activities.

4.1   LMS: Location in the Enterprise Learning Solution

Figure 3 illustrates the multifaceted nature of CEM's Enterprise Learning Solution. Much of the learning material is created and maintained by CEM employees using a variety of products. Content is stored on CEM's own storage repository, on site. In addition, courseware that is created and maintained directly by third parties is stored offsite in the storage repository of the third party organization. Courseware is provided to CEM Corporation by a number of third party learning content suppliers.

The LMS has the capability of managing and tracking both offline activities (e.g. books, 'on the job' training, mentoring, classroom training) and online activities (e.g. video and audio, rich media, web casts, web-based training, virtual classroom training). In the case of online activities, learning content may be accessed and delivered through the LMS either from CEM's repository or from the third party's storage repository. Certain testing is built into the learning content itself, but additional pre-training-testing and post-training-testing may be invoked and this is currently provided by another third party product. While much of the required reporting is provided by the LMS, administrators also use third party software to generate more sophisticated reports.

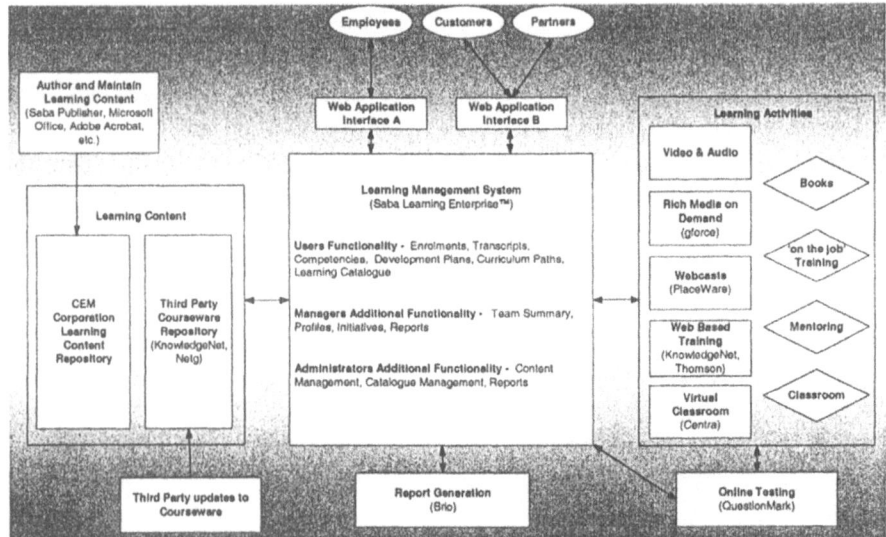

Figure 3. CEM Corporation: Enterprise Learning Solution Components

4.2   LMS: Key Roles

The research framework was tested using the empirical findings and was modified accordingly, see Figure 4. The findings indicate that to some extent, the LMS within this organization fulfills all of the roles suggested in the research framework. Furthermore, the research identified a number of additional roles that were not suggested in the framework, but that are being performed by the LMS. One of the primary roles of the LMS within CEM Corporation to date has been its assistance to Training Managers in **supporting the administration of training**[2] across such a large organization with a variety of training needs. This has had a consequential effect of **increasing the productivity of training.**

----

[2] Bold text within this section indicates that this is a role fulfilled by the Learning Management System.

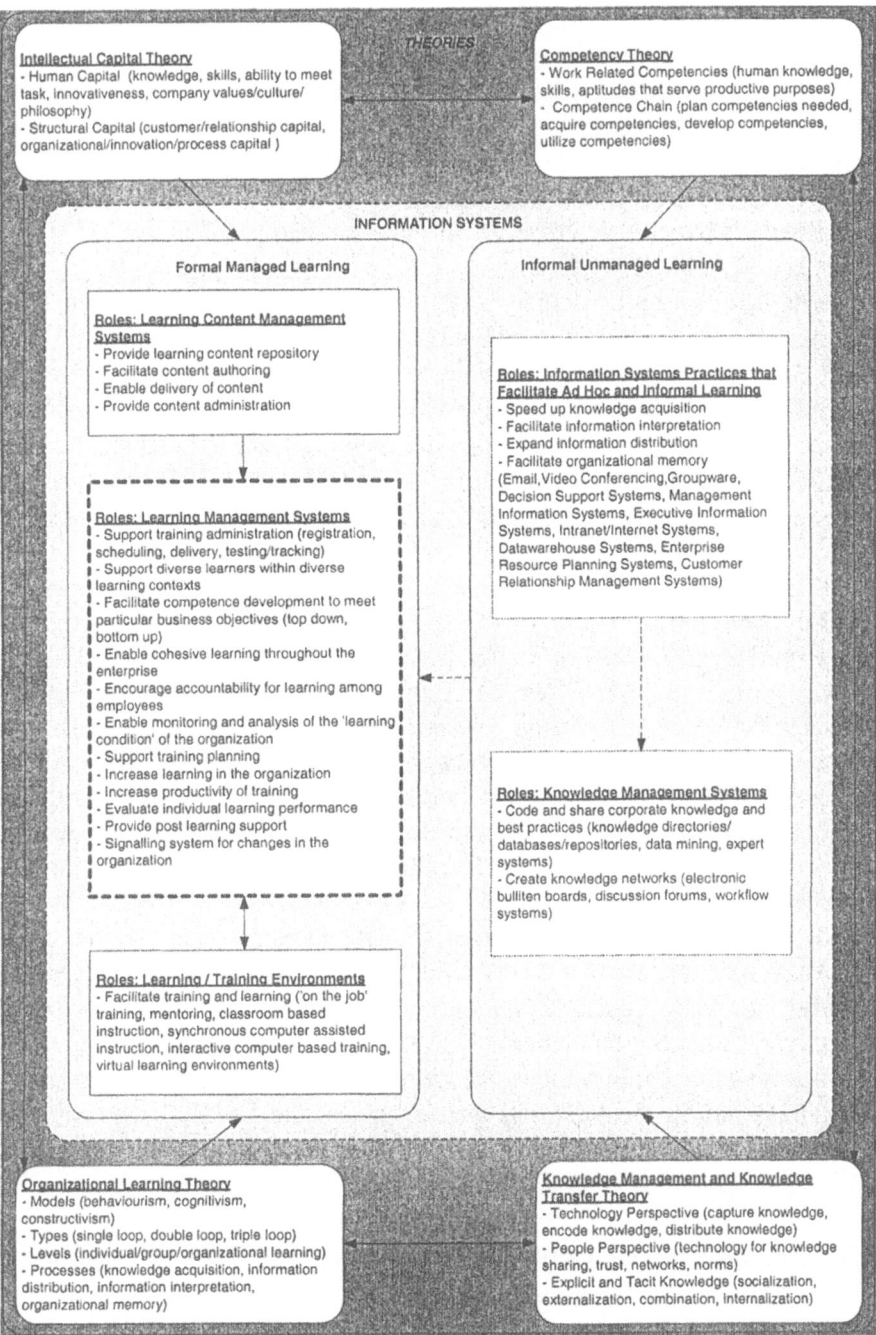

Figure 4. Learning in Organizations: Modified Framework and Summary of Findings

From a learner's perspective, the principal role played by the LMS is that it provides a central repository for a range of learning material in a structured way which enables the system to **support a diverse body of learners within diverse learning contexts.** This leads to the most critical role of all, which is that it increases the use of training and hence, **increases learning in the organization.** Also from the perspective of the learner, two other significant and emerging roles of the LMS were highlighted, namely; the **provision of post learning support;** and the role of the LMS as a **signaling system for changes in the organization,** when new training is made available on the system.

Although role-based competency models have been set up on the LMS for many employees within the organization, CEM Corporation is still in the early stages of constructing workable competency models for all employees. Difficulties were highlighted in relation to certain technical positions which primarily involved specialist skills. Challenges were also emphasized in relation to positions involving varying requirements for depth and breath of knowledge of a particular skill. The extent of investment in time and effort in drawing up these 'technical role-based competency models' is such that the effort is justified only if it is driven by the local business in the context of local business needs. Furthermore, it must be noted that CEM Corporation is also in the early stages of mobilizing the competency assessment process, even in relation to job roles for which there is a standard competency model available. However, in some parts of the organization, employees have already begun to engage with the LMS to carry out competency assessments against the standard competency model related to their own particular role.Subsequently, and usually in consultation with their managers, employees have reviewed the learning options available and devised development plans for acquiring the competencies in which they are deficient. Although this functionality is still being rolled out across CEM Corporation, it is clear that the LMS is beginning to fulfill the vital role of **facilitating competence development to meet particular business objectives,** through a dual approach to learning management (i.e. top down and bottom up). From a top down perspective, Training Managers within CEM Corporation have already started to use the LMS to automate the 'training needs analysis' process which will assist them in the identification of training needs and will **support training planning.** From a bottom up perspective, CEM Corporation is encouraging employees to self manage their own learning using LMS and this has the added benefit of **encouraging accountability for learning among employees.**

The use of competency models for assessing and developing competencies forms the basis of a number of other key roles of the LMS which are beginning to emerge. Through standardizing role-based competency requirements and development options, the LMS is already enabling more consistent and **cohesive learning throughout the enterprise.**

Assessing employees against the standard competency model for their particular role enables the **monitoring and analysis of the 'learning condition' of the organization**. Furthermore, by reviewing progress between one competency assessment and the next, **evaluation of individual learning performance** for any employee is facilitated. This may then form part of the individual's overall performance evaluation. The research findings highlighted the sensitivity that surrounds performance assessments overall and stressed the need for a significant human element in assessing performance. In particular, the need for joint agreement between an employee and their immediate superior on future development plans was emphasized.

4.3   LMS: Relationships to and Influence of Theories

The relationships between the LMS and the other IS that support learning within CEM Corporation were investigated within the context of the research framework. In addition, the impact of leading theories was considered in relation to how they had influenced CEM Corporation in its use of the LMS to manage learning. As can be seen in Figure 4, it was found that while the LMS is fed by a direct link to the LCMS (i.e. The LMS remotely accesses learning content on the LCMS), and has a strong two way relationship with 'learning/training environments', it has only a tenuous link to other Information Systems that support ad hoc or informal learning, including KMS. The link from these systems consists primarily of a need which they generate for formal learning and training programs. The content for this training will often stem from the system itself and the type of environment used will, more than likely, be decided by the nature of the system in question. KMS often store information on problems and solutions relating to other systems that support informal learning and so there is a tentative link between these groups of systems. KMS, while supporting knowledge management in a formal way, only support informal unmanaged learning, as the learning is not delivered in a structured way, nor is it measured or validated. The findings also indicate that the LMS, together with the LCMS and the learning/training environments, all contribute to process of formal managed learning within the organization. On the other hand, IS which facilitate ad hoc or uncontrolled learning together with KMS, support informal unmanaged learning within the organization.

The study found that all of the theories[3] reviewed have influenced CEM Corporation in its use of The LMS. Organizational learning theory, however, has had a greater effect on what types of learning content and on which types of learning environments that CEM Corporation uses to facilitate various forms of learning within the organization, than on how the LMS is employed to manage that learning. Intellectual capital theory played a critical role in the decision to implement an LMS initially and continues to underpin the mission of the initiative which is to maximize intellectual capital within the organization. Competence theory is influencing CEM Corporation on an ongoing basis, in particular, its use of the LMS to manage competency levels across the organization. Knowledge management and knowledge transfer theory has also played a key role in how CEM Corporation uses the LMS to manage all forms of 'knowledge learning'.

The size and complexity of CEM Corporation made it very difficult to establish if the organization as a whole was more influenced by organizational learning theory, and thus more oriented towards a people or community focus, or if it was more influenced by knowledge management theory, and thus more oriented towards a resource or knowledge distribution focus. When questioned about this, the responses of individual interviewees varied considerably and this in itself would indicate that CEM Corporation holds a pluralist approach, and is thus more resource oriented in certain parts of the organization, while it is more people focused in other areas.

5.   LMS: OPPORTUNITIES, CHALLENGES AND CONCLUSIONS

This study illustrates that the introduction of an LMS can replace the isolated and fragmented learning programs in an organization with a systematic means of assessing and raising competency and performance levels throughout the organization (see Geenberg, 2002). For IT organizations, such as CEM Corporation, who operate in a dynamic environment where product sophistication and complexity grows with each generation, and where customer and product support is a competitive advantage, LMS offer a strategic solutions for planning, delivering and managing all learning events, including both online and classroom-based learning. The research findings show that an LMS can play a vital role in increasing learning within the organization. This may be achieved in part by

---

[3] Theories reviewed: organizational learning theory, intellectual capital theory, competence theory, knowledge management/knowledge transfer theory.

increasing the control and management of employee competency levels, but equally importantly, it may also be achieved by empowering employees to be creative in managing their own competency development and learning. The challenge for management is to increase their influence and control over training and learning within the organization, while at the same time encouraging employees to take ownership for their own levels of knowledge and to be committed to ongoing self development.

As outlined earlier, learning is a complex undertaking, especially in organizational contexts. Managing learning and measuring learning outcomes is difficult, but is made even more problematic within complex learning domains, such as those that exist in IT organizations such as CEM Corporation. It is unlikely that an LMS will manage all of the learning in an organization in a 'truly scientific way', though it will assist greatly in managing the diverse and extensive array of learning contexts and learning process which must be supported. Its strengths lie in the new approach and attitude that it will encourage and inspire in the hearts and minds of individuals within the organization, as it enables learning that is more visible within the organization, more structured and more accessible. This stimulation of the hearts and minds is a major contributing factor to learning and is known as 'emotional quotient' (Goleman, 1996). According to Nelson and Winter (1982), an organization's practiced routines determine what it is capable of doing. They propose that the routinization of an activity within an organization constitutes the most important form of storage of the organization's specific operational knowledge. As an LMS facilitates learning structures and learning processes, it thus encourages the routinization of learning within the organization and promotes a learning culture in which formal learning and training are further integrated into everyday work practices.

CEM Corporation still has a way to go to fully exploit the benefits offered by the LMS. Firstly, not all formal training is currently being tracked and managed through the Learning Management System. One of the key organizations within CEM Corporation does not yet use the LMS to manage and track its own technical training. Secondly, role-based competency models have not been drawn up for all roles and this process may be hindered as the construction of competency models for some of the more specialist technical roles may be time consuming and problematic. Even where competency models are available within CEM Corporation, many employees, and indeed managers, have not yet engaged with the competency assessment process. These competency assessments will be critical in establishing if positive learning outcomes have been achieved and will demonstrate if the organization is obtaining a return on its investment in implementing an LMS.

Argyris and Schön (1978, 1996) describe organizational deuterolearning learning as "*a critically important kind of organizational double-loop learning...through which the members of an organization may discover and modify the learning system that conditions prevailing patterns of organizational inquiry.*" They equate this to Bateson's (1972) description of second order learning or '*learning how to learn*'. It is clear that the implementation and utilization of an LMS represents an attempt to fundamentally change how learning is happening in an organization and signifies a concerted effort to achieve deuterolearning. An LMS offers exciting new opportunities to both employees and management in relation to 'learning how to learn'.

5.1   LMS: Research directions

This paper introduces a framework for 'learning in organizations' which incorporates a role for Learning Management Systems (LMS). Researchers maintain that a young field such as IS requires "pre-theory" frameworks to guide research activities while enroute to theory development (Bariff and Ginzberg, 1982; Teng and Galetta, 1991). Crossan, Lane and White (1999) point out that a framework defines the research landscape and directs theory building. The relationships among the framework elements proposed in this paper are described at a high level, however further theory development will expand and deepen these connections and will enable the development of testable hypotheses. Bakos and Treacy (1986) believe that as an area matures, there is an increasing need to move beyond frameworks toward explanatory models of the underlying phenomena. In this way, it becomes possible to build a cumulative tradition and to make normative statements which will guide managerial actions. Hence, a suitable avenue for further research would be to test this study's findings on other similar organizations.

This study supports the contention that learning management methods and tools are well suited to compensate for the deficiencies of current knowledge management techniques (see Hall, 2001; Brennan *et al.*, 2001). Furthermore, new Learning Content Management Systems may be used to store knowledge or course components at object level in relational databases; and are likely to be the closest application yet to bridging knowledge management and learning management. This view is supported by Aldrich (2001), who maintains that using knowledge chunks as the building blocks for electronic learning promotes the convergence of KMS and electronic LMS, in addition to enabling just-in-time-learning.

In its examination of one 'problem-solution oriented' KMS, this study illustrates that there is a potential opportunity for attaching learning activities or specific learning material to knowledge objects within a KMS. Similarly, the findings indicate that there are benefits to be gained by using a KMS to review selected problems and solutions relating to a particular topic area, as supplementary learning material at the end of a learning program. Thus, another interesting area for future research would be to investigate the feasibility and potential advantages of merging Learning Management Systems and practices with Knowledge Management Systems and practices, resulting in an integrated holistic solution to organizational requirements for learning and knowledge management—a 'Knowledge & Learning Management Systems' or 'KLMS'.

REFERENCES

Alavi, M. and Leidner, D. "Knowledge Management Systems: Issues, Challenges and Benefits," In *Communications of the Association of Information Systems,* Volume 1, Issue 2, 1999.

Alavi, M. and Leidner, D. "Review: Knowledge Management and Knowledge Management Systems: Conceptual Foundations and Research Issues," *MIS Quarterly* (25:1), 2001, pp. 107-136.

Aldrich, C. "Can LMSs Survive the Sophisticated buyer?" In *Learning Circuits, American Society for Training and Development,* http://www.learningcircuits.org/2001/nov2001/ttools.html,2001

Alvesson, M. and Kärreman, D. "Odd Couple: Making Sense of the Curious Concept of Knowledge Management," *Journal of Management Studies* (38:7), 2001, pp. 995-1018.

Ambrose, D.W. "The Effects of Hypermedia on Learning: A Literature Review." Educational Technology, December 1991, pp.51-55.

Argyris, C. and Schön, D.A. *Organizational Learning,* Reading, Massachusetts: Addison-Wesley, 1978.

Argyris, C. and Schön, D.A. *Organizational Learning II: Theory, Method and Practice,* Reading, Massachusetts: Addison-Wesley, 1996.

Bakos, J.Y. and Treacy, M.E. "Information Technology and Corporate Strategy: A Research Perspective," *MIS Quarterly* (10:2), 1986, pp. 107-119.

Bariff, M.L and Ginzberg, M.J "MIS and Behavioural Sciences: Research Patterns and Prescriptions," *Data Base,* Fall, 1982, pp. 19-26.

Barron, T. "The LMS Guess," In Learning Circuits, American Society for Training and Development, http://www.learningcircuits.org/apr2000/barron.html, 2000.

Bateson, G. *Steps to an Ecology of Mind,* San Francisco: Chandler, 1972.

Benbasat, I., Goldstein, D.K. and Mead, M. "The Case Research Strategy in Studies of Information Systems," *MIS Quarterly* (11:3), 1987, pp. 369-386.

Borghoff, U.M. and Pareschi, R. *Information Technology for Knowledge Management,* Heidelberg: Springer-Verlag, 1999.

Bornman, H. and Vonsolms S.H. "Hypermedia, Multimedia and Hypertext: Definitions and Overview," Electronic Library 11(4-5), 1993, pp. 259-268.

Brennan, M., Funke, S. and Andersen, C. "The Learning Content Management System: A New eLearning Market Segment Emerges," *An IDC White Paper,* http://www.lcmscouncil.org/resources.html, 2001.

Brown, J. S. and Duguid, P. "Organizational Learning and Communities of Practice: Toward a Unified View of Working, Learning, and Innovation," *Organization Science* (2:1), 1991, pp. 40-57.

Bruner, J. *Toward a Theory of Instruction,* Cambridge, Massachusetts: Harvard University Press, 1966.

Butler, T. "From Data to Knowledge and Back Again: Understanding the Limitations of KMS," *Knowledge and Process Management: The Journal of Corporate Transformation,* Vol. 10, No. 4, 2003, pp. 144-155.

Cairncross, S. and Mannion M. "Interactive Multimedia and Learning," Innovations in Education and Teaching International 38(2), 2001, pp. 156-164.

Chait, L.P. "Creating a Successful Knowledge Management System," *Journal of Business Strategy* (20:2), 1999, pp. 23-26.

Crossan, M., Lane, H and White, R. "An Organizational Learning Framework: From Intuition to Institution," *Academy of Management Review* (24:3), 1999, pp. 522-537.

Cyert, R.M. and March, J.G. *A Behavioral Theory of the Firm,* Englewood Cliffs, NJ: Prentice Hall, 1963.

Easterby-Smith, M. "Disciplines of the Learning Organisation: Contributions and Critiques," *Human Relations* (50:9), 1997, pp. 1085-113.

Easterby-Smith, M., Crossan, M. and Nicolini D. "Organizational Learning: Debates Past, Present and Future," *Journal of Management Studies* (37:6), 2000, pp. 783-796.

Edvinsson, L. and Malone, M.S. "Intellectual Capital. London, HarperCollins," *e-LearningHub.com,*   http://www.e-learninghub.com/articles/learning_management_system.html, 1997.

Eisenhardt, K.M. "Building Theories from Case Study Research," *Academy of Management Review* (14:4), 1989, pp. 532-550.

Fiol, C. M. and M. A. Lyles "Organizational Learning," Academy of Management Review 10(4), 1985, pp. 284-295.

Gagné, R. *The Conditions of Learning,* Fourth Edition, London: Holt, Rinehart and Winston, 1977.

Galliers, R. and Newell S. "Back to the Future: From Knowledge Management to Data Management," In Global Co-Operation in the New Millennium, 9th European Conference on Information Systems, Bled, Slovenia, June 27-29, 2001, pp. 609-615.

Garavelli, A.C., Gorgoglione, M. and Scozzi, B. "Managing Knowledge Transfer by Knowledge Technologies," *Technovation* (22), 2002, pp. 269-279.

Goleman, D. *Emotional Intelligence,* London: Bloomsbury Publishing, 1996.

Goodman, P. and Darr E. "Computer-Aided Systems and Communities: Mechanisms for Organizational Learning in Distributed Environments," MIS Quarterly 22(4), 1998, pp. 417-440.

Greenberg, L. "LMS and LCMS: What's the Difference?" In *Learning Circuits, American Society for Training and Development,* http://www.learningcircuits.org/2002/dec2002/greenberg.htm, 2002.

Hall, B. *Learning Management Systems 2001,* California: brandon-hall.com, 2001.

Harvey, C. and Denton, J. "To come of Age: Antecedents of Organizational Learning," *Journal of Management Studies* (37:7), 1999, pp. 897-918.

Hendriks, P.H. "Many Rivers to cross: from ICT to Knowledge Management Systems," *Journal of Information Technology* (16:2), 2001, pp. 57-72.

Huber, G. P. "Organizational Learning: The Contributing Processes and the Literatures," *Organization Science* (2:1), 1991, pp. 88-115.

Huber, G.P. "Transfer of Knowledge in Knowledge Management Systems: Unexplored Issues and Suggestions," *European Journal of Information Systems* (10:2), 2001, pp. 72-79.

Jacobsen, P. "LMS vs. LCMS," *e-learning magazine*, Advanstar communications, http://www.elearningmag.com/elearning/article/articleDetail.jsp?id=21264, 2002.

Jonassen, D.H. and Grabowski, B.L. *Handbook of Individual Differences, Learning, and Instruction*, New Jersey: Lawrence Erlbaum Associates, 1993.

King, W.R. and Ko, D.G. "Evaluating Knowledge Management and The Learning Organization: An Information/Knowledge Value Chain Approach," In *Communications of the Association for Information Systems*, Volume 5, Article 14, 2001.

Lee, A.S. "A Scientific Methodology for Case Studies," *MIS Quarterly* (13:1), 1989, pp. 33-50.

Leidner, D. E. and Jarvenpaa, S.L. "The Use of Information Technology to Enhance Management School Education, A theoretical View," *MIS Quarterly* (19:3), 1995, pp. 265-291.

Marshall, C. and Rossman B.G. *Designing Qualitative Research*, California, Sage, 1989.

McCombs, B.L. "Assessing the Role of Educational Technology in the Teaching and Learning Process: A Learner Centered Perspective," In *The Secretary's Conference on Educational Technology, US Department of Education*, http://www.ed.gov/Technology/techconf/2000/mccombs_paper.html, 2000.

McDermott, R. "Why Information Technology Inspired, but Cannot Deliver Knowledge Management," *California Management Review* (41:4), 1999, pp. 103-117.

Myers, M.D. "Qualitative Research on Information Systems," *MIS Quarterly* (21:2), 1997, pp. 221-242.

Nelson R.S. and Winter, S.G. *An Evolutionary Theory of Economic Change*, Cambridge, Massachusetts: Harvard University Press, 1982.

Nichani, M. "LCM S = LMS + CMS [RLOs]," *elearningpost*, http://www.elearningpost.com/features/archives/001022.asp, 2001.

Nonaka, I. and Takeuchi, H. *The knowledge creating company*, Oxford, England: Oxford University Press, 1995.

Nordhaug, O. *Human Capital in Organizations*, Scandinavia: University Press, 1994.

Penrose, E. *The Theory of Growth of the Firm*, London: Basil Blackwell, 1959.

Pettigrew, A.M. "Contextualism Research and the Study of Organizational Change Process," In E. Mumford, H. Fitzgerald, H. Klein and A. Wood-Harper (Eds.) *Research Methods in Information Systems*, Holland: Elsevier, 1985.

Picolli, G., Ahmad R. and Ives B. "Web-based Virtual Learning Environments: A Research Framework and a Preliminary Assessment of Effectiveness in Basic IT Skills Training," MIS Quarterly 25(4), 2001, pp. 401-426.

Prahalad, K. and Hamel, G. "The Core Competence of the Corporation," *Harvard Business Review* (68:3), 1990, pp. 79-91.

Robey, D., Boudreau M.C. and Rose G.M. "Information Technology and Organizational Learning: A Review and Assessment of Research," Accounting Management and Information Technologies 10, 2000, pp. 125-155.

Rowley, J. "Knowledge Management in pursuit of learning: the Learning with Knowledge Cycle," *Journal of Information Science* (27:4), 2001, pp. 227-237.

Ruggles, R. "The State of the Notion: Knowledge Management in Practice," California Management Review 40(3), 1998, pp. 80-89.

Scarbrough, H. and Swan, J. "Explaining the Diffusion of Knowledge Management: The Role of Fashion," *British Journal of Management (*12), 2001, pp. 3-12.

Schultze, U. and Boland, R.J. "Knowledge Management Technology and the Reproduction of Knowledge Work Practices," *Journal of Strategic Information Systems* (9:2-3), 2000, pp. 193-212.

Senge, P. M. *The Fifth Discipline: The Art and Practice of The Learning Organization,* New York: Doubleday, 1990.

Skinner, B.F. "Are theories of learning necessary?" *Psychological Review* (57), 1950, pp. 193-216.

Snell, R. and Chak A.M. "The Learning Organization: Learning and Empowerment for Whom?" *Management Learning* (29:3), 1998, pp. 337-364.

Sohn, C. "How Information Systems Provide Competitive Advantage: An Organizational Learning Perspective," In Americas Conference on Information Systems, 1998.

Stake, R.E. "Case Studies," In N.K. Denzin and Y.S. Lincoln *(Eds.) Handbook of Qualitative Research,* California: Sage Publications, 1994.

Stata, R. "Organizational Learning - The Key to Management Innovation," *Sloan Management Review* (30:3), 1989, pp. 63-74.

Stewart, T. *Intellectual Capital,* London: Nicholas Brealey, 1997.

Storey, J and Barnett E. "Knowledge Management Initiatives: Learning from Failure," Journal of Knowledge Management, 4, 2000, pp. 145-156.

Sutton, D.C. "What is Knowledge and can it be managed?" *European Journal of Information Systems* (10:2), 2001, pp. 80-88.

Sveiby, K.E. *The New Organizational Wealth: Managing and Measuring Knowledge Based Assets,* San Francisco: Berrett-Koehler, 1997.

Teece, D.   "Capturing Value from Knowledge Assets: The New Economy, Markets for Know-How and Intangible Assets," California Management Review 40(3), 1998, pp. 55-79.

Teece, D., Pisano, J. and Shuen, A. "Firm Capabilities, Resources and the concept of Strategy," *Working Paper No. 90-8. CA, University of California at Berkley,* 1990.

Teng J.T.C and Galletta, D.F. "MIS Research Directions: A Survey of Researcher's Views," *Data Base* Winter/Spring, 1991, pp. 53-62.

Wagner, E.D. "E-Learning: Where Cognitive Strategies, Knowledge Management, and Information Technology converge," In *Learning without Limits,* Volume 3, Informania Inc, California, http://www.learnativity.com/download/LwoL3.pdf, 2000.

Weick, K.E. Sensemaking in Organizations. Thousand Oaks, CA: Sage, 1995.

Wilson, B.G. Constructivist Learning Environments: Case Studies in Instructional Design. Englewood Cliffs, NJ: Educational Technology Publications, 1996.

Yin, R.K. *Case Study Research, Design and Methods,* Second Edition, Newbury Park: Sage, 1994.

Zeiberg, C. "Ten steps to Successfully Selecting a Learning Management System," In L. Kent, M. Flanagan and C. Hedrick (Eds.), an Lguide publication, http://www.lguide.com/reports/, 2001.

Zuboff, S. In the Age of the Smart Machine: The Future of Work and Power, New York: Basic Books, 1998.

APPENDIX A

Table 1A. Overview of semi-structured interview questions used in the study

Research Question	Semi-Structured Interview Questions	Interviewees
What are the roles of the Learning Management System in managing learning within the organization?	Does LMS support training administration? Does LMS support diverse learners & diverse learning contexts? Does LMS facilitate competence development to meet particular business objectives? Does LMS enable cohesive learning throughout enterprise? Does LMS encourage employee accountability for learning? Does LMS enable monitoring and analysis of 'learning condition' within the organization? What are the other key roles or attributes of the LMS?	HR Training & Dev. Manager / LMS Manager (Irish Operation) HR Manager (Irish Operation) Technical Training Specialist (Manufacturing Organization, EMEA)
What is the relationship between the Learning Management System and other IS that support learning in the organization?	What types of learning/training environments are used in CEM and how does the LMS incorporate them? What Knowledge Management Systems are in use in CEM and how does the LMS incorporate them? What Content Management Systems are in use in CEM and what functionality do they provide to the LMS? Is there any relationship between the LMS and other Information Systems that support ad hoc/informal learning?	Technical Operations Manager (Customer Services Organization, EMEA)
What theories have influenced how the organization uses the Learning Management System to manage learning within the organization?	Does CEM seek to manage work competencies using the LMS? Does CEM seek to manage the competency chain using the LMS? (i.e. planning, acquisition, development and utilization of competencies) Does CEM seek to manage its human capital using the LMS? (i.e. knowledge, innovativeness, company values) Does CEM seek to manage its structural capital using the LMS? (i.e. external customer and supplier relationships, internal innovation and process capital) Does CEM seek to manage knowledge using the LMS? (from a technology perspective or a people perspective) Does CEM seek to transfer/convert knowledge using the LMS? (if so, what forms of conversions are supported) Does CEM seek to support the three key learning models using the LMS? (learn facts/rules, apply concepts/procedures, analysis/exploration) Does CEM seek to support the three types of organizational learning using the LMS? (changes in rules/processes, changes in strategy/assumptions, changes in the way strategy/assumptions are generated)	Training Specialist (Customer Services Organization, EMEA) Training & Knowledge Management Specialist (Customer Services Organization, EMEA) Global Technical Support Training Manager (Customer Services Organization, Asia Pacific)

Research Question	Semi-Structured Interview Questions	Interviewees
	Does CEM seek to support the three levels of organizational learning using the LMS? (individual, group, organizational) Does EMD seek to support the four primary organizational learning processes using the LMS? (acquire knowledge, distribute information, information interpretation, organizational memory) Is CEM's approach to managing learning using the LMS more influenced by organizational Learning Management or Knowledge Management (building learning/mining knowledge, learning as process/knowledge as resource, focus on people and community/knowledge distribution and cognition, organizational projects/IT projects, system/resource based view)	Software Engineer (Software Engineering Organization, EMEA)

WEB-BASED INFORMATION SYSTEMS— INNOVATION OR RE-SPUN EMPEROR'S CLOTHING?

Chris Barry
Department of Accountancy and Finance, National University of Ireland, Galway, Ireland

Abstract: The challenge of developing new systems with Web technologies has led many to take for granted that such Web-based Information Systems (IS) are by their nature, and in their essence, fundamentally innovative and different from conventional IS. This paper questions whether this is in fact the case. Assumptions of Web-related novelty pervade the academic literature, texts and sales literature where impressive claims are made for the potential of e-commerce and e-business information technology (IT) and applications. In this paper a number of closely related aspects of organizational Web-based IS are considered - the business context and the use of Web technologies, systems development and information systems theory. To assess whether Web-based IS are fundamentally innovative, features or aspects of each of these dimensions are critically studied. In doing so the author puts forward a number of revisionist perspectives. The paper concludes that much of what is claimed to be new about Web-based IS is often recycled, re-labelled or simply erroneous.

Key words: Web-based information systems; information technology; e-commerce; information systems development; information systems theory.

1.   INTRODUCTION

In recent years firms have made massive investments in Web-based applications, reaching into almost every aspect of organizational work. The effect of the Web on many firms has been dramatic – new business strategies have been developed based around innovative Web-based technologies; existing business strategies have been re-worked and re-aligned with Web strategies; ways in which business had been done in the past have been changed, such as supply chains, to exploit the potential of the Web; and new business models have been developed. However the economic downturn in all sectors, especially the ICT (Information, Communications and Technology) sector, and the collapse of many Web enterprises in 2001 and 2002 has dampened enthusiasm and sent world stock markets into steep reversals. Some equilibrium is being restored to the market as firms begin to recognize that e-business is really about business and that "market signals" from the early days were heavily distorted (Porter, 2001).

Frenetic Web-based systems development leading up to this period was characterized by processes that were unstructured and usually unsupported by methods or techniques that one might ordinarily expect to ensure high quality or effectiveness. Some poor Websites reflect this lack of formalization in the development process, while others may have been successful precisely because they were not patched together with old methods and techniques. Since little research has been conducted on Web systems development processes it is unclear how some organizations got it right and others got it fabulously wrong. Perhaps luckily for some, much of the chaos in development was masked because many Web projects had investors with deep pockets and systems, perhaps as Ciborra suggests (1999), were assembled as much by improvisation as by design.

There is also uncertainty and confusion within the academic world of Information Systems (IS) about the novelty of Web-based IS, whether we need new ways of developing such systems and whether existing, established IS theories are able to explain and absorb this new "type" of information system. Now, as we emerge from the hype surrounding all things preceded by "Web" and "e-", it gives academics and developers some time for reflection about the nature of Web-based IS.

2.   SO WHAT IS NEW AND WHAT IS NOT?

It is often assumed that the Web has brought "newness" to many aspects related to its development, deployment and use. On the face of it the assumption is reasonable since nothing like the information technology (IT) had been seen before and many, who would never have engaged with IT at any serious level previously, were soon e-mailing, browsing and shopping on-line. The enormous social change that these new technologies brought, served to fuel the notion that all that was Web-related was original and innovative. Some even make claims that it is a "fundamentally a new medium of human communications" (Turoff and Hiltz, 1998) – a phenomenon of superconnectivity. This paper is concerned with information systems issues and evaluates three closely related aspects of Web-based IS:

the business context and the use of Web technologies, systems development and information systems theory. In order to assess whether Web-based IS are fundamentally original, an analysis of each of these dimensions is undertaken to distinctly characterize such systems. The following table illustrates each feature or aspect of the Web and suggests whether some of its novelties are real or whether some degree of revisionism is warranted.

Table 1. Novelty of Web-based Innovation

Web Feature or Aspect	Assumed Novelty	Revisionism?
Business and Technology	New Web business sector	The Web is every business sector!
	New way of selling and advertising	One more distribution channel and advertising medium
	Better direct marketing	Some success, e.g. solicited advertising but damaged by spam
	Redundancy of corresponding conventional businesses	The few successful Web-based competitors were merged or mopped up
	Projects did not demand traditional capital budgeting assessment techniques	Explosive costs went unnoticed until share prices collapsed
	Rapid emergence of new technologies	None - unprecedented rate of technological development in IT
	Cheap alternative to conventional systems development	Growing evidence that large-scale Web-based IS more costly than traditional in-house cousins
	Technology delivers benefits	A proven myth
Systems Development	Early expectation that impact would be limited to small-scale, standalone applications	Explosive growth in use of Web-based technologies demands integration
	Nature of Web-based IS requires alternative approach to development	Yes, since few suggested Web methods (e.g., WebHDM, RMM) are used in practice
	Existing development techniques inappropriate	For the most part correct - but no widely agreed substitutes found
IS Theory	Assumption that Web-based IS are theoretically "new"	IS literature comfortably absorbs them into existing frameworks
	Assumption that Web-based IS are a new type of IS	Analysis reveals non-homogeneity, examples can be found everywhere
	Fertile ground for new theoretical IS models	None found to be widely accepted

Table 1 illustrates how many assumptions and predictions about the Web-based world have been unfounded or unfulfilled. It remains full of promise but it is immature in almost all respects. Each feature or aspect is discussed more fully in the following sections.

3.   BUSINESS ANS TECHNOLOGY DIMENSIONS

As the introduction pointed out, the business world is now more cautious about investing in Web-related enterprises. Little needs to be said about the heightened expectations of that time except that they under-achieved spectacularly. Professionals and academics tried to make sense of the many and varied types of Websites that emerged. Undoubtedly the originality of searching for products on-line and buying them or gathering more information in minutes that you could collect in months by other means grabbed everyone's imagination. But while these features of the Web made everyone sense the emergence of a new age, many such business or technology revolutions have come and gone in the past (for example the railway boom of the nineteenth century or the space age of the twentieth century) - having an impact but ultimately, excessively over-optimistic (Howcroft, 2001). The notion that this was a new business sector was quickly put to rest as it became clear that the Web had tremendous possibilities for every sector. Suggestions that there was a "new economy" where success was measured by factors like growth in customer numbers, increased market share, acquisitions and ultimately company floatation, forged through exciting new Web-delivered businesses, were quickly shelved. As Porter puts it, the idea that the "...Internet rendered all the old rules about companies and competition obsolete ... is a dangerous one." (2001 p. 63). He goes further, suggesting it led many companies to make bad decisions that have eroded their competitive advantage. Profits are now back in fashion. The upheaval has shaken the bad fruit from the tree and left the remaining firms with profitable, sustainable, business models. Perhaps unsurprisingly many of these models closely resemble more old-fashioned ones such as retailing, financial services, education and entertainment.

Below, in Table 2, is an amalgam of several classification schemes for new types of commercial Websites (exceptionally the free-content or library model has been included). It is not necessarily complete but most Websites are characterized by one, and perhaps several, of these models illustrating the non-homogenous nature of Web-based IS. Cleary, the table demonstrates the diverse nature of Websites and while the classifications help to categorize them it does not suggest many fundamentally new business models. Undoubtedly the Web is allowing businesses develop value by leveraging its power (Tapscott, Ticoll and Lowy, 2000) and new business opportunities and markets have emerged - but any suggestion that revenue-based models or banner advertising are important innovations is without merit.

Table 2. Commercial Website Classification

Model	Illustration
The free-content or e-publishing model	Original, philosophically-driven use or a public broadcasting service like bbc.co.uk
The subscription model	The Irish Times at Ireland.com
The mail-order or storefront model	Amazon.com or Tesco.co.uk
The digital delivery model	Most software suppliers / trial versions
The direct marketing model	Solicited and unsolicited mail
The real estate model	Firms sell opportunistic domain names and e-mail addresses
The incentive scheme model	Used by market research firms
The business to business model	For EDI exchange or trade
The free-content, advertising based model	General search engines turned portals like Altavista.com and Yahoo.com
The financial service centre	Web only or new bank channel
Electronic markets	Sector-based markets that extend the value chain
The e-learning model	Formal and informal, paying or free
The on-line entertainment model	Web TV or Music distribution

The not-for-profit ethos of the early days of the Internet seemed to insulate it from commercialization and some of its more unsavoury facets. Now relentless marketing and certain firms have degraded the experience in pursuit of profit. Techniques that have been deployed increasingly frustrate users. Website "stickiness", that makes it difficult to extricate yourself from a site is increasingly common. Similarly, direct marketing that promised great potential for "no-waste", focussed advertising and marketing has become blemished with outrageous techniques of a small number of spammers who are obliterating the medium for legitimate firms. The Web, just like the "real world" of business, is far from insulated from tacky schemes.

A more mature view of the Web is that it is now a new, sophisticated marketing channel. It hold the promise of being potentially better than others because it is possible to collect, in real-time, information about customers as they click their way through pages, buttons and advertisements. It has been proven by some to deliver products and offer services more quickly and more cheaply than through certain traditional channels (e.g., Dell) and to radically re-structure some industries such as Internet based e-ticketing systems (e.g. Ryanair). However this is fundamentally to do with business success, not a technological triumph. There needs to be less concentration on "e" and more on business and business strategy (Brache and Webb, 2000; Porter, 2001; McGrath and Heiens, 2003).

Predictions that conventional businesses like high street banks, retail firms and publishers would melt away as low cost Web-based alternatives attracted customers has not happened. Wild market capitalizations of Web-based firms like lastminute.com (at one time £700 million with £0.6 million turnover) just made no sense. When the bubble burst, even many of the success stories of firms that made profits (or manageable losses) on the Web were unable to survive and were either merged or mopped up by firms with real, "old" money. The business world has come full circle, from bricks to clicks and back to bricks (or bricks and clicks). As Saloner & Spence (2002) put it - "It is now widely accepted that for the vast majority of goods and services, the Internet will supplement rather than replace existing channels."

Without question new Web technologies have moved along at a dizzying pace, producing increasingly sophisticated Websites that extend business services and proffer new ways of conducting business. However there are associated problems, such as compatibility, the organizational cost of change and skills training in education and industry. Often, each new "innovation" is given more credit than it deserves. Exaggerated claims made in the past for new technologies such as 4th generation languages and CASE tools have given those within the IS academic and professional community some healthy scepticism. This is a view rarely shared by those more intimately involved in Web-based development, many of whom began work within that environment, fuelled by youthful enthusiasm and un-lumbered by the legacy of traditional systems and practices.

Most of the technology of the Web is new and has emerged at an unprecedented rate. Where to begin? - Web browsers, HTML, DHTML, XHTML, XML, CGI, ASP, Javascript, PHP, plug-ins, Intranets, Web portals, dedicated e-mail and Web servers, server "farms" to name a few. It was and remains the rapid emergence of new technologies that makes the Web such a hectic place – for businesses, developers and users alike. In the early days, it seemed so simple to put together a few Web pages and declare ones arrival in cyberspace. It was a cheap window on the world. Now the complex technologies of the World Wide Web present immense challenges to systems developers who must ensure that the systems are robust, always-on and universally accessible on multiple platforms. Today it is an expensive proposition to develop large-scale IS using Web technologies, rivalling, if not exceeding, traditional IS in cost and complexity. New requirements or dimensions of a system will naturally impinge on development and maintenance and add to a system's cost. For example, multilingualism and localization, legal issues across national boundaries and the need to regularly update volatile content.

Two schools of thought on competitiveness through IT innovation inform the debate - the technology-driven and the competency-driven approach. The former sees IT as a powerful weapon in achieving competitive advantage and thus business benefits. This view, given weight to and based on the work of Porter (1985), sees the "strategic" use of IT as closely related to an organization's fundamental business strategy (Porter and Dent-Micallef, 1997). The latter view considers any competitive advantage to be organizationally specific, based on its own distinctive resources, from which it may gain benefit. This view hold that the real strength and value of IT is that it can release, support and nurture existing skills and competencies (Clemons, 1991; Booth and Philip, 1996), creating greater organizational flexibility - a hard come-by commodity.

At the outset, a technology-driven version was typically pitched to prospective clients seeking a Web-based system. This was usually so whether or not any clear set of business objectives for the system existed. Firms were encouraged to have a "presence" otherwise some possible competitive advantage would be lost. This contention has long been debunked (Earl, 1992) and the reverse is in fact the case – that a firm must establish the business case first, then other activities such as marketing, financial planning and human resource management, as well as IT investments, would follow. Finally, all going to plan, business benefits will flow. Similarly Howcroft, more recently, noted that investors mistakenly expected that "technology *per se* delivers benefits" (Howcroft, 2001). As Porter suggests we should see the Internet for what it is "...an enabling technology - a powerful set of tools that can be used, wisely or unwisely, in almost any industry and as part of almost any strategy" (2001).

4. SYSTEMS DEVELOPMENT

4.1 What Development Practices are Best?

Does the systems development process for Web-based IS demand an entirely new approach? On the face of it the many distinct characteristics of such systems based on new technologies suggests it should. However it should not be assumed that systems development in a Web-based context differs in every respect. There are still many issues that remain common to all IS projects. These include:
- A business case must be made for a Web-based project
- The feasibility of projects should be evaluated using traditional investment appraisal techniques such as Return on Investment, Net Present Value or Payback Analysis
- The management of Web-based projects is crucial to bring them in on time and within budget
- Some "structure" in the development process is needed (provided by new or existing methods and techniques)
- Systems analysis still needs to be conducted to elicit user needs

– Web-based IS have to be integrated with existing IS

The early, laissez-faire approach to Web-based IS development had led to on-the-fly and ad-hoc development practices. This was to a large extent understandable, as traditional methods seem inappropriate for the development of what were typically small Websites (Barry and Lang, 2001). While the reasons why this might be so may not be fully understood, it should be noted that Web-based IS are far more interactive than traditional systems and aspects such as changing dynamic interfaces, multimedia content and personalized content are rarely modelled by existing IS or software engineering techniques. That traditional methods could further be shown to be unsuitable is evident when the working arrangement between team members is considered. Web-based development teams comprise individuals from quite diverse backgrounds but are nonetheless highly dependent on each other's skills. Now, however, as Web-based IS have grown in scale and include multimedia, data-heavy, business applications, it is essential for practitioners to use comprehensive, easily understood development techniques (Britton, Jones, Myers and Sharif, 1997; Barry and Lang, 2003). The time when small-scale Web-based IS could flaunt more conventional development practice has passed as large Web-projects demand a more managed process. It has also been noted that Web-based IS development differs because they are developed on "internet time" (Iansiti and MacCormack, 1997; Aoyama, 1998; Cusumano and Yoffie, 1999) and that they are characteristically different from other IS because they exhibit properties such as extreme times pressures, vague requirements, a release orientation, parallel development, variable quality and a dependence on quality staff (Baskerville and Pries-Heje, 2002).

So what methods should practitioners be using? Many innovative methods have been put forward by academics (Garzotto, Paolini and Schwabe, 1993; Isakowitz, Stohr and Balasubramanian, 1995; Gellersen, Wicke and Gaedke, 1997) but research indicates that practitioners are not making use of them (Barry and Lang, 2001). Should the IS community be alarmed at this? Not necessarily. Barry and Lang found that many multimedia and Web developers were using some form of in-house method based on a semi-structured SDLC approach and improvising with old, familiar techniques such as ERDs for data design, storyboarding and flowcharting. This improvisational aspect of Web-based IS development is a repeated theme in recent literature (Cusumano and Yoffie, 1999; Vidgen, 2002; Baskerville, Ramesh, Levine, Pries-Heje and Slaughter, 2003) that may help to explain the way methodology, as it is conventionally understood, has become contingent and variable (Fitzgerald, Russo and O'Kane, 2000). This "à la carte" approach to methodology usage is consistent with findings on how multimedia and Web development takes place in Barry and Lang's work. Nevertheless, assuming that without "structured" methods developers are using poorly disciplined approaches to produce Web-based IS is not a safe assumption. Lang's research (2003) reveals that hypermedia and Web-based development is more disciplined than is commonly believed and that talk of a "hypermedia crisis" is not being borne out by research. He found that 84% of respondents used a hypermedia development process that had clear tasks and/or phases and that in half of these organizations the processes were explicitly documented.

This must be a surprise for some academics and many consulting firms that still share a widely held view that development practice remains sloppy and ad-hoc and would benefit by prescribed, engineering oriented methods. More recently, several Web-specific development methods have been proposed (De Troyer and Leune, 1998; Howcroft and Carroll, 2000; Ginige and Murugesan, 2001; Vidgen, Avison, Wood and Wood-Harper, 2003) but it remains to be seen if they will be widely adopted. Development approaches that are emerging from practitioners should make the academic community sit up and take note – they may offer more promise. Some small teams developing software for quick-to-market applications are using agile software development (ASD) approaches that include XP, Scrum, Adaptive Software Development, Feature-Driven Development (FDD) and Dynamic Systems Development Methodology (DSDM). Indeed Web-based IS projects may be well suited to the use of ASD techniques. The first principle of the Agile Software Manifesto states "our highest priority is to satisfy the customer through early and continuous delivery of valuable software." The emphasis is on individuals and interactions rather than processes, tools and project plans. Research by Baskerville et al. has identified practices that characterize Internet speed development and conclude that "agile principles are better suited than traditional software development principles" (Baskerville et al., 2003 p. 70).

4.2   Challenges for the IS Development Community

There are, of course, huge challenges that face the IS community to deal with the rising complexity of Web-based IS. Development methods can be improved, in this author's view by first understanding how such systems are being developed and basing methods and techniques on firmly rooted comprehension. Early assumptions regarding the homogeneity of Web-based projects were misplaced. It soon became apparent that there was no one methodology or set of techniques for developing Web-based IS. This was simply because Web-based IS are as diverse as all other types of IS (see discussion in section 5 below). Different approaches, contingency-style, are needed depending on project type, size, functionality and so on. Indeed differences in the approach to end-user requirements determination are needed depending on whether a Web-based application is being designed for internal rather than external users (Huarng, 2003). The externally oriented systems, where the user base is typically beyond the control of the developer, makes requirements determination and monitoring more difficult. For a particular organization there must also be a match between process maturity and the development approach.

An area that remains as complex as it is unresolved is that of integration. Making Web-based system communicate with, sit alongside or on top of other systems is difficult and hazardous. The most high-tech systems have to be integrated with sometimes the most low-tech, sometimes the most ancient and sometimes proprietary systems that don't like talking to anything. Integration is big business at the moment as consultants capitalize on the plain truth that computers and associated technologies communicate in almost as many languages as we humans do.

Team members need to understand the perspectives and approaches of their colleagues, so that good design decisions are made (Sano, 1996; Rosenfeld and Morville, 1998). Multi-disciplinary collaboration needs to be a vital part of any new approach and development methods and techniques that cater for the differing nature of developer roles must be accounted for. If software engineers are from Mars and graphic designers are from Venus the development environment must productively accommodate both with a "universal" language.

5.   INFORMATION SYSTEMS THEORY

5.1   The Novelty of Web-based IS

Claims for the newness or novelty of all things Web-related have dominated the media and IS press for some years now. While most IS professionals and researchers are well used to exaggerated assertions from software and hardware vendors about new technologies, some of the academic literature has also been guilty of abandonment and adoption in one graceful movement. It is of no surprise then that a good deal of uncertainty and confusion is evident amongst IS students and researchers alike. In a debate on the fundamental novelty of Web-based IS this author recalls a post-graduate student exasperatingly asserting that "...of course Web IS are new, they have just been invented!"

IS researchers do need to ask some essential questions, none the least of which is "are Web-based IS new?" If they are can we construct models to demonstrate their originality? Do Web-based IS represent a novel ensemble of IS components such as people, machines, procedures and activities? Perhaps if they are not entirely new, are they a reasonable extension to some taxonomy, can they usefully extend an existing framework or can they be understood and explained by widely accepted existing IS concepts? There are undoubtedly exciting, productive and new agendas that can be pursued, but overlooking existing research does no service to the field of IS or to the tradition of exemplar research. Anchors in the corpus, oft-cited references, cumulative research – these should be the starting point for sound research. Given the past failings of the discipline in forgoing the establishment of sound theoretical foundations it would be wise for all IS researchers to examine the extant boundaries of the field. We should also as Galliers suggests look for lessons in the reference disciplines (2000).

5.2   Existing IS Definitions and Models

A brief look at some of the longest established, and widely accepted, IS conceptual literature demonstrates that theoretically Web-based IS can be quite comfortably accommodated. If we were to attempt an informal proof of whether Web-based IS are new, a definitional view is an obvious starting point. Definitions of Management Information Systems (MIS) have been generally broad and inclusive. Take one from Whitten and Bentley:

> "An information system is an arrangement of people, data, interfaces and geography that are integrated for the purpose of supporting and improving the day-to-day operations in a business, as well as fulfilling the problem-solving and decision-making information needs of business managers" (Whitten and Bentley, 1998).

At a glance one can think of many types of Web-based IS that neatly fall within this definitional domain such as a distributed group of users of an intranet-based production scheduling system, a Web-based sales reporting system might deliver summarized sales information to Account Managers in support of client sales or an Intranet-based personnel system distributes training opportunities, job postings, task guidelines and newsletters to staff. Definitions of other more recent types of IS such as Decision Support Systems (DSS) or Executive Information Systems (EIS) produce similar outcomes.

> "A DSS is a computer-based information system used to support decision making activities where it is not possible or desirable to have an automated systems perform the entire decision process" (Ginzberg and Stohr, 1981).

"Computer based systems that help decision makers confront ill-structured problems through direct interaction with data and analysis models" (Sprague and Carlson, 1982).

An Intranet-based application that has been designed to support a complex joint venture or a Web-based system that assists managers in forecasting departmental budgeting requirements would both constitute legitimate DSS-like Web-based IS.

EIS, developed to provide executives and senior managers with the internal and external information they need for monitoring operational procedures as well as for strategic decision-making, can be defined as:

"A computerised system that provides executives with easy access to internal and external information that is relevant to their critical success factors" (Watson, Houdeshel and Rainer, 1997).

Once again Web-based IS can be identified as applications of EIS. For example, a browser-based Virtual Private Network might deliver to top executives key information on each production facility such as production schedules, capacity utilization, sales and economic forecasts, and news feeds from the media and industry analysts.

In reality none of this should be surprising if we characterize a Web-based IS as an information system that utilizes Web and other technologies to serve an organizational need. While we can debate the conceptual similarity between the definitions above it is clear that Web-based IS are easily recognizable as applications of each of these types of IS.

5.3   Gorry and Scott Morton's Framework

Looking back at some of the earlier IS literature may help to give researchers and practitioners some perspective. Gorry and Scott Morton's framework (1971) has been widely used to classify information systems. Substituting Gorry and Scott Morton's examples for Web-based IS, Figure 1 demonstrates that such systems can fall above or below a line (shown dashed) that distinguishes structured from semi-structured or even un-structured decision domains. Similarly, such systems can range across managerial activity from operational to strategic.

If one looks for examples of MIS, DSS and EIS defined in the previous section, the retrofitting of Web-based IS into Gorry and Scott Morton's framework works! It would appear that it is possible to find Web-based IS to fall into every part of the framework. There are two possible conclusions that can be drawn from this – either Web-based IS are entirely unexplained by the framework or they are neither unique nor homogeneous. Given the examples of Web-based IS illustrated above it is reasonable to assume the latter explanation.

Decision Structure ⬇	Categories of Managerial Activity		
	Operational Control	Management Control	Strategic Planning
Structured	Web-based Credit Clearance Systems Internet EDI	Intranet-based Project Time Allocation Intranet-based Budget Analysis System	Intranet-based Investment Management System Issue-specific Corporate Public Relations site
Semi-structured	Web-based Taxation Compliance System Web-based Order Tracking Web-based Quality Control Systems	WAP-based Truck Routing System Web-based ERP and CRM Web-enabled Groupware	Web-based Strategic Management System Web-based Corporate Intelligence EIS Web-based Meetingware
Unstructured			

Figure 1. Web-based IS in Gorry and Scott Morton's Framework

5.4   Taxonomic Perspectives

Further evidence that Web-based IS are not new in any theoretical sense, can be found in Mason's seminal paper (1969). In it, he describes a continuum along which assumptions about the decision-making process are added in as you move across from left to right (see Figure 2). At various points (of articulation) in the process the IS stops and the decision-maker takes over. This analysis allows one to demonstrate the relative simplicity or complexity of a system, and to express what remains for the decision-maker to do in order to complete the decision-making process.

	Typical Decision Complexity				
	High		Medium	Low	
Decision-making Activities ⟹	Source	Data	Predictions and Inferences	Values and Choices	Action
Mason's Taxonomy	Databank System		Predictive Information System	Decision-Making System	Decision-Taking System
Types of Information Systems	Executive Information System		Decision Support System	Data Processing	Automated EDI
Simon's decision-making phases	Intelligence		Design		Choice
Web-based IS	Web-based Information Retrieval/Archive		B2B Supply-chain System	Web-based Order Entry	Automated Internet EDI

Figure 2. Web-based IS in Mason's Model

The model also holds a great deal of resilience in classifying new "types" of information systems. For example, a sales forecasting DSS equates to a Predictive Information System and a payroll application equates to a Decision-Making System. It is clear that Web-based IS can be simple information retrieval systems or complex business-to-business supply-chain centres, making it difficult to sustain an argument that Web-based IS are a new "classification" of information system. Indeed the analysis reveals that "Web-based" is little more than an adjective that could easily be replaced by "Client/server" or "GUI-based" and that the "Web-based IS" are just Web-based examples of applications for each of the "Types of Information Systems" above. A similar perspective can be achieved when Alter's taxonomy of decision support systems is analysed (Alter, 1977). More elaborated models such as Sprague's framework for the development of decision support systems (Sprague, 1980) or Scott Morton's three-part taxonomy (Scott Morton, 1985) can also be used to illustrate the ubiquitous nature of Web-based IS.

5.5   Interpretation of the Analysis

The analysis above tells us that Web-based IS are many and varied. They are not homogenous and range across Gorry and Scott Morton's framework and Mason's continuum effortlessly. Some are data processing oriented, some decision support oriented and others are strategically oriented applications. That well established IS theories and frameworks accommodate Web-based IS so well is reassuring. Web-based IS seem to be everywhere within the models because the common denominator is the platform on which they are delivered. Nonetheless many aspects of Web-based IS need to be conceptualized within the context of other types of IS. Indeed, researchers need to develop models and frameworks that are as enlightening as the work of those discussed above.

6. CONCLUSIONS AND KEY ISSUES FOR THE IS COMMUNITY

This paper started out with a simple question – are Web-based IS new and wholly innovative? In the discussion the author has deliberately chosen illustrations that reflect parallels with non-Internet cases. The three dimensions of Web-based IS discussed above all reveal credibility gaps between promise and reality, between fact and fiction and indeed between paradigmatic claims and practice.

On the potential of the Web to deliver new and exciting business applications there have undoubtedly been, and will continue to be, enormous opportunities. While many applications might simply be the Web-based delivery of well-established systems such as libraries, auctions and direct marketing systems, others present new business models such as industry portals, digital distribution and elaborate business-to-business systems. Intranets allow flatter organizations to more easily implement multifunctional teamwork adding hugely to productively. Cairncross has predicted that "the most widespread revolution will come from the rise in collaboration and the decline of the organizational hierarchy" (2002). While these opportunities are of critical interest to business managers and strategists, apart from natural curiosity, the IS development community should remain largely neutral. To put it another way – "Web-based anything" is just another application!

The discussion on systems development on how Web-based IS are actually built does suggest significant transgression from more conventional IS development. While multimedia information systems never achieved the growth anticipated in the early 1990s, Web-based applications have quickly been embedded into mainstream organizational systems. Although the multimedia content of many Websites remains limited, improvements in bandwidth and backend processing may soon remove technical barriers. If as it seems likely, and some research suggests (Barry and Lang, 2001), there will be significantly increased multimedia content in organizational Web-based IS in the near future, the assistance needed by practitioners to develop and implement them successfully will become more urgent. More research such as that conducted by Baskerville et al. (2003) that gets close to developers and delivers fresh insights needs to be conducted.

Perhaps most telling, and of most interest to academics, might be the analysis of the theoretical foundations of Web-based IS. From a theoretical perspective, the IS literature has been able to absorb Web-based IS into the family of information systems without too much difficulty. The nature and characteristics of such systems are satisfactorily explained and therefore Web-based IS are not conceptually new.

It is time for some revisionism. The goldrush is over and the IS community should now be ready for some real debate on the fundamental nature of Web-based IS and innovation, what are their organizational significance and how can improvements be made to the development process? Researchers also need to uncover why there are inadequacies in the use of traditional approaches, how practitioners are developing systems and how they might be further assisted by methods, techniques and tools firmly based on their needs during development? Lastly, in the world of Web-based IS development, the distance between prescriptive methods and usable ones needs to be bridged.

REFERENCES

Alter, S. (1977), A Taxonomy of Decision Support Systems. *Sloan Management Review*, 19(1), 39-56.

Aoyama, M. (1998), Web-based Agile Software Development. *IEEE Software*, Nov/Dec, 56-65.

Barry, C. and Lang, M. (2001), A Survey of Multimedia and Web Development Techniques and Methodology Usage. *IEEE Multimedia*, 8(3), 52-60.

Barry, C. and Lang, M. (2003), A comparison of 'traditional' and multimedia information systems development practices. *Information and Software Technology*, 45(4), 217-227.

Baskerville, R. and Pries-Heje, J. (2002), Information Systems Development @ Internet Speed: A New Paradigm in the Making! Tenth European Conference on Information Systems, 282-291, Gdansk, Poland.

Baskerville, R., Ramesh, B., Levine, L., Pries-Heje, J. and Slaughter, S. (2003), Is Internet-Speed Software Development Different? *IEEE Software*, 20(6), 102-107.

Booth, M. and Philip, G. (1996), Technology driven and competency-driven approaches to competitiveness: Are they reconcilable? *Journal of Information Technology*, 11(2), 143-159.

Brache, A. and Webb, J. (2000), The Eight Deadly Assumptions of e-Business. *Journal of Business Strategy*, 21(3), 13-17.

Britton, C., Jones, S., Myers, M. and Sharif, M. (1997), A Survey of Current Practice in the Development of Multimedia Systems. *Information & Software Technology*, 39(10), 695-705.

Cairncross, F. (2002), The Company of the Future: How the Communications Revolution is Changing Management, Harvard Business School Press.

Ciborra, C. (1999), A Theory of Information Systems Based on Improvisation, in W. Currie & B. Galliers (Eds.), Rethinking Management Information Systems, 136-155, Oxford University Press.

Clemons, E. (1991), Corporate strategies for information technology: A resource-based approach. *Computer*, 24(11), 23-32.

Cusumano, M. and Yoffie, D. (1999), Software Development on Internet Time. *Computer*, 32(10), 60-69.

De Troyer, O. and Leune, C. (1998), WSDM: a user centered design method for Web sites. *Computer Networks and ISDN Systems*, 30, 85-94.

Earl, M. J. (1992), Putting IT in its place: A Polemic for the 1990's. *Journal of Information Technology*, 7(2), 100-108.

Fitzgerald, B., Russo, N. and O'Kane, T. (2000), An Empirical Study of System Development Method Tailoring in Practice. Proceedings of Eighth European Conference on Information Systems, 187-194, Vienna.

Galliers, R. (2000), A Manifesto for the Future of Information Systems as a Topic of Study, in P. Finnegan & C. Murphy (Eds.), Information Systems at the Core: European Perspectives on Deploying and Managing Systems in Business, 13-29, Dublin: Blackhall Publishing.

Garzotto, F., Paolini, P. and Schwabe, D. (1993), HDM - A Model-Based Approach to Hypertext Application Design. *ACM Transactions on Information Systems*, 11(1), 1-26.

Gellersen, H., Wicke, R. and Gaedke, M. (1997), WebComposition: An Object-Oriented Support System for the Web Engineering Lifecycle. Proceedings of the Sixth International WWW Conference, 1429-1437, April, Santa Clara, CA, USA.

Ginige, A. and Murugesan, S. (2001), Web Engineering: A Methodology for Developing Scalable, Maintainable Applications. *Cutter IT Journal*, 14(7), 24-35.

Ginzberg, M. and Stohr, E. (1981), Decision Support Systems: Issues and Perspectives, Amsterdam, The Netherlands: North-Holland.

Gorry, G. and Scott Morton, M. (1971), A Framework for Management Information Systems. *Sloan Management Review*, Fall, 55-70.

Howcroft, D. (2001), After the goldrush: deconstructing the myths of the dot.com market. *Journal of Information Technology*, 16(4), 195-204.

Howcroft, D. and Carroll, J. (2000), A Proposed Methodology for Web Development. Proceedings of the Eighth European Conference on Information Systems, 290-297, Vienna.

Huarng, A. (2003), Web-based Information Systems Requirements Analysis. *Information Systems Management*, Winter, 50-58.

Iansiti, M. and MacCormack, A. (1997), Developing Products on Internet Time. *Harvard Business Review*, 75(5), 108-117.

Isakowitz, T., Stohr, E. and Balasubramanian, P. (1995), RMM: A Methodology for Structured Hypermedia Design. *Communications of the ACM*, 38(8), 34-44.

Lang, M. (2003), Reconsidering the "Software Crisis": A Study of Hypermedia Systems Development. Proceedings of IADIS International WWW/Internet 2003 Conference, November 5-8, Algarve, Portugal.

Mason, R. (1969), Basic Concepts for Designing Management Information Systems. *AIS*, Research paper no. 8.

McGrath, L. and Heiens, R. (2003), Beware the Internet panacea: how tried and true strategy got sidelined. *Journal of Business Strategy*, 24(6).

Porter, M. (1985), How information gives you competitive advantage. *Harvard Business Review*, 65(4), 149-160.

Porter, M. (2001), Strategy and the Internet. *Harvard Business Review*, 79(3), 62-78.

Porter, M. and Dent-Micallef, A. (1997), Information technology for competitive advantage: The role of human, business, and technology resources. *Strategic Management Journal*, 18(5), 375-405.

Rosenfeld, L. and Morville, P. (1998), Information Architecture for the World Wide Web, Sebastopol, CA: O'Reilly & Associates.

Saloner, G. and Spence, M. (2002), Creating and Capturing Value - Perspectives and Cases on Electronic Commerce, New York: John Wiley & Sons.

Sano, D. (1996), Designing Large-scale Websites: A Visual Design Methodology, New York: John Wiley & Sons.

Scott Morton, M. (1985), The State of the Art of Research, in F. McFarlan (Ed), The Information Systems Research Challenge, 13-41, Cambridge, MA: Harvard Business School Press.

Sprague, R. (1980), A Framework for the Development of Decision Support Systems. *MIS Quarterly*, December, 1-26.

Sprague, R. and Carlson, E. (1982), Building Effective Decision Support Systems, Englewood Cliffs, NJ: Prentice-Hall.

Tapscott, D., Ticoll, D. and Lowy, A. (2000), Digital Capital: Harnessing the Power of Business Webs, London: Nicholas Brealey.

Turoff, M. and Hiltz, S. (1998), Superconnectivity. *Communications of the ACM*, 41(7), 116.

Vidgen, R. (2002), What's so different about developing Web-based Information Systems? Tenth European Conference on Information Systems, 262-271, Gdansk, Poland.

Vidgen, R., Avison, D., Wood, B. and Wood-Harper, T. (2003), Developing Web Information Systems, Oxford: Butterworth Heinemann.

Watson, H., Houdeshel, G. and Rainer, R. (1997), Building Executive Information Systems and other Decision Support Applications, New York: John Wiley & Sons.

Whitten, J. and Bentley, L. (1998), Systems Analysis and Design Methods, Boston, MA: Irwin McGraw-Hill.

PART VI

PANELS

ICT INNOVATION: FROM CONTROL TO RISK AND RESPONSIBILITY

Piero Bassetti[1], Claudio Ciborra[2], Edoardo Jacucci[3], and Jannis Kallinikos[4]

[1]*President Bassetti Foundation on Innovation and Responsibility, Milan, Italy;* [2]*Professor of Information Systems, LSE, UK;* [3]*Doctoral Student, Oslo University, Norway;* [4]*Reader of Information Systems, LSE, UK*

The deployment of innovation in ICT, as in other major technologies, has been governed by models and procedures of risk management and control aimed at eliciting the main risks factors of innovation processes and the planning for appropriate control interventions. For example, software engineering has come up with rigorous methods for software projects risks management. But also the management of information systems has developed methods to identify , quantify and control major and minor risks that punctuate the design and implementation of information systems in organizations. What is common to most of these approaches is the quantitative notion of risk linked to the expected value of a possible danger affecting the project. And a series of assumptions concerning: the closed boundaries of the project, the system, the application; and the unified point of decision making responsible for risk management.

The Panel wants to challenge this widely accepted framework and report from a recent research project on the dual nature of risk. Also, it wants to introduce into the debate a large body of the social science literature on risk (Beck; Luhmann; Douglas; Giddens) that has been systematically ignored by the software engineering, IS and strategy scholars in their studies on risk and ICT.

The following areas will be for discussion:

- The multiple, non linear, and reflexive relationships between risk and ICT
- The dynamics of side-effects in ICT infrastructures

- The problematic aspects of the notion of responsibility when risks cannot be fully calculated and governed (and the ensuing impacts on the deployment of innovations)
- The unexpected effects of representing risk as a calculable entity in the management of projects
- The decline of boundary drawing and boundary management (e.g. loose coupling) as forms of containing and managing risks in a connected world.

The broad themes addressed by the panelists regard Control, Knowledge, Responsibility and Life.

1.   CONTROL

The ways by which ICT currently develops (large scale systems, infrastructures, the internet) violates some crucial premises upon which control has traditionally been predicated. An appreciation of the issues that are at stake makes, however, necessary the historical understanding of the traditional forms by which technology has been involved in the regulation of human affairs. Despite significant variations brought about the diffusion of IT-based artefacts, traditional forms of technological control remain operative and are clearly reflected in the way technical artefacts are designed, implemented and managed. Furthermore, the appreciation of the current developments presupposes the adequate understanding of the older system of technological control and the fractures, which grid technologies and the internet afflict upon it. Traditional forms of technological control could be understood in terms of functional simplification and closure (Luhmann). Functional simplification involves the demarcation of an operational domain within which the complexity of the world is reconstructed as a simplified set of causal/instrumental relations or, as in the case of ICT, as a simplified set of procedures. Functional closure implies the construction of a protective cocoon that is placed around the selected causal sequences to safeguard their recurrent unfolding. While possible to gauge in similar terms, the involvement of large-scale information systems in organizations spin a web of technological relations throughout the organization in ways that reconfigures the boundaries of the technological and social domains and blurs the distinctions between referential reality and representation. The traditional forms of technological control, predicated upon the premises of functional simplification and closure, are thereby challenged. These trends are further accentuated by the diffusion of the internet and other grid technologies, and the exit of technology from the secluded world of organizations into the open realm of everyday life.

2.   KNOWLEDGE

Knowledge and risk are closely related at a conceptual level. They are opposites. Risk is what we are confronted with when we do not know what will happen. And in the theory of High Reliability Organization learning from experience in general and from failures in particular is one of the key strategies to avoid accidents and disasters. But there are important barriers and limits to learning which cannot be overcome, and which imply that there are serious limits to the degree we can reduce risks stemming from complex technologies by means of knowledge and learning. Some of these are extrinsic to the knowledge and learning processes, like organizational politics and group interests. Others are intrinsic to knowledge and learning. These limits relate to the partiality of all knowledge, the complexity of the systems of knowledge we are applying related to complex technologies. This implies that there are always unanticipated side-effects of applying any (new) knowledge, conflicting and incompatible systems of knowledge, that a complex system of knowledge gets a character of an independent, autonomous actor. A case study on electronic patient records will exemplify this theme.

3.   RESPONSIBILITY

It is usually understood that ICT as many other innovations can create benefits for organizations operating on markets. What gets overlooked however is that ICT may have an impact on political institutions and hence on the expression of political power. If ICT innovation means "change", if it creates new opportunities as well as risks in the society, who is responsible? In particular who is accountable for the assessment of the social risks of ICT? Often, the political and social institutions do not know how, or are in general unwilling to pick up the responsibilities connected to ICT innovations. It is high time that they consider such a possibility.

4.   LIFE

Finally, the economic and sociological analyses of ICT and risk need to be complemented by a phenomenological/existential one. Life, risk and technology are getting more intimate. The extension of the domain of quantifiable knowledge and representation exposes us to the danger of the further growth of ignorance generated by the new interdependencies and side-effects created by grid technologies. The essence of such a reflexive process needs to be captured by a new notion of risk combined with a new perspective on the question of technology.

The challenge emerging when looking at the next developments in ICT platforms and in risk management is that the unfolding of our life (projects) is conditioned, constrained and enabled by grid technologies. Technology is already there, albeit in an indirect and hidden form when we apply for a loan or seek a life insurance scheme. Next, it will be there in helping us compute whether we should engage in a house move or a career change; whether we can afford a certain course of studies rather than another. For each life choice grid technologies will be able to offer a calculus of probabilities and thus quantify our life projects. Only an exploration of life, risk and technology can offer us some clues to grasp these future developments.

THE DARKER SIDE OF INNOVATION

Frank Land[1], Helga Drummond[2], Philip Vos Fellman[3] (and Roberto Rodriguez), Steve Furnell[4], and Prodromos Tsiavos[5]

[1]*Department of Information Systems, London School of Economics, UK;* [2]*Department of Decision Sciences, Liverpool University, UK;* [3]*School of Management, University of Southern New Hampshire, USA;* [4]*Plymouth University, UK;* [5]*Department of Information Systems, London School of Economics, UK*

Key Words Cybercrime, Open-Source, Peer-to-Peer, Malfeasance, Fraud, Criminal Innovation, Espionage

Abstract: Innovation is widely regarded as one of the crucial capabilities if an enterprise is to be successful and to sustain its success (Kay, 1993). John Simmons, Chief Comptroller of J. Lyons and 'father' of the LEO project said:

Innovation is the lifeblood of successful business management. Without innovation the most successful business will ultimately become of no more value than the once fabulous but now worked-out gold mine. (Simmons, 1962)

However there is an asymmetry in most of the discussion of the field. Innovation is primarily discussed as a force for progress even if, as in the GM debate, the direction of progress is sometimes challenged. In general innovation is regarded as something which benefits society a large.

There is much less discussion, except in the specialist literature, of cybercrime (Furnell, 2001), of the way innovation is a part of the apparatus of fraudsters, criminals, tyrants and the mischievous. Innovation, too, is used to challenge the established system of business values and the powers of the business elite.

The intention is for the panel to debate some of the issues. Panellists are each specialists in some area relevant to the theme. Professor Helga Drummond of Liverpool University (see Drummond, 2003), who has recently written about the Baring Bank, Nick Leeson affair, considers that as businesses systems evolve and change cracks appear in the system, and opportunistic innovators take advantage of these for their own, often nefarious purposes.

Dr Steven Furnell, Plymouth University and Research Co-ordinator of the Network Research Group theme is the ongoing battle between two classes of innovators:

1. Those who attempt to break down the security and integrity of systems partly for mischievous reasons such as some hackers, some to weaken competition by disabling the network, some to wreak vengeance for some slight. Their chosen weapons are viruses worms and spam.

2. Those who try to build defences against the first group. These include not only technical specialists such as those working in the field of encryption, but lawyers drafting new types of regulation.

The third panellist is Prod Tsiavios a Ph.D student at the London School of Economics. He has studied the peer-to-peer and open source movements and his theme is the evolution of approaches which attempt to get around conventional methods of protecting markets via intellectual property laws and regulations.

Finally Professor Phillip Vos Fellman from Southern New Hampshire University and Dr Roberto Rodriguez will together reflect on the dark side of the internet from the perspective of investigators studying the impact of dark innovation.

References

Drummond, H, (2003), Did Nick Leeson have an accomplice? The role of information technology in the collapse of Barings Bank, Journal of Information Technology, Vol. 18, No. 2, pp. 93-101, June

Furnell, S., (2001), Cybercrime: Vandalizing the Information Society, Addison-Wesley, Boston.

Kay, J., (1993), Foundations of Corporate Success: How business strategies add value, Oxford University Press, Oxford.

Simmonds, J.R.M., (1962), 'LEO and the Managers, Macdonald, London.

IT AS A PLATFORM FOR COMPETITIVE AGILITY

V. Sambamurthy
Michigan State University, USA

Abstract: Agility is a fundamental capability for firms to compete in contemporary business environments and underlies the continual launch of innovative competitive moves through speed, surprise, and disruption. Winners in today's economy succeed because of agility in innovative products, markets, channels, and customer segments. Advances in information technologies and the emergence of information management as a superior locus of customer loyalty and value creation have spurred interest in leveraging IT as a platform for competitive agility. Investments in information technology could provide firms with strategic digital options that they can apply to sense opportunities for competitive action and respond with speed and innovation. Therefore, there is a growing interest in understanding how information technologies can be managed as a platform for competitive agility.

Key words: Agility, Organizational Capabilities, Digital Options

This panel develops insights about the linkages between IT and agility through a focus on the following questions:
1. What is competitive agility?
2. How could IT provide a platform for competitive agility?
3. What organizational and IT management practices are likely to position firms to exploit their IT infrastructures and be agile?

The format of the panel will be as follows:

Panel Chair and Moderator: Ritu Agarwal, Tyser Professor of Information Systems, University of Maryland, College Park, ragarwal@rhsmith.umd.edu

Panelists:
1. Jeff Sampler, Professor of Information Systems, London Business School and IESE Business School, Barcelona, Spain, jsampler@london.edu

2. V. Sambamurthy, Eli Broad Professor of Information Technology, Michigan State University, East Lansing, smurthy@msu.edu
3. Robert W. Zmud, Michael Price Chair in MIS, University of Oklahoma, Norman, rzmud@ou.edu

Jeff Sampler will argue that there is need to redefine the role that technology and people must play in organizations. He will explore new models of strategic thinking, technology management, and the vital traits for senior business and technical management in this turbulent world. He will emphasize two roles of technology in the agile organization - being a major driver of organizational turbulence, as well as a potential solution. A simple focus on technology to create agile organizations is a very narrow solution and there is a need to explore the integrative role of strategy, management, and organizational structures.

V. Sambamurthy will focus on two specific concepts: competitive agility and strategic digital options. He will draw upon recent research to describe how these two concepts are vital to rethinking the role of IT as a digital options generator.

Robert Zmud will first describe the different roles that IT can play as a platform for agility. Next, he will focus attention on some organizational capabilities, structures, and processes that are necessary to unleash the digital platforms and engage in agile competitive moves. Implications of this perspective for research and practice will be discussed.

Each panelist will weave together insights from their current research, consulting, and interactions with executives to provide meaningful insights both for research and managerial action.

INNOVATION IN INDUSTRY AND ACADEMIA
A panel that explores innovation in two key areas with invited speakers from industry and academia

Esther Baldwin
Intel Coprporation

Abstract: This Panel will feature invited speakers from industry and academia that will discuss how academia has fostered innovation through research partnerships and incubation centres that have resulted in new businesses, products and other "end values". The panel will explore the techniques used to foster and extract value from innovations in the two environments and discuss the challenges faced by industry and academy in this area. Industry research and development consortiums and internal innovation management practices that help fuel growth will be discussed followed by an open discussion with the audience. How each of these two segments had lead innovation practices, their successes and failures will be discussed.

There are three types of innovation that will be used by the panel to evaluate industry and academia. Radical innovation - the invention of a completely new concept or product that radically affects a key aspect of business or life as we know it. Incremental innovation - the improvement of an existing process, product or technology in order increase its value or differentiate it in the marketplace. Re-applied Innovation - cognitive transfer that allows a product or process to be re-applied in a new application area.

Several problems will be discussed by the panel and the audience will be asked to participate in identifying new solutions.

Innovation in Industry: Industries have grown from the radical innovations of one or two people. The innovators that developed the "firsts" spend considerable time and usually capital in seeing their invention become reality. Seldom have the initial inventors been the people that benefited most financially from the consumption and mass production of a radical innovation. Is this a problem and is anyone working in this area?

Re-applied innovation is seldom seen in industry and usually is part of the diffusion of innovation process that results in people from diverse disciplines making the cognitive transfer that takes innovations to a new field or application. Re-applied innovation is most seen in industry when duress is placed on a company due to major radical or incremental innovations (by someone else) and is used to help them survive these inflection points that may have eliminated their primary customer base. This is a reactive approach. If you consider that re-applied innovation is the area of least activity in both industry and academia and usually results in additional returns on already sunk capital and people investment. It is the area for highest potential to fuel growth. What can be done to increase success in this area will be explored.

The panel members will discuss innovation and how it is managed in industry by companies such as Intel ® Corporation. Charles House (Director of Societal Impact of Technology) will represent innovation in industry. His current research is on Advanced Productivity Initiatives for Intel's IT group. House was recently named to the Electronic Design Hall of Fame to join others such as Hewlett, Packard, Noyce, Moore, Grove, and Hoff from HP and Intel – also Edison, Marconi, Tesla, Edwin Armstrong, &, surprise, Hedy Lamarr.

Also discussed will be innovation in National Research Facilities: Government funded research in an academic and quasi-industrial environment to promote topics of interest to their respective governments and communities. Research and innovations from these environments frequently flow into industry and commercial products result, many from re-applied innovation e.g. ceramic tile technology used on the space shuttle is used in high heat commercial applications.   How good are we at extracting value from innovations at these facilities?

Academia has excelled in the incubation and research of innovative ideas and in recognizing that there is more value in innovation than an end product. There are many examples of Research consortiums and incubation facilities in academia that encourage the development of innovations and even the translation of innovations into viable businesses. Academia does a much more thorough job of protecting these ideas with patents and promoting their ideas through publication.   The President of the National College of Ireland Dr Joyce O'Connor will be one the panelists representing Academia.

Academia is the source of much potential radical innovation today and it is often funded by grants from industry. The partnership that has evolved between industry and academia is leading to innovations reaching the marketplace and financial remuneration for both parties involved. Converting intellectual capital into economic capital is a challenge for both industry and academia and often innovations have languished in the prototype stage without being driven to further end value. Academia has led the two groups in forming incubation centres and partnering with small innovators in order to convert idea into commercial products. Industry is just starting to manage innovation to multiple end values other than products. Intel's Network of University Labs represents a new form of industry-university collaboration that will be explored. How can we as an Innovation community reinforce each other and leverage our strengths?

Key words: Radical Innovation, Incremental Innovation, Re-applied innovation, end value, incubation, diffusion, collaboration, Innovation Management; Fostering a culture of Innovation; Innovation Partnerships and Consortia;

Erratum to: IT Innovation for Adaptability and Competitiveness

Brian Fitzgerald and Eleanor Wynn (eds.)

This book was originally published with a copyright holder in the name of the publisher in error, whereas IFIP International Federation for Information Processing holds the copyright.

The updated original online version for this book can be found at
DOI 10.1007/978-1-4020-8000-5

B. Fitzgerald, et al. (eds.), *IT Innovation for Adaptability and Competitiveness,*
DOI 10.1007/978-1-4020-8000-5_31, © IFIP International Federation for Information Processing, 2017 E1